Springer Texts in Statistics

Series Editors:
Richard DeVeaux
Stephen E. Fienberg
Ingram Olkin

More information about this series at http://www.springer.com/series/417

Stephen B. Vardeman • J. Marcus Jobe

Statistical Methods for Quality Assurance

Basics, Measurement, Control, Capability, and Improvement

Second Edition

Stephen B. Vardeman
Iowa State University
Ames, Iowa, USA

J. Marcus Jobe
Farmer School of Business
Miami University
Oxford, Ohio, USA

Additional material to this book can be downloaded from http://extras.springer.com

ISSN 1431-875X ISSN 2197-4136 (electronic)
Springer Texts in Statistics
ISBN 978-0-387-79105-0 ISBN 978-0-387-79106-7 (eBook)
DOI 10.1007/978-0-387-79106-7

Library of Congress Control Number: 2016942548

2nd edition: © Springer-Verlag New York 2016
1st edition: © Wiley 1999, under the title *Statistical Quality Assurance Methods for Engineers*
This work is subject to copyright. All rights are reserved by the Publisher, whether the whole or part of the material is concerned, specifically the rights of translation, reprinting, reuse of illustrations, recitation, broadcasting, reproduction on microfilms or in any other physical way, and transmission or information storage and retrieval, electronic adaptation, computer software, or by similar or dissimilar methodology now known or hereafter developed.
The use of general descriptive names, registered names, trademarks, service marks, etc. in this publication does not imply, even in the absence of a specific statement, that such names are exempt from the relevant protective laws and regulations and therefore free for general use.
The publisher, the authors and the editors are safe to assume that the advice and information in this book are believed to be true and accurate at the date of publication. Neither the publisher nor the authors or the editors give a warranty, express or implied, with respect to the material contained herein or for any errors or omissions that may have been made.

Printed on acid-free paper

This Springer imprint is published by Springer Nature
The registered company is Springer-Verlag New York

Soli Deo Gloria

PREFACE

This book is a substantial revision of our *Statistical Quality Assurance Methods for Engineers* published in 1999 by Wiley. It has evolved with the Iowa State University course that originally motivated the writing of that first book and fairly accurately represents the current technical content of Industrial Engineering 361 at ISU. (The biggest exception to this is that Sect. 6.2 is not formally part of the course for all students, but does end up being important for a substantial fraction of them as providing tools for their out-of-class process improvement project.)

This book has in Chap. 2 a substantially expanded treatment of measurement and associated statistical methods. This has grown out of the realization that a very large fraction of process improvement projects taken on by ISU IE 361 students ultimately turn out to be problems of measurement (or at least have important measurement components).

There has been some reorganization and extension of the qualitative material that in *Statistical Quality Assurance Methods for Engineers* was split between Chaps. 1 and 9 into this book's Chap. 1. Material from the former text's Sect. 2.2.3 on lower limits of detection, its Chap. 4 on advanced methods of process monitoring, its Sect. 5.5 on balanced hierarchical studies, its Chap. 8 on sampling inspection, and its Sect. 7.3 on mixture studies has been moved out of this book in the interest of concentrating on the most central issues and what can actually be profitably covered in a one-semester course for undergraduates. It is our intention to put this material onto a website as supplemental/optional resources for those interested.

There are currently both .pdf slides for lectures on the material of this text and videos on the material available on the website www.analyticsiowa.com University instructors interested in obtaining laboratories used at ISU since 2006 to teach IE 361 in a "flipped" environment to juniors in industrial engineering are invited to contact the first author. A complete solutions manual for the end-of-chapter exercises is available from the publisher to adopters of the book for university courses.

The authors wish to express their deep gratitude, love, and appreciation for their respective families. These have given form and substance to lives in which the kind of meaningful work represented by this book has been possible and enjoyed. The first author especially thanks his dear wife Jo Ellen Vardeman, their cherished sons Micah and Andrew Vardeman and their wonderful wives and children. The second author especially thanks his beloved parents, John and Caryl Jobe for their guidance, example, and love.

Ames, IA, USA	Stephen B. Vardeman
Oxford, OH, USA	J. Marcus Jobe

ABOUT THE AUTHORS

Stephen B. Vardeman is Professor of Statistics and Industrial Engineering at Iowa State University. He is a Fellow of the American Statistical Association and an (elected) Ordinary Member of the International Statistical Institute. His interests include physical science and engineering applications of statistics, statistics and metrology, quality assurance, business applications of statistics, modern ("big") data analytics, statistical machine learning, directional data analysis, reliability, statistics education, and the development of new statistical theory and methods. He has published textbooks in both quality assurance and engineering statistics and recently developed effective new graduate-level coursework in modern statistical machine learning.

J. Marcus Jobe is Professor of Information Systems and Analytics in the Farmer School of Business at Miami University (Ohio). His current research interests focus on measurement quality, process monitoring, pattern recognition, and oil exploration applications. Professor Jobe has taught extensively in Ukraine and conducted research in Germany. He has won numerous teaching awards at Miami University, including the all-university Outstanding Teaching award and the Richard T. Farmer Teaching Excellence award for senior professors in the school of business. The U.S. Dept. of State twice awarded him the Senior Fulbright Scholar award (1996–1997 and 2005–2006) to Ukraine. Marcus Jobe also works as a consultant and has co-authored two statistics texts with Stephen B. Vardeman.

CONTENTS

Preface vii

1 Introduction 1
   1.1 The Nature of Quality and the Role of Statistics 1
   1.2 Modern Quality Philosophy and Business
       Practice Improvement Strategies 3
       1.2.1 Modern Quality Philosophy and a Six-Step
             Process-Oriented Quality Assurance Cycle 4
       1.2.2 The Modern Business Environment and General
             Business Process Improvement 7
       1.2.3 Some Caveats 10
   1.3 Logical Process Identification and Analysis 12
   1.4 Elementary Principles of Quality Assurance
       Data Collection 15
   1.5 Simple Statistical Graphics and Quality
       Assurance 19
   1.6 Chapter Summary 25
   1.7 Chapter 1 Exercises 25

2 Statistics and Measurement 33
   2.1 Basic Concepts in Metrology and Probability Modeling
       of Measurement 34
   2.2 Elementary One- and Two-Sample Statistical Methods
       and Measurement 40
       2.2.1 One-Sample Methods and Measurement Error 40
       2.2.2 Two-Sample Methods and Measurement Error 45

xii Contents

- 2.3 Some Intermediate Statistical Methods and Measurement 53
  - 2.3.1 A Simple Method for Separating Process and Measurement Variation 54
  - 2.3.2 One-Way Random Effects Models and Associated Inference 56
- 2.4 Gauge R&R Studies 62
  - 2.4.1 Two-Way Random Effects Models and Gauge R&R Studies 62
  - 2.4.2 Range-Based Estimation 65
  - 2.4.3 ANOVA-Based Estimation 68
- 2.5 Simple Linear Regression and Calibration Studies 75
- 2.6 R&R Considerations for Go/No-Go Inspection 80
  - 2.6.1 Some Simple Probability Modeling 81
  - 2.6.2 Simple R&R Point Estimates for 0/1 Contexts 82
  - 2.6.3 Confidence Limits for Comparing Call Rates for Two Operators 84
- 2.7 Chapter Summary 87
- 2.8 Chapter 2 Exercises 87

3 Process Monitoring 107
- 3.1 Generalities About Shewhart Control Charting 108
- 3.2 Shewhart Charts for Measurements/"Variables Data" 113
  - 3.2.1 Charts for Process Location 113
  - 3.2.2 Charts for Process Spread 119
  - 3.2.3 What If $n = 1$? 124
- 3.3 Shewhart Charts for Counts/"Attributes Data" 128
  - 3.3.1 Charts for Fraction Nonconforming 128
  - 3.3.2 Charts for Mean Nonconformities per Unit 132
- 3.4 Patterns on Shewhart Charts and Special Alarm Rules 136
- 3.5 The Average Run Length Concept 144
- 3.6 Statistical Process Monitoring and Engineering Control 150
  - 3.6.1 Discrete Time PID Control 150
  - 3.6.2 Comparisons and Contrasts 157
- 3.7 Chapter Summary 160
- 3.8 Chapter 3 Exercises 160

4 Process Characterization and Capability Analysis 191
- 4.1 More Statistical Graphics for Process Characterization 192
  - 4.1.1 Dot Plots and Stem-and-Leaf Diagrams 192
  - 4.1.2 Quantiles and Box Plots 194
  - 4.1.3 Q-Q Plots and Normal Probability Plots 198

	4.2	Process Capability Measures and Their Estimation 207
	4.3	Prediction and Tolerance Intervals 213
		4.3.1 Intervals for a Normal Process 214
		4.3.2 Intervals Based on Maximum and/or Minimum Sample Values 216
	4.4	Probabilistic Tolerancing and Propagation of Error 220
	4.5	Chapter Summary 229
	4.6	Chapter 4 Exercises 230

5 Experimental Design and Analysis for Process Improvement Part 1: Basics   **251**

- 5.1 One-Way Methods 252
  - 5.1.1 The One-Way Normal Model and a Pooled Estimator of Variance 253
  - 5.1.2 Confidence Intervals for Linear Combinations of Means 257
- 5.2 Two-Way Factorials 261
  - 5.2.1 Graphical and Qualitative Analysis for Complete Two-Way Factorial Data 262
  - 5.2.2 Defining and Estimating Effects 266
  - 5.2.3 Fitting and Checking Simplified Models for Balanced Two-Way Factorial Data 274
- 5.3 2^p Factorials 279
  - 5.3.1 Notation and Defining Effects in p-Way Factorials 280
  - 5.3.2 Judging the Detectability of 2^p Factorial Effects in Studies with Replication 289
  - 5.3.3 Judging the Detectability of 2^p Factorial Effects in Studies Lacking Replication 292
  - 5.3.4 The Yates Algorithm for Computing Fitted 2^p Effects . . . 296
  - 5.3.5 Fitting Simplified Models for Balanced 2^p Data 297
- 5.4 Chapter Summary 301
- 5.5 Chapter 5 Exercises 302

6 Experimental Design and Analysis for Process Improvement Part 2: Advanced Topics   **333**

- 6.1 2^{p-q} Fractional Factorials 334
  - 6.1.1 Motivation and Preliminary Insights 334
  - 6.1.2 Half-Fractions of 2^p Factorials 337
  - 6.1.3 $1/2^q$ Fractions of 2^p Factorials 344
- 6.2 Response Surface Studies 354
  - 6.2.1 Graphics for Understanding Fitted Response Functions . . 355

	6.2.2	Using Quadratic Response Functions 362
	6.2.3	Analytical Interpretation of Quadratic Response Functions 369
	6.2.4	Response Optimization Strategies 372
6.3	Qualitative Considerations in Experimenting for Quality Improvement 378	
	6.3.1	"Classical" Issues 378
	6.3.2	"Taguchi" Emphases 381
6.4	Chapter Summary 384	
6.5	Chapter 6 Exercises 384	

A Tables   **407**

Bibliography   **421**

Index   **425**

CHAPTER 1

INTRODUCTION

This opening chapter first introduces the subject of quality assurance and the relationship between it and the subject of statistics in Sect. 1.1. Then Sect. 1.2 provides context for the material of this book. Standard emphases in modern quality assurance are introduced, and a six-step process-oriented quality assurance cycle is put forward as a framework for approaching this field. Some connections between modern quality assurance and popular business process improvement programs are discussed next. Some of the simplest quality assurance tools are then introduced in Sects. 1.3 through 1.5. There is a brief discussion of process mapping/analysis in Section 1.3, discussion of some simple principles of quality assurance data collection follows in Sect. 1.4, and simple statistical graphics are considered in Sect. 1.5.

1.1 The Nature of Quality and the Role of Statistics

This book's title raises at least two basic questions: "What is 'quality'?" and "What do 'statistical methods' have to do with assuring it?"

Consider first the word "quality." What does it mean to say that a particular good is a quality product? And what does it mean to call a particular service a quality service? In the case of manufactured goods (like automobiles and dishwashers), issues of reliability (the ability to function consistently and effectively across time), appropriateness of configuration, and fit and finish of parts come to mind. In the realm of services (like telecommunications and transportation services) one thinks of consistency of availability and performance, esthetics, and convenience. And in evaluating the "quality" of both goods and services, there is an implicit understanding that these issues will be balanced against corresponding costs to determine overall "value." Here is a popular definition of quality that reflects some of these notions.

Definition 1 *Quality in a good or service is fitness for use. That fitness includes aspects of both product design and conformance to the (ideal) design.*

Quality of design has to do with Appropriateness and the choice and configuration of features that define what a good or service is supposed to be like and is supposed to do. In many cases it is essentially a matter of matching product "species" to an arena of use. One needs different things in a vehicle driven on the dirt roads of the Baja Peninsula than in one used on the German autobahn. Vehicle quality of design has to do with providing the "right" features at an appropriate price. With this understanding, there is no necessary contradiction between thinking of both a Rolls-Royce and a Toyota economy car as quality vehicles. Similarly, both a particular fast-food outlet and a particular four-star restaurant might be thought of as quality eateries.

Quality of conformance has to do with living up to specifications laid down in product design. It is concerned with small variation from what is specified or expected. Variation inevitably makes goods and services undesirable. Mechanical devices whose parts vary substantially from their ideal/design dimensions tend to be noisy, inefficient, prone to breakdown, and difficult to service. They simply don't work well. In the service sector, variation from what is promised/expected is the principal source of customer dissatisfaction. A city bus system that runs on schedule every day that it is supposed to run can be seen as a quality transportation system. One that fails to do so cannot. And an otherwise elegant hotel that fails to ensure the spotless bathrooms its customers expect will soon be without those customers.

This book is concerned primarily with tools for assuring quality of conformance. This is not because quality of design is unimportant. Designing effective goods and services is a highly creative and important activity. But it is just not the primary topic of this text.

Then what does the subject of statistics have to do with the assurance of quality of conformance? To answer this question, it is helpful to have clearly in mind a definition of statistics.

Definition 2 *Statistics is the study of how best to:*

1. *collect data,*

2. *summarize or describe data, and*

3. *draw conclusions or inferences based on data,*

all in a framework that recognizes the reality and omnipresence of variation.

If quality of conformance has to do with small variation and one wishes to assure it, it will be necessary to measure, monitor, find sources of, and seek ways to reduce variation. All of these require data (information on what is happening in a system producing a product) and therefore the tool of statistics. The intellectual framework of the subject of statistics, emphasizing as it does the concept of variation, makes it natural for application in the world of quality assurance. We will see that both simple and also somewhat more advanced methods of statistics have their uses in the quest to produce quality goods and services.

Section 1.1 Exercises

1. "Quality" and "statistics" are related. Briefly explain this relationship, using the definitions of both words.

2. Why is variation in manufactured parts undesirable? Why is variation undesirable in a service industry?

3. If a product or service is designed appropriately, does that alone guarantee quality? Why or why not?

4. If a product or service conforms to design specifications, does that alone guarantee quality? Why or why not?

1.2 Modern Quality Philosophy and Business Practice Improvement Strategies

The global business environment is fiercely competitive. No company can afford to "stand still" if it hopes to stay in business. Every healthy company has explicit strategies for constantly improving its business processes and products.

Over the past several decades, there has been a blurring of distinctions between "quality improvement" and "general business practice improvement." (Formerly, the first of these was typically thought of as narrowly focused on characteristics of manufactured goods.) So there is now much overlap in emphases, language, and methodologies between the areas. The best strategies in both arenas must boil down to good methodical/scientific data-based problem solving.

In this section we first provide a discussion of some elements of modern quality philosophy and an intellectual framework around which we have organized the topics of this book (and that can serve as a road map for approaching quality improvement projects). We then provide some additional discussion and critique of the modern general business environment and its better known process-improvement strategies.

1.2.1 Modern Quality Philosophy and a Six-Step Process-Oriented Quality Assurance Cycle

Modern quality assurance methods and philosophy are focused not (primarily) on products but rather on the **processes** used to produce them. The idea is that if one gets processes to work effectively, resulting products will automatically be good. On the other hand, if one only focuses on screening out or reworking bad product, root causes of quality problems are never discovered or eliminated. The importance of this process orientation can be illustrated by an example.

Example 3 *Process Improvement in a Clean Room. One of the authors of this text once toured a "clean room" at a division of a large electronics manufacturer. Integrated circuit (IC) chips critical to the production of the division's most important product were made in the room, and it was the bottleneck of the whole production process for that product. Initial experience with that (very expensive) facility included 14 % yields of good IC chips, with over 80 people working there trying to produce the precious components.*

Early efforts at quality assurance for these chips centered on final testing and sorting good ones from bad. But it was soon clear that those efforts alone would not produce yields adequate to supply the numbers of chips needed for the end product. So a project team went to work on improving the production process. The team found that by carefully controlling the quality of some incoming raw materials, adjusting some process variables, and making measurements on wafers of chips early in the process (aimed at identifying and culling ones that would almost certainly in the end consist primarily of bad chips), the process could be made much more efficient. At the time of the tour, process-improvement efforts had raised yields to 65 % (effectively quadrupling production capacity with no capital expenditure!), drastically reduced material waste, and cut the staff necessary to run the facility from the original 80 to only eight technicians.

Process-oriented efforts are what enabled this success story. No amount of attention to the yield of the process as it was originally running would have produced these important results.

It is important to note that while process-oriented quality improvement efforts have center stage, product-oriented methods still have their place. In the clean room of Example 3, process-improvement efforts in no way eliminated the need for end-of-the-line testing of the IC chips. Occasional bad chips still needed to be identified and culled. Product-oriented inspection was still necessary, but it alone was not sufficient to produce important quality improvements.

A second important emphasis of modern quality philosophy is its **customer orientation**. This has two faces. First, the final or end user of a good or service is viewed as being supremely important. Much effort is expended by modern corporations in seeing that the "voice of the customer" (the will of the end user) is heard and carefully considered in all decisions involved in product design and production. There are many communication and decision-making techniques (such as "quality function deployment") that are used to see that this happens.

But the customer orientation in modern quality philosophy extends beyond concentration on an end user. All workers are taught to view their efforts in terms of processes that have both "vendors" from whom they receive input and "customers" to whom they pass work. One's most immediate customer need not be the end user of a company product. But it is still important to do one's work in a way that those who handle one's personal "products" are able to do so without difficulties.

A third major emphasis in modern quality assurance is that of **continual improvement**. What is state-of-the-art today will be woefully inadequate tomorrow. Consumers are expecting (and getting!) ever more effective computers, cars, home entertainment equipment, package delivery services, and communications options. Modern quality philosophy says that this kind of improvement must and will continue. This is both a statement of what "ought" to be and a recognition that in a competitive world, if an organization does not continually improve what it does and makes, it will not be long before aggressive competition drives it from the marketplace.

This text presents a wide array of tools for quality assurance. But students do not always immediately see where they might fit into a quality assurance/improvement effort or how to begin a class project in the area. So, it is useful to present an outline for approaching modern quality assurance that places the methods of this book into their appropriate context. Table 1.1 on page 6 presents a six-step process-oriented quality assurance cycle (i.e., the intellectual skeleton of this book) and the corresponding technical tools we discuss.

A sensible first step in any quality improvement project is to attempt to thoroughly understand the current and ideal configurations of the processes involved. This matter of *process mapping* can be aided by very simple tools like the flowcharts and Ishikawa diagrams discussed in Sect. 1.3.

Effective measurement is foundational to efforts to improve processes and products. If one cannot reliably measure important characteristics of what is being done to produce a good or service, there is no way to tell whether design requirements are being met and customer needs genuinely addressed. Chapter 2 introduces some basic concepts of metrology and statistical methodology for quantifying and improving the performance of measurement systems.

When adequate measurement systems are in place, one can begin to collect data on process performance. But there are pitfalls to be avoided in this collection, and if data are to be genuinely helpful in addressing quality assurance issues, they typically need to be summarized and presented effectively. So Sects. 1.4 and 1.5 contain discussions of some elementary principles of quality assurance data collection and effective presentation of such data.

Once one recognizes uniformity as essentially synonymous with quality of conformance (and variation as synonymous with "unquality"), one wants processes to be perfectly consistent in their output. But that is too much to hope for in the real world. Variation is a fact of life. The most that one can expect is that a process be *consistent in its pattern of variation*, that it be describable as physically stable.

TABLE 1.1. A six-step process-oriented quality assurance cycle (and corresponding tools)

Step	Tools
1. Attempt a logical analysis of how a process works (or should work) and where potential trouble spots, sources of variation, and data needs are located	• Flowcharts (Sect. 1.3) • Ishikawa/fishbone/cause-and-effect diagrams (Sect. 1.3)
2. Formulate appropriate (customer-oriented) measures of process performance, and develop corresponding measurement systems	• Basic concepts of measurement/metrology (Chap. 2) • Statistical quantification of measurement precision (Chap. 2) • Regression and calibration (Chap. 2)
3. Habitually collect and summarize process data	• Simple quality assurance data collection principles (Sect. 1.4) • Simple statistical graphics (Sect. 1.5)
4. Assess and work toward process stability	• Control charts (Chap. 3)
5. Characterize current process and product performance	• Statistical graphics for process characterization (Sect. 4.1) • Measures of process capability and performance and their estimation (Sects. 4.2, 4.3) • Probabilistic tolerancing and propagation of error (Sect. 4.4)
6. Work to improve those processes that are unsatisfactory	• Design and analysis of experiments (Chaps. 5, 6)

Control charts are tools for monitoring processes and issuing warnings when there is evidence in process data of physical instability. These essential tools of quality assurance are discussed in Chap. 3.

Even those processes that can be called physically stable need not be adequate for current or future needs. (Indeed, modern quality philosophy views *all* processes as inadequate and in need of improvement!) So it is important to be able to characterize in precise terms what a process is currently doing and to have tools for finding ways of improving it. Chapter 4 of this text discusses a number of methods for quantifying current process and product performance, while Chaps. 5 and 6 deal with methods of experimental design and analysis especially helpful in process-improvement efforts.

The steps outlined in Table 1.1 are a useful framework for approaching most process-related quality assurance projects. They are presented here not only as a road map for this book but also as a list of steps to follow for students wishing to get started on a class project in process-oriented quality improvement.

1.2.2 The Modern Business Environment and General Business Process Improvement

Intense global competition has fueled a search for tools to use in improving all aspects of what modern companies do. At the same time, popular understanding of the realm of "quality assurance" has broadened substantially in the past few decades. As a result, distinctions between what is the improvement of general business practice and what is process-oriented quality improvement have blurred. General business emphases and programs like Total Quality Management (TQM), ISO 9000 certification, Malcolm Baldrige Prize competitions, and Six Sigma programs have much in common with the kind of quality philosophy just discussed.

TQM

Take for example, "TQM," an early instance of the broad business influence of modern quality philosophy. The name *Total Quality Management* was meant to convey the notion that in a world economy, successful organizations will *manage* the *total*ity of what they do with a view toward producing *quality* work. TQM was promoted as appropriate in areas as diverse as manufacturing, education, and government. The matters listed in Table 1.2 came up most frequently when TQM was discussed.

TABLE 1.2. Elements of TQM emphasis

1.	Customer focus
2.	Process/system orientation
3.	Continuous improvement
4.	Self-assessment and benchmarking
5.	Change to flat organizations "without barriers"
6.	"Empowered" people/teams and employee involvement
7.	Management (and others') commitment (to TQM)
8.	Appreciation/understanding of variability

Items 1, 2, and 3 in Table 1.2 are directly related to the emphases of modern quality assurance discussed above. The TQM process orientation in 2 is perhaps a bit broader than the discussion of the previous subsection, as it sees an organization's many processes fitting together in a large **system**. (The billing process needs to mesh with various production processes, which need to mesh with the product-development process, which needs to mesh with the sales process, and so on.) There is much planning and communication needed to see that these work in harmony within a single organization. But there is also recognition that *other* organizations, external suppliers, and customers need to be seen as part of "the system." A company's products can be only as good as the raw materials with which it works. TQM thus emphasized involving a broader and broader "super-organization" (our terminology) in process- and system-improvement efforts.

In support of continual improvement, TQM proponents emphasized knowing what the "best-in-class" practices are for a given business sector or activity. They promoted **benchmarking activities** to find out how an organization's techniques compare to the best in the world. Where an organization was found to be behind, every effort was to be made to quickly emulate the leader's performance. (Where an organization's methodology is state of the art, opportunities for yet another quantum improvement were to be considered.)

It was standard TQM doctrine that the approach could only be effective in organizations that are appropriately structured and properly unified in their acceptance of the viewpoint. Hence, there was a strong emphasis in the movement on **changing corporate cultures and structures** to enable this effectiveness. Proponents of TQM simultaneously emphasized the importance of involving all corporate citizens in TQM activities, beginning with the highest levels of management and at the same time reducing the number of layers between the top and bottom of an organization, making it more egalitarian. Cross-functional project teams composed of employees from various levels of an organization (operating in consensus-building modes, with real authority not only to suggest changes but to see that they were implemented, and drawing on the various kinds of wisdom resident in the organization) were standard TQM fare. One of the corporate evils most loudly condemned was the human tendency to create "little empires" inside an organization that in fact compete with each other, rather than cooperate in ways that are good for the organization as a whole.

In a dimension most closely related to the subject of statistics, the TQM movement placed emphasis on understanding and appreciating the consequences of **variability.** In fact, providing training in elementary statistics (including the basics of describing variation through numerical and graphical means and often some basic Shewhart control charting) was a typical early step in most TQM programs.

TQM had its big names like W.E. Deming, J.M. Juran, A.V. Feigenbaum, and P. Crosby. There were also thousands of less famous individuals, who in some cases provided guidance in implementing the ideas of more famous quality leaders and in others provided instruction in their own modifications of the systems of others. The sets of terminology and action items promoted by these individuals varied consultant to consultant, in keeping with the need for them to have unique products to sell.

Six Sigma

Fashions change and business interest in some of the more managerial emphases of TQM have waned. But interest in business process improvement has not. One particularly popular and long-lived form of corporate improvement emphasis goes under the name "**Six Sigma**." The name originated at Motorola Corporation in the late 1980s. Six Sigma programs at General Electric, AlliedSignal, and Dow Chemical (among other leading examples) have been widely touted as at least partially responsible for important growth in profits and company stock values. So huge interest in Six Sigma programs persists.

The name "Six Sigma" is popularly used in at least three different ways. It refers to

1. a goal for business process performance,

2. a strategy for achieving that performance for all of a company's processes, and

3. an organizational, training, and recognition program designed to support and implement the strategy referred to in 2.

As a goal for process performance, the "Six Sigma" name has a connection to the normal distribution. If a (normal) process mean is set 6σ inside specifications/requirements (even should it inadvertently drift a bit, say by as much as 1.5σ), the process produces essentially no unacceptable results. As a formula for organizing and training to implement universal process improvement, Six Sigma borrows from the culture of the martial arts. Properly trained and effective individuals are designated as "black belts," "master black belts," and so on. These individuals with advanced training and demonstrated skills lead company process-improvement teams.

Here, our primary interest is in item 2 in the foregoing list. Most Six Sigma programs use the acronym DMAIC and the corresponding steps

1. **Define**

2. **Measure**

3. **Analyze**

4. **Improve**

5. **Control**

as a framework for approaching process improvement. The *Define* step involves setting the boundaries of a particular project, laying out the scope of what is to be addressed and bringing focus to a general "we need to work on X" beginning. The *Measure* step requires finding appropriate responses to observe, identifying corresponding measurement systems, and collecting initial process data. The *Analyze* step involves producing data summaries and formal inferences adequate to make clear initial process performance. After seeing how a process is operating, there comes an *Improve*ment effort. Often this is guided by experimentation and additional data collected to see the effects of implemented process changes. Further, there is typically an emphasis on variation reduction (improvement in process consistency). Finally, the Six Sigma five-step cycle culminates in process *Control*. This means process watching/monitoring through the routine collection of and attention to process data. The point is to be sure that improvements made persist over time. Like this book's six-step process-oriented quality assurance cycle in Table 1.1, the Six Sigma five-step DMAIC cycle is full of places where statistics is important. Table 1.3 on page 10 shows where some standard statistical concepts and methods fit into the DMAIC paradigm.

TABLE 1.3. DMAIC and statistics

Element	Statistical topics
Measure	• Measurement concepts • Data collection principles • Regression and linear calibration • Modeling measurement error • Inference in measurement precision studies
Analyze	• Descriptive statistics • Normal plotting and capability indices • Statistical intervals and testing • Confidence intervals and testing
Improve	• Regression analysis and response surface methods • Probabilistic tolerancing • Confidence intervals and testing • Factorial and fractional factorial analysis
Control	• Shewhart control charts

1.2.3 Some Caveats

This book is primarily about technical tools, not philosophy. Nevertheless, some comments about proper context are in order before launching into the technical discussion. It may at first seem hard to imagine anything unhappy issuing from an enthusiastic universal application of quality philosophy and process-improvement methods. Professor G. Box, for example, referred to TQM in such positive terms as "the democratization of science." Your authors are generally supportive of the emphases of quality philosophy and process-improvement *in the realm of commerce*. But it is possible to lose perspective and, by applying them where they are not really appropriate, to create unintended and harmful consequences.

Consider first the matter of "customer focus." To become completely absorbed with what some customers want amounts to embracing them as the final arbiters of what is to be done. And that is a basically amoral (or ultimately immoral) position. This point holds in the realm of commerce but is even more obvious when a customer-focus paradigm is applied in areas other than business.

For example, it is laudable to try to make government or educational systems more efficient. But these institutions deal in fundamentally moral arenas. We should want governments to operate morally, whether or not that is currently in vogue with the majority of (customer) voters. People should want their children to go to schools where serious content is taught, real academic achievement is required, and depth of character and intellect are developed, whether or not it is a "feel-good" experience and popular with the (customer) students, or satisfies the job-training desires of (customer) business concerns. Ultimately, we should fear for a country whose people expect other individuals and all public institutions to immediately gratify their most trivial whims (as deserving customers). The whole of human existence is not economics and commerce. Big words and concepts like "self-sacrifice," "duty," "principle," "integrity," and so on have little relevance in a "customer-driven" world. What "the customer" wants is not always even consistent, let alone moral or wise.

Preoccupation with the analysis and improvement of processes and systems has already received criticism in business circles, as often taking on a life of its own and becoming an end in itself, independent of the fundamental purposes of a company. Rationality is an important part of the human standard equipment, and it is only good stewardship to be moderately organized about how things are done. But enough is enough. The effort and volume of reporting connected with planning (and documentation of that planning) and auditing (what has been done in every conceivable matter) have increased exponentially in the past few years in American business, government, and academia. What is happening in many cases amounts to a monumental triumph of form over substance. In a sane environment, smart and dedicated people will naturally do reasonable things. Process improvement tools are sometimes helpful in thinking through a problem. But slavish preoccupation with the details of how things are done and endless generation of vision and mission statements, strategic plans, process analyses, outcome assessments, and so forth can turn a relatively small task for one person into a big one for a group, with an accompanying huge loss of productivity.

There are other aspects of emphases on the analysis of processes, continuous improvement, and the benchmarking notion that deserve mention. A preoccupation with formal benchmarking has the natural tendency to produce homogenization and the stifling of genuine creativity and innovation. When an organization invests a large effort in determining what others are doing, it is very hard to then turn around and say "So be it. That's not what we're about. That doesn't suit our strengths and interests. We'll go a different way." Instead, the natural tendency is to conform, to "make use" of the carefully gathered data and strive to be like others. And frankly, the tools of process analysis applied in endless staff meetings are not the stuff of which first-order innovations are born. Rather, those almost always come from really bright and motivated people working hard on a problem *individually* and perhaps occasionally coming together for free-form discussions of what they've been doing and what might be possible.

In the end, one has in the quality philosophy and process improvement emphases introduced above a sensible set of concerns, *provided* they are used in limited ways, in appropriate arenas, by ethical and thinking people.

Section 1.2 Exercises

1. A "process orientation" is one of the primary emphases of modern quality assurance. What is the rationale behind this?

2. How does a "customer focus" relate to "quality"?

3. What are motivations for a corporate "continuous improvement" emphasis?

4. Why is effective measurement a prerequisite to success in process improvement?

5. What tools are used for monitoring processes and issuing warnings of apparent process instability?

6. If a process is stable or consistent, is it necessarily producing high-quality goods or services? Why or why not?

1.3 Logical Process Identification and Analysis

Often, simply comparing "what is" in terms of process structure to "what is supposed to be" or to "what would make sense" is enough to identify opportunities for real improvement. Particularly in service industry contexts, the mapping of a process and identification of redundant and unnecessary steps can often lead very quickly to huge reductions in cycle times and corresponding improvements in customer satisfaction. But even in cases where how to make such easy improvements is not immediately obvious, a process identification exercise is often invaluable in locating potential process trouble spots, possibly important sources of process variation, and data collection needs.

The simple **flowchart** is one effective tool in process identification. Figure 1.1 is a flowchart for a printing process similar to one prepared by students (Drake, Lach, and Shadle) in a quality assurance course. The figure gives a high-level view of the work flow in a particular print shop. Nearly any one of the boxes on the chart could be expanded to provide more detailed information about the printing process.

People have suggested many ways of increasing the amount of information provided by a flowchart. One possibility is the use of different shapes for the boxes on the chart, according to some kind of classification scheme for the activities being portrayed. Figure 1.1 uses only three different shapes, one each for input/output, decisions, and all else. In contrast, Kolarik's *Creating Quality: Concepts, Systems, Strategies and Tools* suggests the use of seven different symbols for flowcharting industrial processes (corresponding to operations, transportation, delays, storage, source inspection, SPC charting, and sorting inspection). Of course, many schemes are possible and potentially useful in particular circumstances.

A second way to enhance the analytical value of the flowchart is to make good use of both spatial dimensions on the chart. Typically, top to bottom corresponds at least roughly to time order of activities. That leaves the possibility of using left-to-right positioning to indicate some other important variable. For example, a flowchart might be segmented into several "columns" left to right, each one

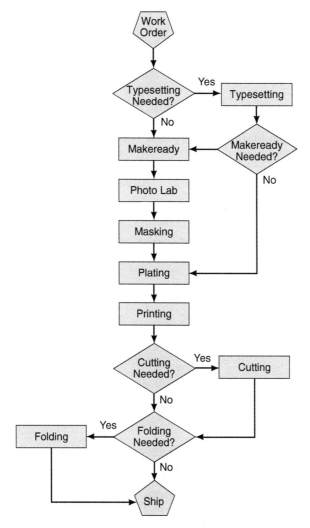

FIGURE 1.1. Flowchart of a printing process

indicating a different physical location. Or the columns might indicate different departmental spheres of responsibility. Such positioning is an effective way of further organizing one's thinking about a process.

Another simple device for use in process identification/mapping activities is the **Ishikawa diagram** (otherwise known as the **fishbone diagram** or **cause-and-effect diagram**). Suppose one has a desired outcome or (conversely) a quality problem in mind and wishes to lay out the various possible contributors to the outcome or problem. It is often helpful to place these factors on a treelike structure, where the further one moves into the tree, the more specific or basic the contributor becomes. For example, if one were interested in quality of an airline flight, general contributors might include on-time performance, baggage handling, in-flight comfort, and so on. In-flight comfort might be further amplified as involving seating, air quality, cabin service, etc. Cabin service could be broken down into components like flight attendant availability and behavior, food quality, entertainment, and so on.

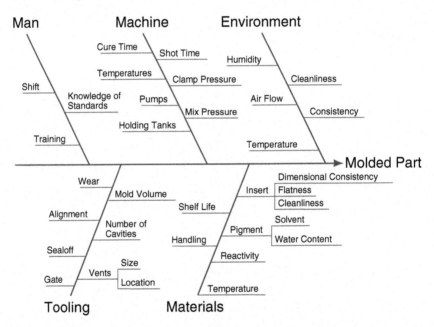

FIGURE 1.2. Cause-and-effect diagram for an injection molding process

Figure 1.2 is part of an Ishikawa diagram made by an industrial team analyzing an injection molding process. Without this or some similar kind of organized method of putting down the various contributors to the quality of the molded parts, nothing like an exhaustive listing of potentially important factors would be possible. The cause-and-effect diagram format provides an easily made and effective organization tool. It is an especially helpful device in group brainstorming sessions, where people are offering suggestions from many different perspectives in an unstructured way, and some kind of organization needs to be provided "on the fly."

Section 1.3 Exercises

1. The top-to-bottom direction on a flowchart usually corresponds to what important aspect of process operation?

2. How might a left-to-right dimension on a flowchart be employed to enhance process understanding?

3. What are other names for an Ishikawa diagram?

4. Name two purposes of the Ishikawa diagram.

1.4 Elementary Principles of Quality Assurance Data Collection

Good (practically useful) data do not collect themselves. Neither do they magically appear on one's desk, ready for analysis and lending insight into how to improve processes. But it sometimes seems that little is said about data collection. And in practice, people sometimes lose track of the fact that no amount of clever analysis will make up for lack of intrinsic information content in poorly collected data. Often, wisely and purposefully collected data will carry such a clear message that they essentially "analyze themselves." So we make some early comments here about general considerations in quality assurance data collection.

A first observation about the collection of quality assurance data is that if they are to be at all helpful, there must be a consistent understanding of exactly how they are to be collected. This involves having **operational definitions** for quantities to be observed and personnel who have been **well trained** in using the definitions and any relevant measurement equipment. Consider, for example, the apparently fairly "simple" problem of measuring "the" diameters of (supposedly circular) steel rods. Simply handed a gauge and told to measure diameters, one would not really know where to begin. Should the diameter be measured at one identifiable end of the rods, in the center, or where? Should the first diameter seen for each rod be recorded, or should perhaps the rods be rolled in the gauge to get maximum diameters (for those cases where rods are not perfectly circular in cross section)?

Or consider a case where one is to collect qualitative data on defects in molded glass automobile windshields. Exactly what constitutes a "defect?" Surely a bubble 1 in in diameter directly in front of the driver's head position is a defect. But would a 10^{-4} in diameter flaw in the same position be a problem? Or what about a 1 in diameter flaw at the very edge of the windshield that would be completely covered by trim molding? Should such a flaw be called a defect? Clearly, if useful data are to be collected in a situation like this, very careful operational definitions need to be developed, and personnel need to be taught to use them.

The importance of consistency of observation/measurement in quality assurance data collection cannot be overemphasized. When, for example, different technicians use measurement equipment in substantially different ways, what looks (in process monitoring data) like a big process change can in fact be nothing more than a change in the person doing the measurement. This is a matter we will consider from a more technical perspective Chap. 2. But here we can make the qualitative point that if operator-to-operator variation in measuring is of the same magnitude as important physical effects, and multiple technicians are going to make measurements, operator differences must be reduced through proper training and practice before there is reason to put much faith in data that are collected.

A second important point in the collection of quality assurance data has to do with **when and where** they are gathered. The closer in time and space that data are taken to an operation whose performance they are supposed to portray, the better. The ideal here is typically for well-trained workers actually doing the work or running the equipment in question to do their own data collection. There are several reasons for this. For one thing, it is such people who are in a position (after being trained in the interpretation of process monitoring data and given the authority to act on them) to react quickly and address any process ills suggested by the data that they collect. (Quick reaction to process information can prevent process difficulties from affecting additional product and producing unnecessary waste.) For another, it is simply a fact of life that data collected far away in time and space from a process rarely lead to important insights into "what is going on." Your authors have seen many student groups (against good advice) take on company projects of the variety "Here are some data we've been collecting for the past three years. Tell us what they mean." These essentially synthetic postmortem examinations never produce anything helpful for the companies involved. Even if an interesting pattern is found in such data, it is very rare that root causes can be identified completely after the fact.

If one accepts that much of the most important quality assurance data collection will be done by people whose primary job is not data collection but rather working in or on a production process, a third general point comes into focus. That is, that routine data collection should be made as **convenient** as possible, and where at all feasible, the methods used should make the data **immediately useful**. These days, quality assurance data are often entered as they are collected (sometimes quite automatically) into computer systems that produce real-time displays intended to show those who gathered them their most important features.

Whether automatic or pencil-and-paper data recording methods are used, thought needs to go into the making of the forms employed and displays produced. There should be no need for transfer to another form or medium before using the data. Figure 1.3 is a so-called two-variable **check sheet**. Rather than making a list of (x, y) pairs and later transferring them to a piece of graph paper or a computer program for making a scatterplot, the use of a pencil-and-paper form like this allows immediate display of any relationship between x and y.

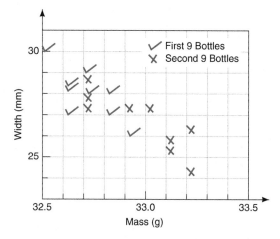

FIGURE 1.3. Check sheet for bottle mass and width of bottom piece for 18 PVC bottles

(Note that the use of different symbols or even colors can carry information on variables besides x and y, like time order of observation.) The point here is that if one's goal is process improvement, data are for using, and their collection and immediate display needs to be designed to be practically effective.

A fourth general principle of quality assurance data collection regards adequate **documentation**. One typically collects process data hoping to locate (and subsequently eliminate) possible sources of variation. If this is to be done, care needs to be taken to keep track of conditions associated with each data point. One needs to know not only that a measured widget diameter was 1.503 mm but also the machine on which it was made, who was running the machine, what raw material lot was used, when it was made, what gauge was used to do the measuring, who did the measuring, and so on. Without such information, there is, for example, no way to ever discover consistent differences between two machines that contribute significantly to overall variation in widget diameters. A sheet full of numbers without their histories is of little help in quality assurance.

Several additional important general points about the collection of quality assurance data have to do with the **volume** of information one is to handle. In the first place, a small or moderate amount of carefully collected (and immediately used) data will typically be worth much more than even a huge amount that is haphazardly collected (or never used). One is almost always better off trying to learn about a process based on a small data set collected with specific purposes and questions in mind than when rummaging through a large "general purpose" database assembled without the benefit of such focus.

Further, when trying to answer the question "How much data do I need to...?," one needs at least a qualitative understanding (hopefully gained in a first course in statistics) of what things govern the information content of a sample. For one thing (even in cases where one is gathering data from a particular finite lot of objects rather than from a process), it is the absolute (and not relative) size of a sample that governs its information content. So blanket rules like "Take a 10%

sample" are not rational. Rather than seeking to choose sample sizes in terms of some fraction of a universe of interest, one should think instead in terms of (1) the size of the unavoidable background variation and of (2) the size of an effect that is of practical importance. If there is no variation at all in a quantity of interest, a sample of size $n = 1$ will characterize it completely! On the other hand, if substantial variation is inevitable and small overall changes are of practical importance, huge sample sizes will be needed to illuminate important process behavior.

A final general observation is that one must take careful account of **human nature, psychology, and politics** when assigning data collection tasks. If one wants useful information, he or she had better see that those who are going to collect data are convinced that doing so will genuinely aid (and *not* threaten) them and that accuracy is more desirable than "good numbers" or "favorable results." People who have seen data collected by themselves or colleagues used in ways that they perceive as harmful (for instance, identifying one of their colleagues as a candidate for termination) will simply not cooperate. Nor will people who see nothing coming of their honest efforts at data collection cooperate. People who are to collect data need to believe that these can help them do a better job and help their organization be successful.

Section 1.4 Exercises

1. Why is it more desirable to have data that provide a true picture of process behavior than to obtain "good numbers" or "favorable results?"

2. What personnel issues can almost surely guarantee that a data collection effort will ultimately produce nothing useful?

3. Why is it important to have agreed upon operational definitions for characteristics of interest before beginning data collection?

4. Making real use of data collected in the past by unnamed others can be next to impossible. Why?

5. How can the problem alluded to in question 4 be avoided?

6. A check sheet is a simple but informative tool. How many variables of potential interest can a form like this portray?

7. What is another virtue of a well-designed check sheet (besides that alluded to in question 6)?

8. Is a large volume of data necessarily more informative than a moderate amount? Explain.

1.5 Simple Statistical Graphics and Quality Assurance

The old saying "a picture is worth a thousand words" is especially true in the realm of statistical quality assurance. Simple graphical devices that have the potential to be applied effectively by essentially all workers have a huge potential impact. In this section, the usefulness of simple histograms, Pareto charts, scatterplots, and run charts in quality assurance efforts is discussed. This is done with the hope that readers will see the value of routinely using these simple devices as the important data organizing and communication tools that they are.

Essentially every elementary statistics book has a discussion of the making of a **histogram** from a sample of measurements. Most even provide some terminology for describing various histogram shapes. That background will not be repeated here. Instead we will concentrate on the interpretation of patterns sometimes seen on histograms in quality assurance contexts and on how they can be of use in quality improvement efforts.

Figure 1.4 is a bimodal histogram of widget diameters.

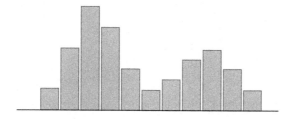

FIGURE 1.4. A bimodal histogram

Observing that the histogram has two distinct "humps" is not in and of itself particularly helpful. But asking the question "Why is the data set bimodal?" begins to be more to the point. Bimodality (or multimodality) in a quality assurance data set is a strong hint that there are two (or more) effectively different versions of something at work in a process. Bimodality might be produced by two different workers doing the same job in measurably different ways, two parallel machines that are adjusted somewhat differently, and so on. The systematic differences between such versions of the same process element produce variation that often can and should be eliminated, thereby improving quality. Viewing a plot like Fig. 1.4, one can hope to identify and eliminate the physical source of the bimodality and effectively be able to "slide the two humps together" so that they coincide, thereby greatly reducing the overall variation.

The modern trend toward reducing the size of supplier bases and even "single sourcing" has its origin in the kind of phenomenon pictured in Fig. 1.4. Different suppliers of a good or service will inevitably do some things slightly differently.

As a result, what they supply will inevitably differ in systematic ways. Reducing a company's number of vendors then has two effects. Variation in the products that it makes from components or raw materials supplied by others is reduced, and the costs (in terms of lost time and waste) often associated with switchovers between different material sources are also reduced.

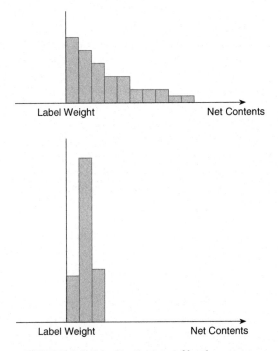

FIGURE 1.5. Two distributions of bottle contents

Other shapes on histograms can also give strong clues about what is going on in a process (and help guide quality improvement efforts). For example, sorting operations often produce distinctive truncated shapes. Figure 1.5 shows two different histograms for the net contents of some containers of a liquid. The first portrays a distribution that is almost certainly generated by culling those containers (filled by an imprecise filling process) that are below label weight. The second looks as if it might be generated by a very precise filling process aimed only slightly above the labeled weight. The histograms give both hints at how the guaranteed minimum contents are achieved in the two cases and also a pictorial representation of the waste produced by imprecision in filling. A manufacturer supplying a distribution of net contents like that in the first histogram must both deal with the rework necessitated by the part of the first distribution that has been "cut off" and also suffer the "give away cost" associated with the fact that much of the truncated distribution is quite a bit above the label value.

Figure 1.6 is a histogram for a very interesting set of data from *Engineering Statistics and Quality Control* by I.W. Burr. The very strange shape of the data set almost certainly also arose from a sorting operation. But in this case, it appears

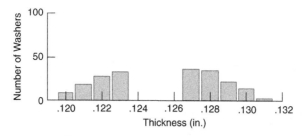

FIGURE 1.6. Thicknesses of 200 mica washers (specifications $1.25 \pm .005$ in)

that the *center* part of the distribution is missing. In all probability, one large production run was made to satisfy several orders for parts of the same type. Then a sorting operation graded those parts into classes depending upon how close actual measurements were to nominal. Customers placing orders with tight specifications probably got (perhaps at a premium price) parts from the center of the original distribution, while others with looser specifications likely received shipments with distributions like the one in Fig. 1.6.

Marking engineering specifications on a histogram is a very effective way of communicating to even very nonquantitative people what is needed in the way of process improvements. Figure 1.7 on page 22 shows a series of three histograms with specifications for a part dimension marked on them. In the first of those three histograms, the production process seems quite "capable" of meeting specifications for the dimension in question (in the sense of having adequate intrinsic precision) but clearly needs to be "re-aimed" so that the mean measurement is lower. The second histogram portrays the output of a process that is properly aimed but incapable of meeting specifications. The intrinsic precision is not good enough to fit the distribution between the engineering specifications. The third histogram represents data from a process that is both properly aimed *and* completely capable of meeting specifications.

Capability of a Process to Meet Specifications

Another kind of bar chart that is quite popular in quality assurance contexts is the so-called **Pareto diagram**. This tool is especially useful as a political device for getting people to prioritize their efforts and focus first on the biggest quality problems an organization faces. One makes a bar chart where problems are listed in decreasing order of frequency, dollar impact, or some other measure of importance. Often, a broken line graph indicating the cumulative importance of the various problem categories is also added to the display. Figure 1.8 on page 23 shows a Pareto diagram of assembly problems identified on a production run of 100 pneumatic hand tools. By the measure of frequency of occurrence, the most important quality problem to address is that of leaks.

The name "Pareto" is that of a mathematician who studied wealth distributions and concluded that most of the money in Italy belonged to a relatively few people. His name has become associated with the so-called Pareto principle or 80–20 principle. This states that "most" of anything (like quality problems or hot dog consumption) is traceable to a relatively few sources (like root causes of quality problems or avid hot dog eaters). Conventional wisdom in modern quality

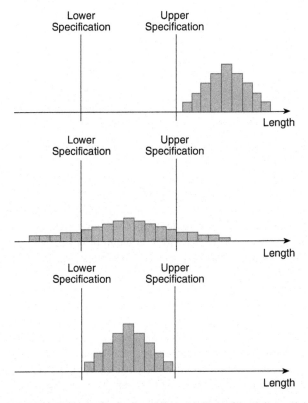

FIGURE 1.7. Three distributions of a critical machined dimension

assurance is that attention to the relatively few major causes of problems will result in huge gains in efficiency and quality.

Discovering *relationships* between variables is often important in discovering means of process improvement. An elementary but most important start in looking for such relationships is often the making of simple **scatterplots** (plots of (x, y) pairs). Consider Fig. 1.9. This consists of two scatterplots of the numbers of occurrences of two different quality problems in lots of widgets. The stories told by the two scatterplots are quite different. In the first, there seems to be a positive correlation between the numbers of problems of the two types, while in the second, no such relationship is evident. The first scatterplot suggests that a single root cause may be responsible for both types of problems and that in looking for it, one can limit attention to causes that could possibly produce both effects. The second scatterplot suggests that two different causes are at work and one will need to look for them separately. It is true, of course, that one can use numerical measures (like the sample correlation) to investigate the extent to which two variables are related. But a simple scatterplot can be understood and used even by people with little quantitative background. Besides, there are things that can be seen in plots (like nonlinear relationships) that will be missed by looking only at numerical summary measures.

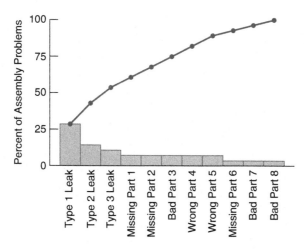

FIGURE 1.8. Pareto chart of assembly problems

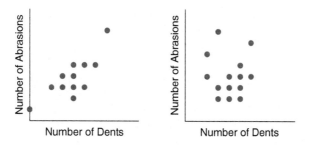

FIGURE 1.9. Two scatterplots of numbers of occurrences of manufacturing defects

The habit of plotting data is one of the best habits a quality engineer can develop. And one of the most important ways of plotting is in a scatterplot against time order of observation. Where there is only a single measurement associated with each time period and one connects consecutive plotted points with line segments, it is common to call the resulting plot a **run chart**. Figure 1.10 on page 24 is a run chart of some data studied by a student project group (Williams and Markowski). Pictured are 30 consecutive outer diameters of metal parts turned on a lathe.

Investigation of the somewhat strange pattern on the plot led to a better understanding of how the turning process worked (and could have led to appropriate compensations to eliminate much of the variation in diameters seen on the plot). The first 15 diameters generally decrease with time, and then there is a big jump in diameter, after which diameters again decrease. Checking production records, the students found that the lathe in question had been shut down and allowed to cool off between parts 15 and 16. The pattern seen on the plot is likely related to the dynamics of the lathe hydraulics. When cold, the hydraulics did not push the cutting tool into the workpiece as effectively as when they were warm. Hence

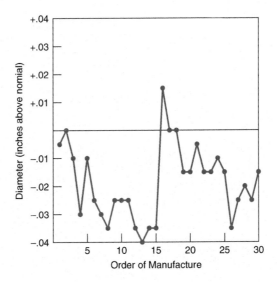

FIGURE 1.10. A run chart for 30 consecutive outer diameters turned on a lathe

the diameters tended to decrease as the lathe warmed up. (The data collection in question did not cover a long enough period to see the effects of tool wear, which would have tended to increase part diameters as the length of the cutting tool decreased.) If one knows that this kind of phenomenon exists, it is possible to compensate for it (and increase part uniformity) by setting artificial target diameters for parts made during a warm-up period below those for parts made after the lathe is warmed up.

Section 1.5 Exercises

1. In what ways can a simple histogram help in understanding process performance?

2. What aspect(s) of process performance cannot be pictured by a histogram?

3. The run chart is a graphical representation of process data that is not "static"; it gives more than a snapshot of process performance. What about the run chart that makes it an improvement over the histogram for monitoring a process?

4. Consider Fig. 1.7. The bottom histogram appears "best" with respect to being completely within specification limits and reasonably mound shaped. Describe run charts for two different scenarios that could have produced this "best" histogram and yet reflect undesirable situations, i.e., an unstable process.

5. What is the main use of a Pareto diagram?

6. What is the rationale behind the use of a Pareto diagram?

1.6 Chapter Summary

Modern quality assurance is concerned with quality of design and quality of conformance. Statistical methods, dealing as they do with data and variation, are essential tools for producing quality of conformance. Most of the tools presented in this text are useful in the process-oriented approach to assuring quality of conformance that is outlined in Table 1.1. After providing general background on modern quality and business process-improvement emphases, this chapter has introduced some simple tools. Section 1.3 considered elementary tools for the use in process mapping. Important qualitative principles of engineering and quality assurance data collection were presented in Sect. 1.4. And Sect. 1.5 demonstrated how effective simple methods of statistical graphics can be when wisely used in quality improvement efforts.

1.7 Chapter 1 Exercises

1. An engineer observes several values for a variable of interest. The average of these measurements is exactly what the engineer desires for any single response. Why should the engineer be concerned about variability in this context? How does the engineer's concern relate to product quality?

2. What is the difference between quality of conformance and quality of design?

3. Suppose 100 % of all brake systems produced by an auto manufacturer have been inspected and met safety standards. What type of quality is this? Why?

4. Describe how a production process might be characterized as exhibiting quality of conformance, but potential customers are wisely purchasing a competitor's version of the product.

5. In Example 3, initial experience at an electronics manufacturing facility involved 14 % yields of good IC chips.

    (a) Explain how this number (14 %) was probably obtained.

    (b) Describe how the three parts of Definition 2 are involved in your answer for part (a).

6. The improved yield discussed in Example 3 came as a result of improving the chip production process. Material waste and the staff necessary to run the facility were reduced. What motivation do engineers have to improve processes if improvement might lead to their own layoff? Discuss the issues this matter raises.

7. Suppose an engineer must choose among vendors 1, 2, and 3 to supply tubing for a new product. Vendor 1 charges $20 per tube, vendor 2 charges $19 per tube, and vendor 3 charges $18 per tube. Vendor 1 has implemented the six-step process-oriented quality assurance cycle (and corresponding tools) in Table 1.1. As a result, only one tube in a million from vendor 1 is nonconforming. Vendor 2 has just begun implementation of the six steps and is producing 10 % nonconforming tubes. Vendor 3 does not apply quality assurance methodology and has no idea what percent of its tubing is nonconforming. What is the price per *conforming* item for vendors 1, 2, and 3?

8. The following matrix (suggested by Dr. Brian Joiner) can be used to classify production outcomes. Good result of production means there is a large proportion of product meeting engineering specifications. (Bad result of production means there is a low proportion of product meeting requirements.) Good method of production means that quality variables are consistently near their targets. (Bad method of production means there is considerable variability about target values.)

		Result of production	
		Good	Bad
Method of production	Good	1	2
	Bad	3	4

Describe product characteristics for items produced under circumstances corresponding to each of cells 1, 2, 3, and 4.

9. **Plastic Packaging.** Hsiao, Linse, and McKay investigated the production of some plastic bags, specifically hole positions on the bags. Production of these bags is done on a model 308 poly bag machine using preprinted, prefolded plastic film delivered on a roll. The plastic film is drawn through a series of rollers to punches that make holes in the bag lips. An electronic eye scans the film after it is punched and triggers heated sills which form the seals on the bags. A conveyor transports the bags to a machine operator who counts and puts them onto wickets (by placing the holes of the bags over 6- in. metal rods) and then places them in boxes. Discuss how this process and its output might variously fall into the cells 1, 2, 3, or 4 in problem 8.

10. Consider again the **plastic packaging** case of problem 9.

    (a) Who is the immediate customer of the hole-punching process?

(b) Is it possible for the hole-punching process to produce hole locations with small variation and yet still produce a poor-quality bag? Why or why not?

(c) After observing that 100 out of 100 sampled bags fit over the two 6-in wickets, an analyst might conclude that the hole-punching process needs no improvement. Is this thinking correct? Why or why not?

(d) Hsiao, Linse, and McKay used statistical methodologies consistent with steps 1, 2, 3, and 4 of the six-step process-oriented quality assurance cycle and detected unacceptable variation in hole location. Would it be advisable to pursue step 6 in Table 1.1 in an attempt to improve the hole-punching process? Why or why not?

11. **Hose Skiving.** Siegler, Heches, Hoppenworth, and Wilson applied the six-step process-oriented quality assurance cycle to a skiving operation. Skiving consists of taking rubber off the ends of steel-reinforced hydraulic hose so that couplings may be placed on these ends. A crimping machine tightens the couplings onto the hose. If the skived length or diameter are not as designed, the crimping process can produce an unacceptable finished hose.

    (a) What two variables did the investigators identify as directly related to product quality?

    (b) Which step in the six-step cycle was probably associated with identifying these two variables as important?

    (c) The analysts applied steps 3 and 4 of the six-step cycle and found that for a particular production line, aim and variation in skive length were satisfactory. (Unfortunately, outside diameter data were not available, so study of the outside diameter variable was not possible.) In keeping with the doctrine of continual improvement, steps 5 and 6 were considered. Was this a good idea? Why or why not?

12. Engineers at an aircraft engine manufacturer have identified several "givens" regarding cost of quality problems. Two of these are "making it right the first time is always cheaper than doing it over" and "fixing a problem at the source is always cheaper than fixing it later." Describe how the six-step process-oriented quality assurance cycle in Table 1.1 relates to the two givens.

13. A common rule of thumb for the cost of quality problems is the "rule of 10." This rule can be summarized as follows (in terms of the dollar cost required to fix a nonconforming item).

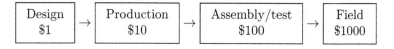

Cost history of nonconforming parts for an aircraft engine manufacturer has been roughly as follows.

Nonconforming item found	Cost to find and fix
At production testing	$200
At final inspection	$260
At company rotor assembly	$20,000
At company assembly teardown	$60,000
In customer's airplane	$200,000
At unscheduled engine removal	$1,200,000

(a) For each step following "at production testing," calculate the ratios of "costs to find and fix" to "cost to find and fix at production testing."

(b) How do the ratios in (a) compare to the rule of 10 summarized in the four-box schematic?

(c) What does your response to (b) suggest about implementation of step 3 of the six-step cycle of Table 1.1?

14. The following quotes are representative of some engineering attitudes toward quality assurance efforts. "Quality control is just a police function." "The quality control people are the ones who come in and shoot the wounded." "Our machinists will do what's easiest for them, so we'll start out with really tight engineering specifications on that part dimension."

    (a) How might an engineer develop such attitudes?

    (b) How can quality engineers personally avoid these attitudes and work to change them in others?

15. **Brush Ferrules.** Adams, Harrington, Heemstra, and Snyder did a quality improvement project concerned with the manufacture of some paint brushes. Bristle fibers are attached to a brush handle with a so-called ferrule. If the ferrule is too thin, bristles fall out. If the ferrule is too thick, brush handles are damaged and can fall apart. At the beginning of the study, there was some evidence that bristle fibers were falling out. "Crank position," "slider position," and "dwell time" are three production process variables that may affect ferrule thickness.

    (a) What feature should analysts measure on each brush in this kind of problem?

    (b) Suggest how an engineer might evaluate whether the quality problem is due to poor conformance or to poor design.

    (c) From the limited information given above, what seems to have motivated the investigation?

(d) The students considered plotting the variable identified in (a) versus the time at which the corresponding brush was produced. One of the analysts suggested first sorting the brushes according to the different crank position, slider position, and dwell-time combinations and then plotting the variable chosen in (a) versus time of production *on separate graphs*. The others argued that no insight into the problem would be gained by having separate graphs for each combination. What point of view do you support? Defend your answer.

16. **Window Frames.** Christenson, Hutchinson, Mechem, and Theis worked with a manufacturing engineering department in an effort to identify cause(s) of variation and possibly reduce the amount of offset in window frame corner joints. (Excessive offset had previously been identified as the most frequently reported type of window nonconformity.)

    (a) How might the company have come to know that excessive offset in corner joints was a problem of prime importance?

    (b) What step in the six-step cycle corresponds to your answer in (a)?

    (c) The team considered the following six categories of factors potentially contributing to unacceptable offset: (1) measurements, (2) materials, (3) workers, (4) environment, (5) methods, and (6) machines. Suggest at least one possible cause in each of these categories.

    (d) Which step in the six-step cycle of Table 1.1 is most clearly related to the kind of categorization of factors alluded to in part (c)?

17. **Machined Steel Slugs.** Harris, Murray, and Spear worked with a plant that manufactures steel slugs used to seal a hole in a certain type of casting. The group's first task was to develop and write up a standard operating procedure for data collection on several critical dimensions of these slugs. The slugs are turned on a South Bend Turrett Lathe using 1018 cold rolled steel bar stock. The entire manufacturing process is automated by means of a CNC (computer numerical control) program and only requires an operator to reload the lathe with new bar stock. The group attempted to learn about the CNC lathe program. It discovered that it was possible for the operator to change the finished part dimensions by adjusting the offset on the lathe.

    (a) What benefit is there to having a standard data collection procedure in this context?

    (b) Why was it important for the group to learn about the CNC lathe program? Which step of the six-step cycle is directly affected by their knowledge of the lathe program?

18. **Cutoff Machine.** Wade, Keller, Sharp, and Takes studied factors affecting tool life for carbide cutting inserts. The group discovered that "feed rate" and "stop delay" were two factors known by production staff to affect tool life. Once a tool wears a prescribed amount, the tool life is over.

(a) What steps might the group have taken to independently verify that feed rate and stop delay impact tool life?

(b) What is the important response variable in this problem?

(c) How would you suggest that the variable in (b) be measured?

(d) Suggest why increased tool life might be attractive to customers using the inserts.

19. **Potentiometers.** Chamdani, Davis, and Kusumaptra worked with personnel from a potentiometer assembly plant to improve the quality of finished trimming potentiometers. The 14 wire springs fastened to the potentiometer rotor assemblies (produced elsewhere) were causing short circuits and open circuits in the final potentiometers. Engineers suspected that the primary cause of the problems was a lack of symmetry on metal strips holding these springs. Of concern were the distance from one edge of the metal strip to the first spring and the corresponding distance from the last spring to the other end of the strip.

    (a) Suggest how the assembly plant might have discovered the short and open circuits.

    (b) Suggest how the plant producing the rotor assemblies perhaps became aware of the short and open circuits (the production plant does not test every rotor assembly). (Hint: Think about one of the three important emphases of modern quality philosophy. How does your response relate to the six-step cycle in Table 1.1?)

    (c) If "lack of symmetry" is the cause of quality problems, what should henceforth be recorded for each metal strip inspected?

    (d) Based on your answer to (c), what measurement value corresponds to perfect symmetry?

20. "Empowerment" is a term frequently heard in organizations in relation to process improvement. Empowerment concerns moving decision-making authority in an organization down to the lowest appropriate levels. Unfortunately, the concept is sometimes employed only until a mistake is made, then a severe reprimand occurs, and/or the decision-making privilege is moved back up to a higher level.

    (a) Name two things that are lacking in an approach to quality improvement like that described above. (Consider decision-making resulting from empowerment as a process.)

    (b) How does real, effective (and consistent) empowerment logically fit into the six-step quality improvement cycle?

21. **Lab Carbon Blank.** The following data were provided by L. A. Currie of the National Institute of Standards and Technology (NIST). The data are

preliminary and exploratory but real. The unit of measure is "instrument response" and is approximately equal to one microgram of carbon. (That is, 5.18 corresponds to 5.18 instrument units of carbon and about 5.18 micrograms of carbon.) The responses come from consecutive tests on "blank" material generated in the lab.

Test number	1	2	3	4	5	6	7
Measured carbon	5.18	1.91	6.66	1.12	2.79	3.91	2.87

Test number	8	9	10	11	12	13	14
Measured carbon	4.72	3.68	3.54	2.15	2.82	4.38	1.64

(a) Plot measured carbon content versus order of measurement.

(b) The data are ordered in time, but (as it turns out) time intervals between measurements were not equal (an appropriate plan for data collection was not necessarily in place). What feature of the plot in (a) might still have meaning?

(c) If one treats the measurement of lab-generated blank material as repeat measurements of a single blank, what does a trend on a plot like that in (a) suggest regarding variation of the measurement process? (Assume the plot is made from data equally spaced in time and collected by a single individual.)

(d) Make a frequency histogram of these data with categories 1.00–1.99, 2.00–2.99, etc.

(e) What could be missed if only a histogram was made (and one didn't make a plot like that in (a)) for data like these?

CHAPTER 2

STATISTICS AND MEASUREMENT

Good measurement is fundamental to quality assurance. That which cannot be measured cannot be guaranteed to a customer. If Brinell hardness 220 is needed for certain castings and one has no means of reliably measuring hardness, there is no way to provide the castings. So successful companies devote substantial resources to the development and maintenance of good measurement systems. In this chapter, we consider some basic concepts of measurement and discuss a variety of statistical tools aimed at quantifying and improving the effectiveness of measurement.

The chapter begins with an exposition of basic concepts and introduction to probability modeling of measurement error. Then elementary one- and two-sample statistical methods are applied to measurement problems in Sect. 2.2. Section 2.3 considers how slightly more complex statistical methods can be used to quantify the importance of sources of variability in measurement. Then Sect. 2.4 discusses studies conducted to evaluate the sizes of unavoidable measurement variation and variation in measurement chargeable to consistent differences between how operators use a measurement system. Section 2.5 considers statistical treatment of the measurement calibration problem. Finally, in Sect. 2.6 the chapter concludes with a brief section on contexts where "measurements" are go/no-go calls on individual items.

2.1 Basic Concepts in Metrology and Probability Modeling of Measurement

Metrology is the science of measurement. Measurement of many physical quantities (like lengths from inches to miles and weights from ounces to tons) is so commonplace that we think little about basic issues involved in metrology. But often engineers are forced by circumstances to leave the world of off-the-shelf measurement technology and devise their own instruments. And frequently because of externally imposed quality requirements for a product, one must ask "Can we even measure that?" Then the fundamental issues of **validity**, **precision**, and **accuracy** come into focus.

Definition 4 *A measurement or measuring method is said to be valid if it usefully or appropriately represents the feature of the measured object or phenomenon that is of interest.*

Definition 5 *A measurement system is said to be precise if it produces small variation in repeated measurement of the same object or phenomenon.*

Definition 6 *A measurement system is said to be accurate (or sometimes unbiased) if on average it produces the true or correct values of quantities of interest.*

Validity is the first concern when developing a measurement method. Without it, there is no point in proceeding to consider precision or accuracy. The issue is whether a method of measurement will faithfully portray the quantity of interest. When developing a new pH meter, one wants a device that will react to changes in acidity, not to changes in temperature of the solution being tested or to changes in the amount of light incident on the container holding the solution. When looking for a measure of customer satisfaction with a new model of automobile, one needs to consider those things that are important to customers. (For example, the number of warranty service calls per vehicle is probably a more valid measure of customer satisfaction or aggravation with a new car than warranty dollars spent per vehicle by the manufacturer.)

Precision of measurement has to do with getting similar values every time a particular measurement is done. A bathroom scale that can produce any number between 150 lb and 160 lb when one gets on it repeatedly is really not very useful. After establishing that a measurement system produces valid measurements, consistency of those measurements is needed. Figure 2.1 portrays some hardness measurements made by a group of students (Blad, Sobotka, and Zaug) on a single metal specimen with three different hardness testers. The figure shows that the dial Rockwell tester produced the most consistent results and would therefore be termed the most precise.

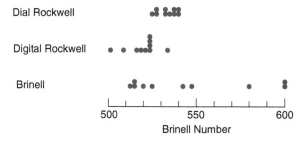

FIGURE 2.1. Brinell hardness measurements made on three different machines

Precision is largely an intrinsic property of a measurement method or device. There is not really any way to "adjust" for poor precision or to remedy it except to (1) overhaul or replace measurement technology or to (2) average multiple measurements. In this latter regard, the reader should be familiar with the fact from elementary statistics that if $y_1, y_2, \ldots, y_n$ can be thought of as independent measurements of the same quantity, each with some mean μ and standard deviation σ, then the sample mean, $\bar{y}$, has expected or average value μ and standard deviation $\sigma/\sqrt{n}$. So people sometimes rely on multiple measurements and averaging to reduce an unacceptable precision of individual measurement (quantified by σ) to an acceptable precision of average measurement (quantified by $\sigma/\sqrt{n}$).

But even validity and precision together don't tell the whole story regarding the usefulness of real-world measurements. This can be illustrated by again considering Fig. 2.1. The dial Rockwell tester is apparently the most precise of the three testers. But it is *not* obvious from the figure what the truth is about "the" real Brinell hardness of the specimen. That is, the issue of accuracy remains. Whether any of the three testers produces essentially the "right" hardness value on average is not clear. In order to assess that, one needs to reference the testers to an accepted standard of hardness measurement.

The task of comparing a measurement method or device to a standard one and, if necessary, working out conversions that will allow the method to produce "correct" (converted) values on average is called **calibration.** In the United States, the National Institute of Standards and Technology (NIST) is responsible for maintaining and disseminating consistent standards for calibrating measurement equipment. One could imagine (if the problem were important enough) sending the students' specimen to NIST for an authoritative hardness evaluation and using the result to calibrate the testers represented in Fig. 2.1. Or more likely, one might test some other specimens supplied by NIST as having known hardnesses and use those to assess the accuracy of the testers in question (and guide any recalibration that might be needed).

An analogy that is sometimes helpful in remembering the difference between accuracy and precision of measurement is that of target shooting. Accuracy in target shooting has to do with producing a pattern centered on the bull's eye (the ideal). Precision has to do with producing a tight pattern (consistency). Figure 2.2 on page 36 illustrates four possibilities for accuracy and precision in target shooting.

36   Chapter 2. Statistics and Measurement

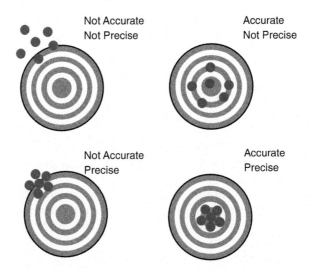

FIGURE 2.2. Measurement/target-shooting analogy

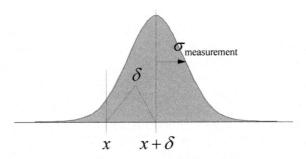

FIGURE 2.3. The distribution of a measurement y of a quantity x where measurement bias is δ and standard deviation of measurement is $\sigma_{\text{measurement}}$. Modified from "Elementary Statistical Methods and Measurement Error" by S.B. Vardeman et al., 2010, *The American Statistician*, 64(1), 47. © 2010 Taylor & Francis. Adapted with permission

Probability theory provides a helpful way to describe measurement error/ variation. If a fixed quantity x called the **measurand** is to be measured with error (as all real-world quantities are), one might represent what is actually observed as

$$y = x + \epsilon \tag{2.1}$$

where ϵ is a random variable, say with mean δ and standard deviation $\sigma_{\text{measurement}}$. Model (2.1) says that the mean of what is observed is

$$\mu_y = x + \delta. \tag{2.2}$$

If $\delta = 0$, the measurement of x is accurate or unbiased. If δ is not 0, it is called the **measurement bias**. The standard deviation of y is (for fixed x) the standard deviation of ϵ, $\sigma_{\text{measurement}}$. So $\sigma_{\text{measurement}}$ quantifies measurement precision in model (2.1). Figure 2.3 pictures the probability distribution of y and the elements x, δ, and $\sigma_{\text{measurement}}$.

Ideally, δ is 0 (and it is the work of calibration to attempt to make it 0). At a minimum, measurement devices are designed to have a **linearity** property. This means that over the range of measurands a device will normally be used to evaluate, if its bias is not 0, it is at least constant (i.e., δ does not depend upon x). This is illustrated in Fig. 2.4 (where we assume that the vertical and horizontal scales are the same).

Device "Linearity"

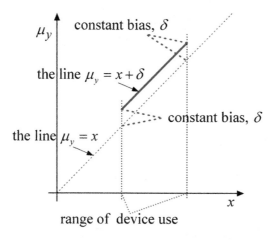

FIGURE 2.4. Measurement device "linearity" is bias constant in the measurand

Thinking in terms of model (2.1) is especially helpful when the measurand x itself is subject to variation. For instance, when parts produced on a machine have varying diameters x, one might think of model (2.1) as applying separately to each individual part diameter. But then in view of the reality of manufacturing variation, it makes sense to think of diameters as random, say with mean μ_x and standard deviation σ_x, independent of the measurement errors. This combination of assumptions then implies (for a linear device) that the mean of what is observed is

$$\mu_y = \mu_x + \delta \tag{2.3}$$

and the standard deviation of what is observed is

$$\sigma_y = \sqrt{\sigma_x^2 + \sigma_{\text{measurement}}^2} . \tag{2.4}$$

Standard Deviation of Observations Subject to Measurement Error

A nonzero δ is still a measurement bias, but now observed variation across parts is seen to include one component due to variation in x and another due to measurement error. The relationships (2.3) and (2.4) between the distributions of measurement error (ϵ) and item-to-item variation in the measurand (x) and the distribution of the observed measurements (y) are pictured in Fig. 2.5.

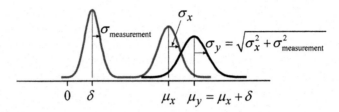

FIGURE 2.5. Random measurement error (*maroon*) and part variation (*maroon*) combine to produce observed variation (*black*). Modified from "Elementary Statistical Methods and Measurement Error" by S.B. Vardeman et al., 2010, *The American Statistician*, 64(1), 47. © 2010 Taylor & Francis. Adapted with permission

Left to right on Fig. 2.5, the two distributions in maroon represent measurement error (with bias $\delta > 0$) and measurand variation that combine to produce variation in y represented by the distribution in black. It is the middle (maroon) distribution of x that is of fundamental interest, and the figure indicates that measurement error will both tend to shift location of that distribution and flatten it in the creation of the distribution of y. It is only this last distribution (the black one) that can be observed directly, and only when both δ and $\sigma_{\text{measurement}}$ are negligible (close to 0) are the distributions of x and y essentially the same.

Observe that Eq. (2.4) implies that

$$\sigma_x = \sqrt{\sigma_y^2 - \sigma_{\text{measurement}}^2}.$$

This suggests a way of estimating σ_x alone. If one has (single) measurements y for several parts that produce a sample standard deviation s_y and several measurements on a single part that produce a sample standard deviation s, then a plausible estimator of σ_x is

Estimator of Process or Part Variation Excluding Measurement Error

$$\hat{\sigma}_x = \sqrt{\max\left(0, s_y^2 - s^2\right)}. \tag{2.5}$$

In the next sections, we will explore the use of reasoning like this, formulas like (2.5), and the application of elementary confidence interval methods to quantify various aspects of measurement precision and bias.

Section 2.1 Exercises

1. In a calibration study one compares outputs of a measurement device to "known" or "standard" values. What purpose does this serve?

2. **Pellet Densification**. Crocfer, Downey, Rixner, and Thomas studied the densification of Nd_2O_3. Pellets of this material were fired at 1,400 °C for various lengths of time and the resulting densities measured (in g/cc). In this context, what are the measurand (x), y, ϵ, and δ?

3. Suppose that in the context of problem 2, five pellets were each fired (for the same length of time), and their densities were each measured using a single device. Further, assume the measurement device has constant bias. How many measurands (xs), ys, ϵs, and δs are there in this setting?

4. In the study of problem 2, the purpose was to evaluate the *effect* of firing on pellet density. Each of the pellets fired had different original densities (that were not recorded). Does the measurement protocol described in problem 2 provide data that track what is of primary interest, i.e., does it produce a valid measure of firing effect? What additional data should have been collected? Why?

5. In the context of problem 2, the density of a single pellet was repeatedly measured five times using a single device. How many measurands (xs), ys, ϵs, and δs are there in this setting?

6. In the context of problem 2 suppose that the standard deviation of densities from repeated measurements of the same pellet with the same device is $\sqrt{2.0}$. Suppose further that the standard deviation of actual densities one pellet to the next (the standard deviation of measurands) is $\sqrt{1.4}$. What should one then expect for a standard deviation of measured density values pellet to pellet?

7. Consider the five pellets mentioned in problem 3. Density measurements similar to the following were obtained by a single operator using a single piece of equipment with a standard protocol under fixed physical circumstances:

    $$6.5, 6.6, 4.9, 5.1, \text{ and } 5.4 .$$

    (a) What is the sample standard deviation of the $n = 5$ density measurements?

    (b) In the notation of this section, which of σ_y, σ_x, or $\sigma_{\text{measurement}}$ is legitimately estimated by your sample standard deviation in (a)?

8. Again consider the five pellets of problem 3 and the five density values recorded in problem 7.

    (a) Compute the average measured density.

    (b) Assuming an appropriate model and using the notation of this section, what does your sample average estimate?

2.2 Elementary One- and Two-Sample Statistical Methods and Measurement

Elementary statistics courses provide basic inference methods for means and standard deviations based on one and two normal samples. (See, e.g., Sects. 6.3 and 6.4 of Vardeman and Jobe's *Basic Engineering Data Collection and Analysis*.) In this section we use elementary one- and two-sample confidence interval methods to study (in the simplest contexts possible) (1) how measurement error influences what can be learned from data and (2) how basic properties of that measurement error can be quantified. Subsequent sections will introduce more complicated data structures and statistical methods, but the basic modeling ideas and conceptual issues can most easily be understood by first addressing them without unnecessary (and tangential) complexity.

2.2.1 One-Sample Methods and Measurement Error

"Ordinary" confidence interval formulas based on a model that says that $y_1, y_2, \ldots, y_n$ are a sample from a normal distribution with mean μ and standard deviation σ are

| Confidence Limits for a Normal Mean |

mean $\qquad \bar{y} \pm t \dfrac{s}{\sqrt{n}}$ for estimating μ (2.6)

and

| Confidence Limits for a Normal Standard Deviation |

stdev $\qquad \left(s\sqrt{\dfrac{n-1}{\chi^2_{\text{upper}}}},\, s\sqrt{\dfrac{n-1}{\chi^2_{\text{lower}}}} \right)$ for estimating σ. (2.7)

These are mathematically straightforward, but little is typically said in basic courses about the practical meaning of the parameters μ and σ. So a first point to make here is that sources of physical variation (and in particular, sources of measurement error and item-to-item variation) interact with data collection plans to give practical meaning to "μ" and "σ." This in turn governs what of practical importance can be learned from application of formulas like (2.6) and (2.7).

Two Initial Applications

Figures 2.6 and 2.7 are schematic representations of two different ways that a single "sample" of n observed values y might arise. These are:

1. Repeat measurements on a single measurand made using the same device, and

2. Single measurements made on multiple measurands coming from a stable process made using the same (linear) device.

Notice that henceforth we will use the language "device" as shorthand for a fixed combination of physical measurement equipment, operator identity, measurement procedure, and surrounding physical circumstances (like time of day, temperature, etc.). We will also use the shorthand "y_i's $\sim$ ind(μ, σ)" for the model statement that observations are independent with mean μ and standard deviation σ. And in schematics like Figs. 2.6 and 2.7, the rulers will represent generic measurement devices, the spheres generic measurands, and the factories generic processes.

This Book's Use of the Word "Device"

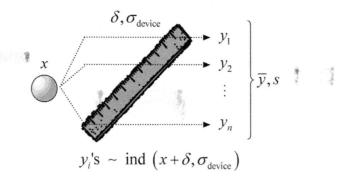

FIGURE 2.6. A single sample derived from n repeat measurements made with a fixed device on a single measurand. Modified from "Elementary Statistical Methods and Measurement Error" by S.B. Vardeman et al., 2010, *The American Statistician*, 64(1), 48. © 2010 Taylor & Francis. Adapted with permission

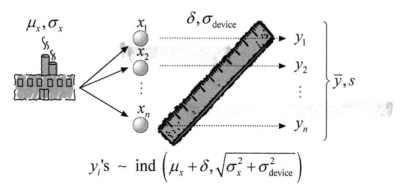

FIGURE 2.7. A single sample derived from single measurements made with a fixed (linear) device on each of n different measurands from a physically stable process. Modified from "Elementary Statistical Methods and Measurement Error" by S.B. Vardeman et al., 2010, *The American Statistician*, 64(1), 48. © 2010 Taylor & Francis. Adapted with permission

The case represented in Fig. 2.6 also corresponds to Fig. 2.3 (where "measurement" variation is simply that inherent in the reuse of the device to evaluate a given measurand). The case represented in Fig. 2.7 also corresponds to Fig. 2.5 (where again "measurement" variation is a variation inherent in the "device" and

42   Chapter 2. Statistics and Measurement

now real part-to-part variation is represented by σ_x). Consider what formulas (2.6) and (2.7) provide in the two situations.

First, if as in Fig. 2.6 n **repeat measurements of a single measurand**, $y_1, y_2, \ldots, y_n$, have sample mean $\bar{y}$ and sample standard deviation s, applying the t confidence interval for a mean, one gets an inference for

$$x + \delta = \text{measurand plus bias} .$$

So:

1. In the event that the measurement device is known to be well-calibrated (one is sure that $\delta = 0$, there is no systematic error), the limits $\bar{y} \pm ts/\sqrt{n}$ based on $\nu = n - 1$ df are limits for x,

2. In the event that what is being measured is a *standard for which x is known*, one may use the limits

$$(\bar{y} - x) \pm t \frac{s}{\sqrt{n}}$$

(once again based on $\nu = n - 1$ df) to estimate the device bias, δ.

Further, applying the χ^2 confidence interval for a standard deviation, one has an inference for the size of the device "noise," σ_{device}.

Next consider what can be inferred from **single measurements made on n different measurands** $y_1, y_2, \ldots, y_n$ from a stable process with sample mean $\bar{y}$ and sample standard deviation s as illustrated in Fig. 2.7. Here:

1. The limits $\bar{y} \pm ts/\sqrt{n}$ (for t based on $\nu = n - 1$ df) are limits for

$$\mu_x + \delta = \text{the mean of the distribution of true values plus (the constant) bias},$$

2. The quantity s estimates $\sigma_y = \sqrt{\sigma_x^2 + \sigma_{\text{device}}^2}$ that we met first in display (2.4) and have noted really isn't of fundamental interest. So there is little point in direct application of the χ^2 confidence limits (2.7) in this context.

Example 7 *Measuring Concentration.* Below are $n = 5$ consecutive concentration measurements made by a single analyst on a single physical specimen of material using a particular assay machine (the real units are not available, so for the sake of example, let's call them "moles per liter," mol/l):

$$1.0025, .9820, 1.0105, 1.0110, .9960$$

These have mean $\bar{y} = 1.0004 \, \text{mol}/\text{l}$ *and* $s = .0120 \, \text{mol}/\text{l}$. *Consulting an* χ^2 *table like Table A.3 using* $\nu = 5 - 1 = 4$ *df, we can find* 95 % *confidence limits for* σ_{device} *(the size of basic measurement variability) as*

$$.0120\sqrt{\frac{4}{11.143}} \text{ and } .0120\sqrt{\frac{4}{.484}} \text{ i.e., } .0072 \, \text{mol}/\text{l} \text{ and } .0345 \, \text{mol}/\text{l} .$$

(One moral here is that ordinary small sample sizes give very wide confidence limits for a standard deviation.) Consulting a t table like Table A.2 also using 4 df, we can find 95 % confidence limits for the measurand plus instrument bias $(x + \delta)$ *to be*

$$1.0004 \pm 2.776 \frac{.0120}{\sqrt{5}}, \text{ i.e., } 1.0004\,\text{mol}/\text{l} \pm .0149\,\text{mol}/\text{l}.$$

Note that if the measurements in question were done on a standard material "known" to have actual concentration $1.0000\,\text{mol}/\text{l}$, *these limits then correspond to limits for device bias of*

$$0.0004\,\text{mol}/\text{l} \pm .0149\,\text{mol}/\text{l}.$$

Finally, suppose that subsequently samples from $n = 20$ *different batches are analyzed and* $\overline{y} = .9954$ *and* $s_y = .0300$. *The 95 % t confidence limits*

$$.9954 \pm 2.093 \frac{.0300}{\sqrt{20}}, \text{ i.e., } .9954\,\text{mol}/\text{l} \pm .0140\,\text{mol}/\text{l}.$$

are for $\mu_x + \delta$, *the process mean plus any device bias/systematic error.*

Application to a Sample Consisting of Single Measurements of a Single Measurand Made Using Multiple Devices (From a Large Population of Such Devices)

The two cases illustrated by Figs. 2.6 and 2.7 do not begin to exhaust the ways that the basic formulas (2.6) and (2.7) can be applied. We present two more applications of the one-sample formulas, beginning with an application where single measurements of a single measurand are made using multiple devices (from a large population of such devices).

There are contexts in which an organization has many "similar" measurement devices that could potentially be used to do measuring. In particular, a given piece of equipment might well be used by any of a large number of operators. Recall that we are using the word "device" to describe a particular combination of equipment, people, procedures, etc. used to produce a measurement. So, in this language, different operators with a fixed piece of equipment are different "devices." A way to compare these devices would be to use some (say n of them) to measure a single measurand. This is illustrated in Fig. 2.8 on page 44.

In this context, a measurement is of the form

$$y = x + \epsilon,$$

where $\epsilon = \delta + \epsilon^*$, for δ the (randomly selected) bias of the device used and ϵ^* a measurement error with mean 0 and standard deviation σ_{device} (representing a repeat measurement variability for any one device). So one might write

$$y = x + \delta + \epsilon^*.$$

44   Chapter 2. Statistics and Measurement

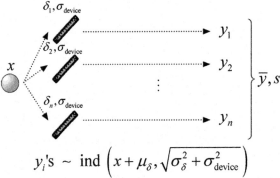

$$y_i\text{'s} \sim \text{ind}\left(x+\mu_\delta, \sqrt{\sigma_\delta^2 + \sigma_{\text{device}}^2}\right)$$

FIGURE 2.8. A single sample consisting of n single measurements of a fixed measurand made with each of n devices (from a large population of such devices with a common precision). Modified from "Elementary Statistical Methods and Measurement Error" by S.B. Vardeman et al., 2010, *The American Statistician*, 64(1), 49. © 2010 Taylor & Francis. Adapted with permission

Thinking of x as fixed and δ and ϵ^* as independent random variables (δ with mean μ_δ, the average device bias, and standard deviation σ_δ measuring variability in device biases), the laws of mean and variance from elementary probability then imply that

$$\mu_y = x + \mu_\delta + 0 = x + \mu_\delta \tag{2.8}$$

and

$$\sigma_y = \sqrt{0 + \sigma_\delta^2 + \sigma_{\text{device}}^2} = \sqrt{\sigma_\delta^2 + \sigma_{\text{device}}^2} \tag{2.9}$$

as indicated on Fig. 2.8. The theoretical average measurement is the measurand plus the average bias, and the variability in measurements comes from both the variation in device biases and the intrinsic imprecision of any particular device.

In a context where a schematic like Fig. 2.8 represents a study where several operators each make a measurement on the same item using a fixed piece of equipment, the quantity

$$\sqrt{\sigma_\delta^2 + \sigma_{\text{device}}^2}$$

is a kind of overall measurement variation that is sometimes called "$\sigma_{\text{R\&R}}$," the first "R" standing for **repeatability** and referring to σ_{device} (a variability for fixed operator on the single item) and the second "R" standing for **reproducibility** and referring to σ_δ (a between-operator variability).

With μ_y and σ_y identified in displays (2.8) and (2.9), it is clear what the one-sample confidence limits (2.6) and (2.7) estimate in this context. Of the two, interval (2.7) for "σ" is probably the most important, since σ_y is interpretable in the context of an R&R study, while μ_y typically has little practical meaning. It is another question (that we will address in future sections with more complicated methods) how one might go about separating the two components of σ_y to assess the relative sizes of repeatability and reproducibility variation.

Repeatability and Reproducibility

Application to a Sample Consisting of Differences in Measurements on Multiple Measurands Made Using Two Linear Devices

Another way to create a single sample of numbers is this. With two devices available and n different measurands, one might measure each once with both devices and create n differences between device 1 and device 2 measurements. This is a way of potentially comparing the two devices and is illustrated in Fig. 2.9.

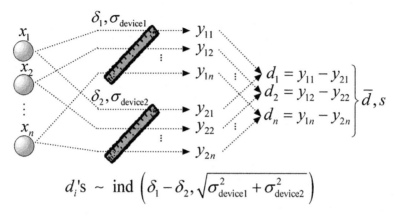

FIGURE 2.9. A single sample consisting of n differences of single measurements of n measurands made using two devices (assuming device linearity). Modified from "Elementary Statistical Methods and Measurement Error" by S.B. Vardeman et al., 2010, *The American Statistician,* 64(1), 49. © 2010 Taylor & Francis. Adapted with permission

In this context, a difference is of the form

$$d = y_1 - y_2 = (x + \epsilon_1) - (x + \epsilon_2) = \epsilon_1 - \epsilon_2$$

and (again applying the laws of mean and variance from elementary probability) it follows that

$$\mu_d = \delta_1 - \delta_2 \text{ and } \sigma_d = \sqrt{\sigma_{\text{device1}}^2 + \sigma_{\text{device2}}^2}$$

as indicated on Fig. 2.9. So applying the t interval for a mean (2.6), the limits

$$\bar{d} \pm t \frac{s}{\sqrt{n}} \qquad (2.10)$$

Confidence Limits for a Mean Difference

provide a way to estimate $\delta_1 - \delta_2$, the difference in device biases.

2.2.2 Two-Sample Methods and Measurement Error

Parallel to the one-sample formulas are the two-sample formulas of elementary statistics. These are based on a model that says that

$$y_{11}, y_{12}, \ldots, y_{1n_1} \text{ and } y_{21}, y_{22}, \ldots, y_{2n_2}$$

are independent samples from normal distributions with respective means μ_1 and μ_2 and respective standard deviations σ_1 and σ_2. In this context, the so-called Satterthwaite approximation gives limits

Confidence Limits for a Difference in Normal Means

$$\bar{y}_1 - \bar{y}_2 \pm \hat{t}\sqrt{\frac{s_1^2}{n_1} + \frac{s_2^2}{n_2}} \text{ for estimating } \mu_1 - \mu_2, \quad (2.11)$$

where appropriate "approximate degrees of freedom" for $\hat{t}$ are

Satterthwaite Degrees of Freedom for Formula (2.11)

$$\hat{\nu} = \frac{\left(\dfrac{s_1^2}{n_1} + \dfrac{s_2^2}{n_2}\right)^2}{\dfrac{s_1^4}{(n_1-1)n_1^2} + \dfrac{s_2^4}{(n_2-1)n_2^2}}. \quad (2.12)$$

(This method is one that you may not have seen in an elementary statistics course, where often only methods valid when one assumes that $\sigma_1 = \sigma_2$ are presented. We use this method not only because it requires less in terms of model assumptions than the more common formula but also because we will have other uses for the Satterthwaite idea in this chapter, so it might as well be met first in this simple context.) It turns out that $\min((n_1 - 1), (n_2 - 1)) \leq \hat{\nu}$, so that a simple conservative version of this method uses degrees of freedom

Conservative Simplification of Formula (2.12)

$$\hat{\nu}^* = \min((n_1 - 1), (n_2 - 1)). \quad (2.13)$$

Further, in the two-sample context, there are elementary confidence limits

Confidence Limits for a Ratio of Two Normal Standard Deviations

$$\frac{s_1}{s_2} \cdot \frac{1}{\sqrt{F_{(n_1-1),(n_2-1),\text{upper}}}} \text{ and } \frac{s_1}{s_2} \cdot \frac{1}{\sqrt{F_{(n_1-1),(n_2-1),\text{lower}}}} \text{ for } \frac{\sigma_1}{\sigma_2} \quad (2.14)$$

(and be reminded that $F_{(n_1-1),(n_2-1),\text{lower}} = 1/F_{(n_2-1),(n_1-1),\text{upper}}$ so that standard F tables giving only upper percentage points can be employed).

Application to Two Samples Consisting of Repeat Measurements of a Single Measurand Made Using Two Different Devices

One way to create "two samples" of measurements is to measure the same item repeatedly with two different devices. This possibility is illustrated in Fig. 2.10.

Direct application of the two-sample confidence interval formulas here shows that the two-sample Satterthwaite approximate t interval (2.11) provides limits for

$$\mu_1 - \mu_2 = (x + \delta_1) - (x + \delta_2) = \delta_1 - \delta_2$$

(the difference in device biases), while the F interval (2.14) provides a way of comparing device standard deviations σ_{device1} and σ_{device2} through direct estimation of

$$\frac{\sigma_{\text{device1}}}{\sigma_{\text{device2}}}.$$

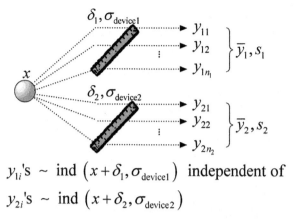

y_{1i}'s ~ ind $(x+\delta_1, \sigma_{device1})$ independent of
y_{2i}'s ~ ind $(x+\delta_2, \sigma_{device2})$

FIGURE 2.10. Two samples consisting of n_1 and n_2 measurements of a single measurand with two devices. Modified from "Elementary Statistical Methods and Measurement Error" by S.B. Vardeman et al., 2010, *The American Statistician*, 64(1), 49. © 2010 Taylor & Francis. Adapted with permission

This data collection plan thus provides for straightforward comparison of the basic characteristics of the two devices.

Example 8 *Measuring Polystyrene "Packing Peanut" Size.* In an in-class measurement exercise, two students used the same caliper to measure the "size" of a single polystyrene "packing peanut" according to a class-standard measurement protocol. Some summary statistics from their work follow.

Student 1	Student 2
$n_1 = 4$	$n_2 = 6$
$\bar{y}_1 = 1.42\,\text{cm}$	$\bar{y}_2 = 1.44\,\text{cm}$
$s_1 = .20\,\text{cm}$	$s_2 = .40\,\text{cm}$

In this context, the difference in the two measurement "devices" is the difference in "operators" making the measurements. Consider quantifying how this difference affects measurement.

To begin, note that from formula (2.12)

$$\hat{\nu} = \frac{\left(\frac{(.20)^2}{4} + \frac{(.40)^2}{6}\right)^2}{\frac{(.20)^4}{(4-1)(4)^2} + \frac{(.40)^4}{(6-1)(6)^2}} \approx 7.7,$$

or using the more conservative display (2.13), one gets

$$\hat{\nu}^* = \min\left((4-1),(6-1)\right) = 3.$$

So (rounding the first of these down to 7) one should use either 7 or 3 degrees of freedom with formula (2.11). For the sake of example, using $\hat{\nu}^* = 3$ degrees of freedom, the upper 2.5% point of the t distribution with 3 df is 3.182. So 95% confidence limits for the difference in biases for the two operators using this caliper are

$$1.42 - 1.44 \pm 3.182 \sqrt{\frac{(.20)^2}{4} + \frac{(.40)^2}{6}},$$

i.e.,

$$-.02 \text{ cm} \pm .61 \text{ cm}.$$

The apparent difference in biases is small in comparison with the imprecision associated with that difference.

Then, since the upper 2.5% point of the $F_{3,5}$ distribution is 7.764 and the upper 2.5% point of the $F_{5,3}$ distribution is 14.885, 95% confidence limits for the ratio of standard deviations of measurement for the two operators are

$$\frac{.20}{.40} \cdot \frac{1}{\sqrt{7.764}} \quad \text{and} \quad \frac{.20}{.40} \cdot \sqrt{14.885},$$

i.e.,

$$.19 \text{ and } 1.93.$$

Since this interval covers values both smaller and larger than 1.00, there is in the limited information available here no clear indicator of which of these students is the most consistent in his or her use of the caliper in this measuring task.

Comparing Devices When Measurement is Destructive

Application to Two Samples Consisting of Single Measurements Made with Two Devices on Multiple Measurands from a Stable Process (Only One Device Being Used for a Given Measurand)

There are quality assurance contexts in which measurement is **destructive** (and cannot be repeated for a single measurand) and nevertheless one needs to somehow compare two different devices. In such situations, the only thing that can be done is to take items from some large pool of items or from some stable process and (probably after randomly assigning them one at a time to one or the other of the devices) measure them and try to make comparisons based on the resulting samples. This possibility is illustrated in Fig. 2.11. This is a schematic for two samples consisting of single measurements made with two devices on multiple measurands from a stable process (only one device used for a given measurand).

Direct application of the two-sample Satterthwaite approximate t interval (2.11) provides limits for

$$\mu_1 - \mu_2 = (\mu_x + \delta_1) - (\mu_x + \delta_2) = \delta_1 - \delta_2$$

(the difference in device biases). So, even in contexts where measurement is destructive, it is possible to compare device biases. It's worth contemplating, however, the difference between the present scenario and the immediately preceding one (represented by Fig. 2.10).

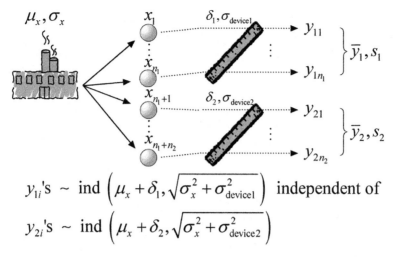

y_{1i}'s $\sim$ ind $\left(\mu_x + \delta_1, \sqrt{\sigma_x^2 + \sigma_{device1}^2}\right)$ independent of

y_{2i}'s $\sim$ ind $\left(\mu_x + \delta_2, \sqrt{\sigma_x^2 + \sigma_{device2}^2}\right)$

FIGURE 2.11. Two samples consisting of single measurements made on $n_1 + n_2$ measurands from a stable process, n_1 with device 1 and n_2 with device 2. Modified from "Elementary Statistical Methods and Measurement Error" by S.B. Vardeman et al., 2010, *The American Statistician,* 64(1), 50. © 2010 Taylor & Francis. Adapted with permission

The measurements y in Fig. 2.10 on page 47 are less variable than are the measurements y here in Fig. 2.11. This is evident in the standard deviations shown on the figures and follows from the fact that in the present case (unlike the previous one), measurements are affected by unit-to-unit/measurand-to-measurand variation. So all else being equal, one should expect limits (2.11) applied in the present context to be wider/less informative than when applied to data collected as in the last application. That should be in accord with intuition. One should expect to be able to learn more useful to comparing devices when the same item(s) can be remeasured than when it (they) cannot be remeasured.

Notice that if the F limits (2.14) are applied here, one winds up with only an indirect comparison of $\sigma_{device1}$ and $\sigma_{device2}$, since all that can be easily estimated (using the limits (2.14)) is the ratio

$$\frac{\sqrt{\sigma_x^2 + \sigma_{device1}^2}}{\sqrt{\sigma_x^2 + \sigma_{device2}^2}}$$

and NOT the (more interesting) ratio $\sigma_{device1}/\sigma_{device2}$.

Application to Two Samples Consisting of Repeat Measurements Made with One Device on Two Measurands

A basic activity of quality assurance is the comparison of nominally identical items. Accordingly, another way to create two samples is to make repeated measurements on two measurands with a single device. This is illustrated in Fig. 2.12 on page 50.

In this context,

$$\mu_1 - \mu_2 = (x_1 + \delta) - (x_2 + \delta) = x_1 - x_2$$

so that application of the two-sample Satterthwaite approximate t interval (2.11) provides limits for the *difference in the measurands* and a direct way of comparing the measurands. The device bias affects both samples in the same way and "washes out" when one takes a difference. (This, of course, assumes that the device is linear, i.e., that the bias is constant.)

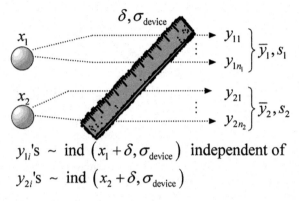

FIGURE 2.12. Two samples consisting of repeat measurements made with one device on two measurands. Modified from "Elementary Statistical Methods and Measurement Error" by S.B. Vardeman et al., 2010, *The American Statistician*, 64(1), 50. © 2010 Taylor & Francis. Adapted with permission

Application to Two Samples Consisting of Single Measurements Made Using a Single Linear Device on Multiple Measurands Produced by Two Stable Processes

Another basic activity of quality assurance is the comparison of nominally identical *processes*. Accordingly, another way to create two samples is to make single measurements on samples of measurands produced by two processes. This possibility is illustrated in Fig. 2.13.

In this context,

$$\mu_1 - \mu_2 = (\mu_{x1} + \delta) - (\mu_{x2} + \delta) = \mu_{x1} - \mu_{x2}$$

so that application of the two-sample Satterthwaite approximate t interval (2.11) provides limits for the *difference in the process mean measurands* and hence a direct way of comparing the processes. Again, the device bias affects both samples in the same way and "washes out" when one takes a difference (still assuming that the device is linear, i.e., that the bias is constant).

If the F limits (2.14) are applied here, one winds up with only an indirect comparison of σ_{x1} and σ_{x2}, since what can be easily estimated is the ratio

$$\frac{\sqrt{\sigma_{x1}^2 + \sigma_{device}^2}}{\sqrt{\sigma_{x2}^2 + \sigma_{device}^2}}$$

and not the practically more interesting σ_{x1}/σ_{x2}.

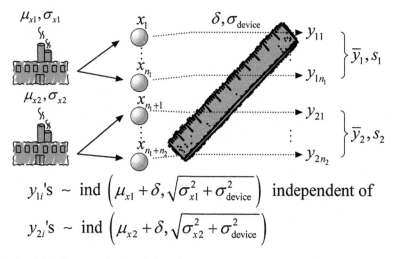

FIGURE 2.13. Two samples consisting of single measurements made using a single device on multiple measurands produced by two stable processes. Modified from "Elementary Statistical Methods and Measurement Error" by S.B. Vardeman et al., 2010, *The American Statistician*, 64(1), 50. © 2010 Taylor & Francis. Adapted with permission

---

Section 2.2 Exercises

1. Consider again the **Pellet Densification** case of problem 7 in Sect. 2.1. Suppose the five data values $6.5, 6.6, 4.9, 5.1$, and 5.4 were measured densities for a single pellet produced by five different operators using the same piece of measuring equipment (or by the same operator using five different pieces of equipment—the two scenarios are conceptually handled in the same way). Use the notation of this section ($x, \delta, \mu_\delta, \sigma_\delta$, and σ_{device}) below.

    (a) What does the sample average of these five data values estimate?

    (b) What does the sample standard deviation of these five data values estimate?

    (c) Which of the two estimates in (a) and (b) is probably more important? Why?

2. Return again to the context of problem 7 of Sect. 2.1. Suppose the original set of five data values $6.5, 6.6, 4.9, 5.1$, and 5.4 was obtained from five different pellets by operator 1 using piece of equipment 1. Using a second piece of equipment, operator 1 recorded densities $6.6, 5.7, 5.9, 6.2$, and 6.3 for the same five pellets. So, for pellet 1, "device 1" produced measurement 6.5 and "device 2" produced 6.6; for pellet 2, "device 1" produced measurement 6.6 and "device 2" produced 5.7 and so on.

(a) Give the five differences in measured densities (device 1 minus device 2). Calculate the sample average difference. What does this estimate? (Hint: Consider δs .)

(b) Calculate the sample standard deviation of the five differences (device 1 minus device 2). What does this estimate? (Hint: Consider the σ_{device}s.)

(c) Find 90 % confidence limits for the average difference in measurements from the two devices.

3. Suppose the two sets of five measurements referred to in problems 1 and 2 actually came from one pellet, i.e., operator 1 measured the same pellet five times with piece of equipment 1 and then measured that same pellet five times with piece of equipment 2.

   (a) Find a 95 % confidence interval for the ratio of the two device standard deviations ($\sigma_{\text{device1}}/\sigma_{\text{device2}}$). What do your limits indicate about the consistency of measurements from device 1 compared to that of measurements from device 2?

   (b) Find a 95 % two-sample Satterthwaite approximate t interval for the difference in the two device averages (device 1 minus device 2). If your interval were to include 0, what would you conclude regarding device biases 1 and 2?

4. Consider now the same ten data values referred to in problems 2 and 3, but a different data collection plan. Suppose the first five data values were measurements on five different pellets by operator 1 using piece of equipment 1 and the second set of data values was for another set of pellets by operator 1 using piece of equipment 2. Assume both sets of pellets came from the same physically stable process.

   (a) What does the sample standard deviation from the first set of five data values estimate?

   (b) What does the sample standard deviation from the second set of five data values estimate?

   (c) What does the difference in the two-sample average densities estimate?

5. Reflect on problems 3 and 4. Which data-taking approach is better for estimating the difference in device biases? Why?

6. In the same **Pellet Densification** context considered in problems 1 through 5, suppose one pellet was measured five times by operator 1 and a different pellet was measured five times by operator 1 (the same physical equipment was used for the entire set of ten observations). What is estimated by the difference in the two sample averages?

7. Once again in the context of problems 1 through 6, suppose the first five data values were measurements on five different pellets made by operator 1 using piece of equipment 1 and the second five were measurements of a different set of pellets by operator 1 using piece of equipment 1. Assume the two sets of pellets come from different firing methods (method 1 and method 2). Assume the two firing processes are physically stable.

   (a) Find the two-sided 95 % two-sample Satterthwaite approximate t interval for the difference in the process mean measurands (method 1 minus method 2).

   (b) In words, what does the interval in (a) estimate? In symbols, what does the interval in (a) estimate?

   (c) With this approach to data taking, can either device bias be estimated directly? Why or why not?

8. Still in the context of problems 1 through 7, density measurements 6.5, 6.6, 4.9, 5.1, and 5.4 were obtained for five different pellets by a single operator using a single piece of measuring equipment under a standard protocol and fixed physical circumstances. Use the t confidence interval for a mean, and give 95 % confidence limits for the mean of the distribution of true densities plus measurement bias.

9. Suppose the five measurements in problem 8 are repeat measurements from only one pellet, not from five different pellets.

   (a) Use the χ^2 confidence limits for a standard deviation (from elementary statistics), and give a 95 % confidence interval for $\sigma_{\text{measurement}}$.

   (b) Use the t confidence interval formula for a mean from elementary statistics and give 95 % confidence limits for the (single) true pellet density plus measurement bias.

2.3 Some Intermediate Statistical Methods and Measurement

Through reference to familiar elementary one- and two-sample methods of statistical inference, Sect. 2.2 illustrated the basic insight that:

> How sources of physical variation interact with a data collection plan governs what of practical importance can be learned from a data set, and in particular, how measurement error is reflected in the data set.

In this section we consider some computationally more complicated statistical methods and what they provide in terms of quantification of the impact of measurement variation on quality assurance data.

2.3.1 A Simple Method for Separating Process and Measurement Variation

In Sect. 2.1 we essentially observed that:

1. Repeated measurement of a single measurand with a single device allows one to estimate device variability,

2. Single measurements made on multiple measurands from a stable process allow one to estimate a combination of process and measurement variability,

and remarked that these facts suggest formula (2.5) as a way to estimate a process standard deviation alone. Our first objective in this section is to elaborate a bit on this thinking.

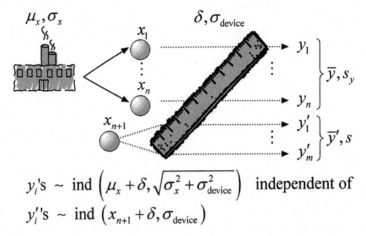

y_i's $\sim$ ind $\left(\mu_x + \delta, \sqrt{\sigma_x^2 + \sigma_{\text{device}}^2}\right)$ independent of

y_i''s $\sim$ ind $\left(x_{n+1} + \delta, \sigma_{\text{device}}\right)$

FIGURE 2.14. Schematic of a data collection plan that allows evaluation of σ_x without inflation by measurement variation

Figure 2.14 is a schematic of a data collection plan that combines elements 1 and 2 above. Here we use the notation y for the single measurements on n items from the process and the notation y' for the m repeat measurements on a single measurand. The sample standard deviation of the ys, s_y, is a natural empirical approximation for $\sigma_y = \sqrt{\sigma_x^2 + \sigma_{\text{device}}^2}$ and the sample standard deviation of the y''s, s, is a natural empirical approximation for σ_{device}. That suggests that one estimates the process standard deviation with

Estimator of Process Standard Deviation not Inflated by Measurement Variability

$$\hat{\sigma}_x = \sqrt{\max\left(0, s_y^2 - s^2\right)} \qquad (2.15)$$

as indicated in display (2.5). (The maximum of 0 and $s_y^2 - s^2$ under the root is there simply to ensure that one is not trying to take the square root of a negative number in the rare case that s exceeds s_y.) $\hat{\sigma}_x$ is not only a sensible single-number estimate of σ_x, but can also be used to make approximate confidence limits for the process standard deviation. The so-called Satterthwaite approximation suggests that one uses

$$\hat{\sigma}_x\sqrt{\frac{\hat{\nu}}{\chi^2_{\text{upper}}}} \quad \text{and} \quad \hat{\sigma}_x\sqrt{\frac{\hat{\nu}}{\chi^2_{\text{lower}}}} \qquad (2.16)$$

Satterthwaite Approximate Confidence Limits for a Process Standard Deviation

as limits for σ_x, where appropriate approximate degrees of freedom $\hat{\nu}$ to be used finding χ^2 percentage points are

$$\hat{\nu} = \frac{\hat{\sigma}_x^4}{\frac{s_y^4}{n-1} + \frac{s^4}{m-1}} \qquad (2.17)$$

Satterthwaite Approximate df for Use With Limits (2.16)

Example 9 *(Example 7 Revisited.)* *In Example 7, we considered $m = 5$ measurements made by a single analyst on a single physical specimen of material using a particular assay machine that produced $s = .0120\,\text{mol}/\text{l}$. Subsequently, specimens from $n = 20$ different batches were analyzed and $s_y = .0300\,\text{mol}/\text{l}$. Using formula (2.15), an estimate of real process standard deviation uninflated by measurement variation is*

$$\hat{\sigma}_x = \sqrt{\max\left(0, (.0300)^2 - (.0120)^2\right)} = .0275\,\text{mol}/\text{l}$$

and this value can be used to make confidence limits. By formula (2.17) approximate degrees of freedom are

$$\hat{\nu} = \frac{(.0275)^4}{\frac{(.0300)^4}{19} + \frac{(.0120)^4}{4}} = 11.96\ .$$

So rounding down to $\hat{\nu} = 11$, since the upper 2.5 % point of the χ^2_{11} distribution is 21.920 and the lower 2.5 % point is 3.816, by formula (2.16) approximate 95 % confidence limits for the real process standard deviation (σ_x) are

$$.0275\sqrt{\frac{11}{21.920}} \quad \text{and} \quad .0275\sqrt{\frac{11}{3.816}},$$

i.e.,

$$.0195\,\text{mol}/\text{l} \ \ \text{and} \ \ .0467\,\text{mol}/\text{l}\ .$$

2.3.2 One-Way Random Effects Models and Associated Inference

One of the basic models of intermediate statistical methods is the so-called one-way random effects model for I samples of observations

$$y_{11}, y_{12}, \ldots, y_{1n_1}$$
$$y_{21}, y_{22}, \ldots, y_{2n_2}$$
$$\vdots$$
$$y_{I1}, y_{I2}, \ldots, y_{In_I}$$

This model says that the observations may be thought of as

One-Way Random Effects Model

$$y_{ij} = \mu_i + \epsilon_{ij}$$

where the ϵ_{ij} are independent normal random variables with mean 0 and standard deviation σ, while the I values μ_i are independent normal random variables with mean μ and standard deviation σ_μ (independent of the ϵs). (One can think of I means μ_i drawn at random from a normal distribution of μ_is and subsequently observations y generated from I different normal populations with those means and a common standard deviation.) In this model, the three parameters are σ (the "within-group" standard deviation), σ_μ (the "between-group" standard deviation), and μ (the overall mean). The squares of the standard deviations are called "variance components" since for any particular observation, the laws of expectation and variance imply that

$$\mu_y = \mu + 0 = \mu \text{ and } \sigma_y^2 = \sigma_\mu^2 + \sigma^2$$

(i.e., σ_μ^2 and σ^2 are components of the variance of y).

Two quality assurance contexts where this model can be helpful are where

1. Multiple measurands from a stable process are each measured multiple times using the same device,

2. A single measurand is measured multiple times using multiple devices.

These two scenarios and the accompanying parameter values are illustrated in Figs. 2.15 and 2.16.

There are well-established (but not altogether simple) methods of inference associated with the one-way random effects model that can be applied to make confidence intervals for the model parameters (and inferences of practical interest in metrological applications). Some of these are based on so-called ANOVA methods and the one-way ANOVA identity that says that with

$$\bar{y}_i = \frac{1}{n_i} \sum_j y_{ij}, n = \sum_i n_i, \text{ and } \bar{y} = \frac{1}{n} \sum_i n_i \bar{y}_i,$$

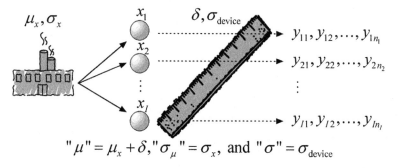

$$"\mu" = \mu_x + \delta, "\sigma_\mu" = \sigma_x, \text{ and } "\sigma" = \sigma_{\text{device}}$$

FIGURE 2.15. Multiple measurands from a stable process each measured multiple times using the same device

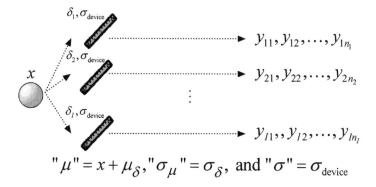

$$"\mu" = x + \mu_\delta, "\sigma_\mu" = \sigma_\delta, \text{ and } "\sigma" = \sigma_{\text{device}}$$

FIGURE 2.16. A single measurand measured multiple times using multiple devices

it is the case that

$$\sum_{i,j}(y_{ij} - \overline{y})^2 = \sum_i n_i(\overline{y}_i - \overline{y})^2 + \sum_{i,j}(y_{ij} - \overline{y}_i)^2 \qquad (2.18)$$

One-Way ANOVA Identity

or in shorthand "sum of squares" notation

$$SSTot = SSTr + SSE \qquad (2.19)$$

One-Way ANOVA Identity in Sum of Squares Notation

$SSTot$ is a measure of overall raw variability in the whole data set. $SSTot$ is $n-1$ times the overall sample variance computed ignoring the boundaries between samples. SSE is a measure of variability left unaccounted for after taking account of the sample boundaries and is a multiple of a weighted average of the I sample variances. $SSTr$ is a measure of variation in the sample means $\overline{y}_{i\cdot}$ and is most simply thought of as the difference $SSTot - SSE$. The "sums of squares" SSE and $SSTr$ have respective associated degrees of freedom $n - I$ and $I - 1$. The ratios of sums of squares to their degrees of freedom are called "mean squares" and symbolized as

$$MSE = \frac{SSE}{n - I} \text{ and } MSTr = \frac{SSTr}{I - 1}. \qquad (2.20)$$

Confidence limits for the parameter σ^2 of the one-way random effects model can be built on the error mean square. A single-number estimate of σ is

One-Way ANOVA Estimator of σ

$$\hat{\sigma} = \sqrt{MSE} \qquad (2.21)$$

and confidence limits for σ are

One-Way ANOVA-Based Confidence Limits for σ

$$\hat{\sigma}\sqrt{\frac{n-I}{\chi^2_{\text{upper}}}} \quad \text{and} \quad \hat{\sigma}\sqrt{\frac{n-I}{\chi^2_{\text{lower}}}} \qquad (2.22)$$

where the appropriate degrees of freedom are $\nu = n - I$. Further, in the case that all n_is are the same, i.e., $n_i = m$ for all i, the Satterthwaite approximation can be used to make fairly simple approximate confidence limits for σ_μ. That is, a single-number estimator of σ_μ is

One-Way ANOVA-Based Estimator for σ_μ

$$\hat{\sigma}_\mu = \sqrt{\frac{1}{m} \max\left(0, MSTr - MSE\right)}, \qquad (2.23)$$

and with approximate degrees of freedom

Satterthwaite Approximate df for Use with Limits (2.25)

$$\hat{\nu} = \frac{m^2 \cdot \hat{\sigma}_\mu^4}{\frac{MSTr^2}{I-1} + \frac{MSE^2}{n-I}} \qquad (2.24)$$

approximate confidence limits for σ_μ are

One-Way ANOVA-Based Confidence Limits for σ_μ

$$\hat{\sigma}_\mu \sqrt{\frac{\hat{\nu}}{\chi^2_{\text{upper}}}} \quad \text{and} \quad \hat{\sigma}_\mu \sqrt{\frac{\hat{\nu}}{\chi^2_{\text{lower}}}}. \qquad (2.25)$$

Operationally, the mean squares implicitly defined in displays (2.18) through (2.20) are rarely computed "by hand." And given that statistical software is going to be used, rather than employ the methods represented by formulas (2.21) through (2.25), more efficient methods of confidence interval estimation can be used. High-quality statistical software (like the open-source command line-driven R package or the commercial menu-driven JMP package) implements the best known methods of estimation of the parameters σ, σ_μ, and μ (based not on ANOVA methods, but instead on computationally more difficult REML methods) and prints out confidence limits directly.

Example 10 *Part Hardness.* *Below are $m = 2$ hardness values (in mm) measured on each of $I = 9$ steel parts by a single operator at a farm implement manufacturer.*

Part	1	2	3	4	5	6	7	8	9
	3.30	3.20	3.20	3.25	3.25	3.30	3.15	3.25	3.25
	3.30	3.25	3.30	3.30	3.30	3.30	3.20	3.20	3.30

This is a scenario of the type illustrated in Fig. 2.15. Either working "by hand" with formulas (2.18) through (2.20) or reading directly off a report from a statistical package,

$$MSE = .001389 \text{ and } MSTr = .003368$$

So using formulas (2.21) and (2.22) (here $n = mI = 18$ so that error degrees of freedom are $n - I = 18 - 9 = 9$), 95 % confidence limits for σ_{device} (= σ here) are

$$\sqrt{.001389}\sqrt{\frac{9}{19.023}} \text{ and } \sqrt{.001389}\sqrt{\frac{9}{2.700}},$$

i.e.,

$$.026 \text{ mm } and \text{ } .068 \text{ mm }.$$

Further, using formulas (2.23) through (2.25), Satterthwaite degrees of freedom for $\hat{\sigma}_\mu$ are

$$\hat{\nu} = \frac{(2^2)\left(\frac{1}{2}(.003368 - .001389)\right)^2}{\frac{(.003368)^2}{9-1} + \frac{(.001389)^2}{18-9}} \approx 2.4$$

and rounding down to 2 degrees of freedom, approximate 95 % confidence limits for σ_x (= σ_μ here) are

$$\sqrt{\frac{1}{2}(.003368 - .001389)}\sqrt{\frac{2}{7.378}} \text{ and } \sqrt{\frac{1}{2}(.003368 - .001389)}\sqrt{\frac{2}{.051}},$$

i.e.,

$$.016 \text{ mm } and \text{ } .197 \text{ mm }.$$

The JMP package (using REML methods instead of the Satterthwaite approximation based on ANOVA mean squares) produces limits for σ_x:

$$0 \text{ mm } and \text{ } \sqrt{.0027603} = .053 \text{ mm }.$$

These more reliable limits at least confirm that the simpler methods "get into the right ballpark" in this example.

What is clear from this analysis is that this is a case where part-to-part variation in hardness (measured by σ_x) is small enough and poorly determined enough in comparison with basic measurement noise (measured by σ_{device} estimated as $.03726 = \sqrt{.001389}$) that it is impossible to really tell its size.

Example 11 *Paper Weighing.* *Below are $m = 3$ measurements of the weight (in g) of a single* $20 \text{ cm} \times 20 \text{ cm}$ *piece of* 20 lb *bond paper made by each of $I = 5$ different technicians using a single balance.*

Operator	1	2	3	4	5
	3.481	3.448	3.485	3.475	3.472
	3.477	3.472	3.464	3.472	3.470
	3.470	3.470	3.477	3.473	3.474

This is a scenario of the type illustrated in Fig. 2.16 and further illustrates the concepts of repeatability (fixed device) variation and reproducibility (here, device-to-device, i.e., operator-to-operator) variation first discussed on page 44. Use of the JMP *statistical package (and REML estimation) with these data produces 95 % confidence limits for the two standard deviations σ_δ ($= \sigma_\mu$ here) and σ_{device} ($= \sigma$ here). These place*

$$0 < \sigma_\delta < \sqrt{4.5 \times 10^{-5}} = .0067\,\text{g}$$

and

$$.0057\,\text{g} = \sqrt{3.2 \times 10^{-5}} < \sigma_{device} < \sqrt{.0002014} = .0142\,\text{g}$$

with 95 % confidence. This is a case where repeatability variation is clearly larger than reproducibility (operator-to-operator) variation in weight measuring. If one doesn't like the overall size of measurement variation, it appears that some fundamental change in equipment or how it is used will be required. Simple training of the operators aimed at making how they use the equipment more uniform (and reduction of differences between their biases) has far less potential to improve measurement precision.

Section 2.3 Exercises

1. **Fiber Angle**. Grunig, Hamdorf, Herman, and Potthoff studied a carpet-like product. Fiber angles (to the backing) were of interest. Operator 1 obtained the values 19, 20, 20, and 23 (in degrees) from four measurements of fiber angle for a single specimen. This same operator then measured fiber angles once each for three other specimens of the "carpet" and obtained the values 20, 15, and 23.

    (a) Using the methods of this section, give an estimate of the specimen-to-specimen standard deviation of fiber angle.

    (b) Give the appropriate "approximate degrees of freedom" associated with your estimate from (a). Then find a 95 % confidence interval for the specimen-to-specimen fiber angle standard deviation.

2. Continue with the **Fiber Angle** case of problem 1. Operator 2 obtained the fiber angle measurements 20, 25, 17, and 22 from the first specimen mentioned in problem 1 and operator 3 obtained the values 20, 19, 15, and 16. (Fiber angle for the same specimen was measured four times by each of the three operators.) As before, all measurements were in degrees. The data summaries below are from the use of the JMP statistical package with these $n = 12$ measurements of fiber angle for this specimen. Use them to answer (a) through (c). (The estimates and confidence intervals in the

second table are for variances, not standard deviations. You will need to take square roots to get inferences for standard deviations.)

ANOVA table

Source	SS	df	MS
Operator	28.66	2	14.33
Error	60	9	6.66
Total	88.66	11	

REML variance component analysis

Random effect	VarComponent	95 % lower	95 % upper
Operator	1.92	−5.27	9.11
Error	6.66	3.15	22.22

(a) Give an appropriate single-number estimate of $\sigma_{\text{repeatability}}$. Determine 95 % confidence limits for device (repeatability) standard deviation, $\sigma_{\text{repeatability}}$.

(b) From the computer output, give the appropriate estimate of $\sigma_{\text{reproducibility}}$. Give 95 % confidence limits for $\sigma_{\text{reproducibility}}$.

(c) Based on your answers to (a) and (b), where would you focus measurement improvement efforts?

3. Continuing with the **Fiber Angle** case, in addition to the repeat measurements $19, 20, 20$, and 23 made by operator 1 on specimen 1, this person also measured angles on two other specimens. Four angle measurements on specimen 2 were $15, 17, 20$, and 20, and four angle measurements on specimen 3 were $23, 20, 22$, and 20. The data summaries below are from the use of the JMP statistical package with these $n = 12$ measurements for these three specimens. Use them to answer (a) through (c). (The estimates and confidence intervals in the second table are for variances, not standard deviations. You will need to take square roots to get inferences for standard deviations.)

ANOVA table

Source	SS	df	MS
Specimen	23.17	2	11.58
Error	33.75	9	3.75
Total	56.92	11	

REML variance component analysis

Random effect	VarComponent	95 % lower	95 % upper
Specimen	1.96	3.78	7.69
Error	3.75	1.77	12.5

(a) Give an appropriate single-number estimate of σ_{device}. Determine 95 % confidence limits for device variation, σ_{device}.

(b) From the computer output, give an appropriate estimate of σ_x. Give 95 % confidence limits for σ_x.

(c) Based on your answers to (a) and (b), does it seem possible to determine fiber angle for a fixed specimen with acceptable precision? (Hint: Consider the sizes of the estimated σ_{device} and σ_x.)

2.4 Gauge R&R Studies

We have twice made some discussion of "gauge R&R," first on page 44 in the context of comparison of two operators and then in Example 11, where three operators were involved. In both cases, only a single part (or measurand) was considered. In a typical industrial gauge R&R study, each of J operators uses the same gauge or measurement system to measure each of I parts (common to all operators) a total of m different times. Variation in measurement typical of that seen in the m measurements for a particular operator on a particular part is called the **repeatability** variation of the gauge. Variation which can be attributed to differences between the J operators is called **reproducibility** variation of the measurement system.

This section considers the analysis of such full-blown gauge R&R studies involving a total of mIJ measurements. We begin with a discussion of the two-way random effects model that is commonly used to support analyses of gauge R&R data. Then primarily for ease of exposition and making connections to common analyses of gauge R&R studies, we discuss some range-based statistical methods. Finally, we provide what are really superior analyses, based on ANOVA calculations.

2.4.1 Two-Way Random Effects Models and Gauge R&R Studies

Typical industrial gauge R&R data are conveniently thought of as laid out in the cells of a table with I rows corresponding to parts and J columns corresponding to operators.

Example 12 *Gauge R&R for a 1-Inch Micrometer Caliper. Heyde, Kuebrick, and Swanson conducted a gauge R&R study on a certain micrometer caliper as part of a class project. Table 2.1 shows data that the $J = 3$ (student) operators obtained, each making $m = 3$ measurements of the heights of $I = 10$ steel punches.*

Notice that even for a given punch/student combination, measured heights are not exactly the same. Further, it is possible to verify that averaging the 30 measurements made by student 1, a mean of about .49853 in is obtained, while corresponding means for students 2 and 3 are, respectively, about .49813 in and .49840 in. Student 1 may tend to measure slightly higher than students 2 and

TABLE 2.1. Measured heights of ten steel punches in 10^{-3} in

Punch	Student 1	Student 2	Student 3
1	496, 496, 499	497, 499, 497	497, 498, 496
2	498, 497, 499	498, 496, 499	497, 499, 500
3	498, 498, 498	497, 498, 497	496, 498, 497
4	497, 497, 498	496, 496, 499	498, 497, 497
5	499, 501, 500	499, 499, 499	499, 499, 500
6	499, 498, 499	500, 499, 497	498, 498, 498
7	503, 499, 502	498, 499, 499	500, 499, 502
8	500, 499, 499	501, 498, 499	500, 501, 499
9	499, 500, 499	500, 500, 498	500, 499, 500
10	497, 496, 496	500, 494, 496	496, 498, 496

3. That is, by these rough "eyeball" standards, there is some hint in these data of both repeatability and reproducibility components in the overall measurement imprecision.

To this point in our discussions of R&R, we have not involved more than a single measurand. Effectively, we have confined attention to a single row of a table like Table 2.1. Standard industrial gauge R&R studies treat multiple parts (partially as a way of making sure that reliability of measurement doesn't obviously vary wildly across parts). So here we consider the kind of multiple-part case represented in Table 2.1.

The model most commonly used in this context is the so-called two-way random effects model that can be found in many intermediate-level statistical method texts. (See, e.g., Section 8.4 of Vardeman's *Statistics for Engineering Problem Solving*.) Let

y_{ijk} = the kth measurement made by operator j on part i .

The model is

$$y_{ijk} = \mu + \alpha_i + \beta_j + \alpha\beta_{ij} + \epsilon_{ijk} , \qquad (2.26)$$

Two-Way Random Effects Model

where the μ is an (unknown) constant, the α_i are normal random variables with mean 0 and variance σ_α^2, the β_j are normal random variables with mean 0 and variance σ_β^2, the $\alpha\beta_{ij}$ are normal random variables with mean 0 and variance $\sigma_{\alpha\beta}^2$, the ϵ_{ijk} are normal random variables with mean 0 and variance σ^2, and all of the αs, βs, $\alpha\beta$s, and ϵs are independent. In this model, the unknown constant μ is an average (over all possible operators and all possible parts) measurement, the αs are (random) effects of different parts, the βs are (random) effects of different operators, the $\alpha\beta$s are (random) joint effects peculiar to particular part×operator combinations, and the ϵs are (random) measurement errors. The variances $\sigma_\alpha^2, \sigma_\beta^2, \sigma_{\alpha\beta}^2$, and σ^2 are called "variance components" and their sizes govern how much variability is seen in the measurements y_{ijk}.

64   Chapter 2. Statistics and Measurement

Consider a hypothetical case with $I = 2$, $J = 2$, and $m = 2$. Model (2.26) says that there is a normal distribution with mean 0 and variance σ_α^2 from which α_1 and α_2 are drawn. And there is a normal distribution with mean 0 and variance σ_β^2 from which β_1 and β_2 are drawn. And there is a normal distribution with mean 0 and variance $\sigma_{\alpha\beta}^2$ from which $\alpha\beta_{11}$, $\alpha\beta_{12}$, $\alpha\beta_{21}$, and $\alpha\beta_{22}$ are drawn. And there is a normal distribution with mean 0 and variance σ^2 from which eight ϵs are drawn. Then these realized values of the random effects are added to produce the eight measurements as indicated in Table 2.2.

TABLE 2.2. Measurements in a hypothetical gauge R&R study

	Operator 1	Operator 2
Part 1	$y_{111} = \mu + \alpha_1 + \beta_1 + \alpha\beta_{11} + \epsilon_{111}$	$y_{121} = \mu + \alpha_1 + \beta_2 + \alpha\beta_{12} + \epsilon_{121}$
	$y_{112} = \mu + \alpha_1 + \beta_1 + \alpha\beta_{11} + \epsilon_{112}$	$y_{122} = \mu + \alpha_1 + \beta_2 + \alpha\beta_{12} + \epsilon_{122}$
Part 2	$y_{211} = \mu + \alpha_2 + \beta_1 + \alpha\beta_{21} + \epsilon_{211}$	$y_{221} = \mu + \alpha_2 + \beta_2 + \alpha\beta_{22} + \epsilon_{221}$
	$y_{212} = \mu + \alpha_2 + \beta_1 + \alpha\beta_{21} + \epsilon_{212}$	$y_{222} = \mu + \alpha_2 + \beta_2 + \alpha\beta_{22} + \epsilon_{222}$

Repeatability Standard Deviation in the Two-Way Model

Either directly from Eq. (2.26) or as illustrated in Table 2.2, according to the two-way random effects model, the only differences between measurements for a fixed part×operator combination are the measurement errors ϵ. And the variability of these is governed by the parameter σ. That is, σ is a measure of repeatability variation in this model, and one objective of an analysis of gauge R&R data is to estimate it.

Then, if one looks at a fixed "part i" (row i), the quantity $\mu + \alpha_i$ is common across the row. In the context of a gauge R&R study this can be interpreted as the value of the ith measurand (these vary across parts/rows because the α_i vary). Then, still for a fixed part i, it is the values $\beta_j + \alpha\beta_{ij}$ that vary column/operator to column/operator. So in this gauge R&R context, this quantity functions as a kind of *part-i-specific operator bias*. (More on the qualifier "part i specific" in a bit.) According to model (2.26), the variance of $\beta_j + \alpha\beta_{ij}$ is $\sigma_\beta^2 + \sigma_{\alpha\beta}^2$, so an appropriate measure of reproducibility variation in this model is

Reproducibility Standard Deviation in the Two-Way Model

$$\sigma_{\text{reproducibility}} = \sqrt{\sigma_\beta^2 + \sigma_{\alpha\beta}^2}. \tag{2.27}$$

According to the model, this is the standard deviation that would be experienced by many operators making a single measurement on the same part *assuming that there is no repeatability component to the overall variation*. Another way to say the same thing is to recognize this quantity as the standard deviation that would be experienced computing with long-run average measurements for many operators on the same part. That is, the quantity (2.27) is a measure of variability in operator bias for a fixed part in this model.

As long as one confines attention to a single row of a standard gauge R&R study, the one-way random effects model and analysis of Sect. 2.3 are relevant. The quantity $\sigma_{\text{reproducibility}}$ here is exactly σ_δ from application of the one-way model to a single-part gauge R&R study. (And the present σ is exactly σ_{device}.) What is new and at first perhaps a bit puzzling is that in the present context of multiple parts and display (2.27), the reproducibility variation has two components, σ_β and $\sigma_{\alpha\beta}$. This is because for a given part i, the model says that bias for operator

j has both components β_j and $\alpha\beta_{ij}$. The model terms $\alpha\beta_{ij}$ allow "operator bias" to change part to part/measurand to measurand (an issue that simply doesn't arise in the context of a single-part study). As such, they are *a measure of nonlinearity* (bias nonconstant in the measurand) in the overall measurement system. Two-way data like those in Table 2.1 allow one to estimate all of $\sigma_{\text{reproducibility}}, \sigma_\beta$ and $\sigma_{\alpha\beta}$, and all else being equal, cases where the $\sigma_{\alpha\beta}$ component of $\sigma_{\text{reproducibility}}$ is small are preferable to those where it is large.

The quantity

$$\sigma_{\text{R\&R}} = \sqrt{\sigma_\beta^2 + \sigma_{\alpha\beta}^2 + \sigma^2} = \sqrt{\sigma_{\text{reproducibility}}^2 + \sigma^2} \qquad (2.28)$$

Combined R&R Standard Deviation

is the standard deviation implied by the model (2.26) for many operators each making a single measurement on the same part. That is, quantity (2.28) is a measure of the combined imprecision in measurement attributable to *both* repeatability and reproducibility sources. And one might think of

$$\frac{\sigma^2}{\sigma_{\text{R\&R}}^2} = \frac{\sigma^2}{\sigma_\beta^2 + \sigma_{\alpha\beta}^2 + \sigma^2} \quad \text{and} \quad \frac{\sigma_{\text{reproducibility}}^2}{\sigma_{\text{R\&R}}^2} = \frac{\sigma_\beta^2 + \sigma_{\alpha\beta}^2}{\sigma_\beta^2 + \sigma_{\alpha\beta}^2 + \sigma^2} \qquad (2.29)$$

as the fractions of total measurement variance due, respectively, to repeatability and reproducibility. If one can produce estimates of σ and $\sigma_{\text{reproducibility}}$, estimates of these quantities (2.28) and (2.29) follow in straightforward fashion.

It is common to treat some multiple of $\sigma_{\text{R\&R}}$ (often the multiplier is six, but sometimes 5.15 is used) as a kind of uncertainty associated with a measurement made using the gauge or measurement system in question. And when a gauge is being used to check conformance of a part dimension or other measured characteristics to engineering specifications (say, some lower specification L and some upper specification U), this multiple is compared to the spread in specifications. **Specifications** U and L are numbers set by product design engineers that are supposed to delineate what is required of a measured dimension in order that the item in question be functional. The hope is that measurement uncertainty is at least an order of magnitude smaller than the spread in specifications. Some organizations go so far as to call the quantity

Engineering Specifications

$$GCR = \frac{6\sigma_{\text{R\&R}}}{U - L} \qquad (2.30)$$

Gauge Capability Ratio

a **gauge capability (or precision-to-tolerance) ratio** and require that it be no larger than .1 (and preferably as small as .01) before using the gauge for checking conformance to such specifications. (In practice, one will only have an estimate of $\sigma_{\text{R\&R}}$ upon which to make an empirical approximation of a gauge capability ratio.)

2.4.2 Range-Based Estimation

Because range-based estimation (similar to, but not exactly the same as, what follows) is in common use for the analysis of gauge R&R studies and is easy to describe, we will treat it here. In the next subsection, better methods based on ANOVA calculations (and REML methods) will be presented.

Consider first the estimation of σ. Restricting attention to any particular part×operator combination, say part i and operator j, model (2.26) says that observations obtained for that combination differ only by independent normal random measurement error with mean 0 and variance σ^2. That suggests that a measure of variability for the ij sample might be used as the basis of an estimator of σ. Historical precedent and ease of computation suggest measuring variability using a range (instead of a sample standard deviation or variance).

So let R_{ij} be the range of the m measurements on part i by operator j. The expected value of the range of a sample from a normal distribution is a constant (depending upon m) times the standard deviation of the distribution being sampled. The constants are well known and called d_2. (We will write $d_2(m)$ to emphasize their dependence upon m and note that values of $d_2(m)$ are given in Table A.5.) It then follows that

$$ER_{ij} = d_2(m)\sigma,$$

which in turn suggests that the ratio

$$\frac{R_{ij}}{d_2(m)}$$

is a plausible estimator of σ. Better yet, one might average these over all $I \times J$ part×operator combinations to produce the range-based estimator of σ:

Range-Based Estimator for Repeatability Standard Deviation

$$\hat{\sigma}_{\text{repeatability}} = \frac{\overline{R}}{d_2(m)}. \quad (2.31)$$

Example 13 (Example 12 continued.) Subtracting the smallest measurement for each part×operator combination in Table 2.1 from the largest for that combination, one obtains the ranges in Table 2.3. These have mean $\overline{R} = 1.9$. From Table A.5, $d_2(3) = 1.693$. So using expression (2.31), an estimate of σ, the repeatability standard deviation for the caliper used by the students, is

$$\hat{\sigma}_{\text{repeatability}} = \frac{\overline{R}}{d_2(3)} = \frac{1.9}{1.693} = 1.12 \times 10^{-3} \text{ in}.$$

(Again, this is an estimate of the (long-run) standard deviation that would be experienced by any particular student measuring any particular punch many times.)

Consider now the standard deviation (2.27) representing the reproducibility portion of the gauge imprecision. It will be convenient to have some additional notation. Let

$$\overline{y}_{ij} = \text{the (sample) mean measurement made on part } i \text{ by operator } j \quad (2.32)$$

and

$$\Delta_i = \max_j \overline{y}_{ij} - \min_j \overline{y}_{ij}$$
$$= \text{the range of the mean measurements made on part } i\,.$$

TABLE 2.3. Ranges of 30 part×operator samples of measured punch Heights

Punch	Student 1	Student 2	Student 3
1	3	2	2
2	2	3	3
3	0	1	2
4	1	3	1
5	2	0	1
6	1	3	0
7	4	1	3
8	1	3	2
9	1	2	1
10	1	6	2

Notice that with the obvious notation for the sample average of the measurement errors ϵ, according to model (2.26),

$$\overline{y}_{ij} = \mu + \alpha_i + \beta_j + \alpha\beta_{ij} + \overline{\epsilon}_{ij}\,.$$

Thus, for a fixed part i these means $\overline{y}_{ij}$ vary only according to independent normal random variables $\beta_j + \alpha\beta_{ij} + \overline{\epsilon}_{ij}$ that have mean 0 and variance $\sigma_\beta^2 + \sigma_{\alpha\beta}^2 + \sigma^2/m$. Thus their range, Δ_i, has mean

$$E\Delta_i = d_2(J)\sqrt{\sigma_\beta^2 + \sigma_{\alpha\beta}^2 + \sigma^2/m}\,.$$

This suggests $\Delta_i/d_2(J)$ or, better yet, the average of these over all parts i, $\overline{\Delta}/d_2(J)$, as an estimator of $\sqrt{\sigma_\beta^2 + \sigma_{\alpha\beta}^2 + \sigma^2/m}$. This in turn suggests that one can estimate $\sigma_\beta^2 + \sigma_{\alpha\beta}^2 + \sigma^2/m$ with $(\overline{\Delta}/d_2(J))^2$. Then remembering that $\overline{R}/d_2(m) = \hat{\sigma}_{\text{repeatability}}$ is an estimator of σ, an obvious estimator of $\sigma_\beta^2 + \sigma_{\alpha\beta}^2$ becomes

$$\left(\frac{\overline{\Delta}}{d_2(J)}\right)^2 - \frac{1}{m}\left(\frac{\overline{R}}{d_2(m)}\right)^2. \tag{2.33}$$

The quantity (2.33) is meant to approximate $\sigma_\beta^2 + \sigma_{\alpha\beta}^2$, which is nonnegative. But the estimator (2.33) can on occasion give negative values. When this happens, it is sensible to replace the negative value by 0 and thus expression (2.33) by

$$\max\left(0, \left(\frac{\overline{\Delta}}{d_2(J)}\right)^2 - \frac{1}{m}\left(\frac{\overline{R}}{d_2(m)}\right)^2\right). \tag{2.34}$$

TABLE 2.4. Part×operator means and ranges of such means for the punch height data

Punch(i)	$\bar{y}_{i1}$	$\bar{y}_{i2}$	$\bar{y}_{i3}$	Δ_i
1	497.00	497.67	497.00	.67
2	498.00	497.67	498.67	1.00
3	498.00	497.33	497.00	1.00
4	497.33	497.00	497.33	.33
5	500.00	499.00	499.33	1.00
6	498.67	498.67	498.00	.67
7	501.33	498.67	500.33	2.67
8	499.33	499.33	500.00	.67
9	499.33	499.33	499.67	.33
10	496.33	496.67	496.67	.33

So finally, an estimator of the reproducibility standard deviation can be had by taking the square root of expression (2.34). That is, one may estimate the quantity (2.27) with

Range-Based Estimator for Reproducibility Standard Deviation

$$\hat{\sigma}_{\text{reproducibility}} = \sqrt{\max\left(0, \left(\frac{\overline{\Delta}}{d_2(J)}\right)^2 - \frac{1}{m}\left(\frac{\overline{R}}{d_2(m)}\right)^2\right)}. \quad (2.35)$$

Example 14 *(Examples 12 and 13 continued.) Table 2.4 organizes $\bar{y}_{ij}$ and Δ_i values for the punch height measurements of Table 2.1. Then $\overline{\Delta} = 8.67/10 = .867$, and since $J = 3$, $d_2(J) = d_2(3) = 1.693$. So using Eq. (2.35),*

$$\begin{aligned}
\hat{\sigma}_{\text{reproducibility}} &= \sqrt{\max\left(0, \left(\frac{.867}{1.693}\right)^2 - \frac{1}{3}\left(\frac{1.9}{1.693}\right)^2\right)}, \\
&= \sqrt{\max(0, -.158)}, \\
&= 0.
\end{aligned}$$

This calculation suggests that this is a problem where σ appears to be so large that the reproducibility standard deviation cannot be seen above the intrinsic "noise" in measurement conceptualized as the repeatability component of variation. Estimates of the ratios (2.29) based on $\hat{\sigma}_{\text{repeatability}}$ and $\hat{\sigma}_{\text{reproducibility}}$ would attribute fractions 1 and 0 of the overall variance in measurement to, respectively, repeatability and reproducibility.

2.4.3 ANOVA-Based Estimation

The formulas of the previous subsection are easy to discuss and use, but they are not at all the best available. Ranges are not the most effective tools for estimating normal standard deviations. And the range-based methods have no corresponding way for making confidence intervals. More effective (and computationally more demanding) statistical tools are available, and we proceed to discuss some of them.

TABLE 2.5. A generic gauge R&R two-way ANOVA table

Source	SS	df	MS
Part	SSA	$I-1$	$MSA = SSA/(I-1)$
Operator	SSB	$J-1$	$MSB = SSB/(J-1)$
Part × Operator	$SSAB$	$(I-1)(J-1)$	$MSAB = SSAB/(I-1)(J-1)$
Error	SSE	$IJ(m-1)$	$MSE = SSE/IJ(m-1)$
Total	$SSTot$	$IJm-1$	

An $I \times J \times m$ data set of y_{ijk}s like that produced in a typical gauge R&R study is often summarized in a so-called two-way ANOVA table. Table 2.5 is a generic version of such a summary. Any decent statistical package will process a gauge R&R data set and produce such a summary table. As in a one-way ANOVA, "mean squares" are essentially sample variances (squares of sample standard deviations). MSA is essentially a sample variance of part averages, MSB is essentially a sample variance of operator averages, MSE is an average of within-cell sample variances, and $MSTot$ isn't typically calculated, but is a grand sample variance of all observations.

For purposes of being clear (and not because they are typically used for "hand calculation") we provide formulas for sums of squares. With cell means $\bar{y}_{ij}$ as in display (2.32), define row and column averages and the grand average of these

$$\bar{y}_{i.} = \frac{1}{J}\sum_j \bar{y}_{ij} \text{ and } \bar{y}_{.j} = \frac{1}{I}\sum_i \bar{y}_{ij} \text{ and } \bar{y}_{..} = \frac{1}{IJ}\sum_{ij} \bar{y}_{ij}.$$

Then the sums of squares are

$$SSTot = \sum_{ijk}(y_{ijk} - \bar{y}_{..})^2,$$

$$SSE = \sum_{ijk}(y_{ijk} - \bar{y}_{ij})^2,$$

$$SSA = mJ\sum_i(\bar{y}_{i.} - \bar{y}_{..})^2,$$

$$SSB = mI\sum_j(\bar{y}_{.j} - \bar{y}_{..})^2, \text{ and}$$

$$SSAB = m\sum_{ij}(\bar{y}_{ij} - \bar{y}_{i.} - \bar{y}_{.j} + \bar{y}_{..})^2$$

$$= SSTot - SSE - SSA - SSB$$

TABLE 2.6. Data from a small in-class gauge R&R study

Part	Operator 1	Operator 2	Operator 3
1	.52, .52	.54, .53	.55, .55
2	.56, .55	.54, .54	.55, .56
3	.57, .56	.55, .56	.57, .57
4	.55, .55	.54, .55	.56, .55

Corresponding degrees of freedom and mean squares are

$$dfE = (m-1)IJ \text{ and } MSE = SSE/(m-1)IJ,$$
$$dfA = I-1 \text{ and } MSA = SSA/(I-1),$$
$$dfB = J-1 \text{ and } MSB = SSB/(J-1), \text{ and}$$
$$dfAB = (I-1)(J-1) \text{ and } MSAB = SSAB/(I-1)(J-1).$$

Example 15 *In-Class Gauge R&R Study. The data in Table 2.6 on page 69 were collected in an in-class gauge R&R exercise where $I = 4$ polystyrene packing peanuts were measured for size (in in) by $J = 3$ students $m = 2$ times apiece using the same inexpensive caliper. The* JMP *statistical package produces the sums of squares*

$$SSA = .00241250, SSB = .00080833, SSAB = .00072500,$$
$$SSE = .00035000, \text{ and } SSTot = .00429583.$$

for these data that can be used as raw material for making important inferences for the R&R study based on model (2.26). Corresponding mean squares are

$$MSE = .00035000/(2-1)(4)(3) = .00002917,$$
$$MSA = .00241250/(4-1) = .00080417,$$
$$MSB = .00080833/(3-1) = .00040417, \text{ and}$$
$$MSAB = .00072500/(4-1)(3-1) = .00012083.$$

High-quality statistical software (like JMP or R) will automatically produce REML-based estimates and confidence intervals for the variance components $\sigma_\alpha^2, \sigma_\beta^2, \sigma_{\alpha\beta}^2$, and σ^2. As the quantities $\sigma_{\text{reproducibility}}^2$ and $\sigma_{\text{R\&R}}^2$ are a bit specialized (being of interest in our R&R application of the two-way random effects model, but not in other common applications), inferences for them are not automatically available. It is possible, but usually not convenient, to use the output of REML analyses to make inferences for these more specialized quantities. So here we will provide formulas for ANOVA-based estimators of $\sigma, \sigma_{\text{reproducibility}}$, and $\sigma_{\text{R\&R}}$ and appropriate Satterthwaite approximate degrees of freedom for making confidence limits. (Where readers know how to obtain REML-based estimates and intervals, our recommendation is to use them in preference to ANOVA-based estimators that follow.)

Single-number estimators for the quantities of most interest in a gauge R&R study are

ANOVA-Based Estimator for Repeatability Standard Deviation

$$\hat{\sigma}_{\text{repeatability}} = \hat{\sigma} = \sqrt{MSE}, \tag{2.36}$$

$$\hat{\sigma}_{\text{reproducibility}} = \sqrt{\max\left(0, \frac{MSB}{mI} + \frac{(I-1)}{mI}MSAB - \frac{1}{m}MSE\right)}, \quad (2.37)$$

ANOVA-Based Estimator for Reproducibility Standard Deviation

and

$$\hat{\sigma}_{\text{R\&R}} = \sqrt{\frac{1}{mI}MSB + \frac{I-1}{mI}MSAB + \frac{m-1}{m}MSE}. \quad (2.38)$$

ANOVA-Based Estimator for $\sigma_{\text{R\&R}}$

Confidence limits based on any of these estimators are of the generic form (already used several times in this chapter)

$$\text{``}\hat{\sigma}\text{''}\sqrt{\frac{\text{``}\hat{\nu}\text{''}}{\chi^2_{\text{upper}}}} \quad \text{and} \quad \text{``}\hat{\sigma}\text{''}\sqrt{\frac{\text{``}\hat{\nu}\text{''}}{\chi^2_{\text{lower}}}} \quad (2.39)$$

Generic Confidence Limits for a Standard Deviation

where "$\hat{\sigma}$" is one of the estimators, "$\hat{\nu}$" is a corresponding (exact or "Satterthwaite approximate") degrees of freedom, and the χ^2 percentage points are based on "$\hat{\nu}$." So it only remains to record formulas for appropriate degrees of freedom. These are

$$\nu_{\text{repeatability}} = IJ(m-1), \quad (2.40)$$

Degrees of Freedom for Use with Formulas (2.36) and (2.39)

$$\hat{\nu}_{\text{reproducibility}} = \frac{\hat{\sigma}^4_{\text{reproducibility}}}{\frac{\left(\frac{MSB}{mI}\right)^2}{J-1} + \frac{\left(\frac{(I-1)MSAB}{mI}\right)^2}{(I-1)(J-1)} + \frac{\left(\frac{MSE}{m}\right)^2}{IJ(m-1)}}$$

$$= \frac{\hat{\sigma}^4_{\text{reproducibility}}}{\frac{1}{m^2}\left(\frac{MSB^2}{I^2(J-1)} + \frac{(I-1)MSAB^2}{I^2(J-1)} + \frac{MSE^2}{IJ(m-1)}\right)},$$

Degrees of Freedom for Use with Formulas (2.37) and (2.39)

$$(2.41)$$

and

Degrees of Freedom for Use with Formulas (2.38) and (2.39)

$$\hat{\nu}_{R\&R} = \frac{\hat{\sigma}_{R\&R}^4}{\left(\frac{MSB}{mI}\right)^2 + \left(\frac{(I-1)MSAB}{mI}\right)^2 + \left(\frac{(m-1)MSE}{m}\right)^2}$$

$$= \frac{\hat{\sigma}_{R\&R}^4}{\frac{1}{m^2}\left(\frac{MSB^2}{I^2(J-1)} + \frac{(I-1)MSAB^2}{I^2(J-1)} + \frac{(m-1)MSE^2}{IJ}\right)}$$

(2.42)

Formulas (2.37), (2.41), (2.38), and (2.42) are tedious (but hardly impossible) to use with a pocket calculator. But a very small program, MathCAD worksheet, or spreadsheet template can be written to evaluate the estimates of standard deviations and approximate degrees of freedom from the sums of squares, m, I, and J.

Example 16 *(Example 15 continued.) A two-way random effects analysis of the data of Table 2.6 made using the* JMP *statistical package produces REML-based confidence limits of*

$$0 \text{ and } \sqrt{.0001359}, \text{ i.e., } 0 \text{ in and } .012 \text{ in for } \sigma_\beta$$

and

$$0 \text{ and } \sqrt{.0001152}, \text{ i.e., } 0 \text{ in and } .011 \text{ in for } \sigma_{\alpha\beta}.$$

There is thus at least the suggestion that a substantial part of the reproducibility variation in the data of Table 2.6 is a kind of nonconstant bias on the part of the student operators measuring the peanuts.

Using formulas (2.36), (2.37), and (2.38) it is possible to verify that in this problem

$$\hat{\sigma}_{repeatability} = \hat{\sigma} = .005401 \text{ in},$$
$$\hat{\sigma}_{reproducibility} = .009014 \text{ in}, \text{ and}$$
$$\hat{\sigma}_{R\&R} = .011 \text{ in}.$$

Using formulas (2.40), (2.41), and (2.42), these have corresponding degrees of freedom

$$\nu_{repeatability} = (4)(3)(2-1) = 12,$$
$$\hat{\nu}_{reproducibility} = 4.04, \text{ and}$$
$$\hat{\nu}_{R\&R} = 7.45.$$

So (rounding degrees of freedom down in the last two cases) using the limits (2.39), 95 % confidence limits for $\sigma_{repeatability}$ are

$$.005401\sqrt{\frac{12}{23.337}} \text{ and } .005401\sqrt{\frac{12}{4.404}}$$

i.e.,

$$.0039\,\text{in} \text{ and } .0089\,\text{in},$$

approximate 95 % confidence limits for $\sigma_{reproducibility}$ are

$$.009014\sqrt{\frac{4}{11.143}} \text{ and } .009014\sqrt{\frac{4}{.484}}$$

i.e.,

$$.0054\,\text{in} \text{ and } .0259\,\text{in},$$

and approximate 95 % confidence limits for $\sigma_{R\&R}$ are

$$.011\sqrt{\frac{7}{16.013}} \text{ and } .011\sqrt{\frac{7}{1.690}}$$

i.e.,

$$.0073\,\text{in} \text{ and } .0224\,\text{in}.$$

These intervals show that none of these standard deviations are terribly well determined (degrees of freedom are small and intervals are wide). If better information is needed, more data would have to be collected. But there is at least some indication that $\sigma_{repeatability}$ and $\sigma_{reproducibility}$ are roughly of the same order of magnitude. The caliper used to make the measurements was a fairly crude one, and there were detectable differences in the way the student operators used that caliper.

Suppose, for the sake of example, that engineering requirements on these polystyrene peanuts were that they be of size $.50\,\text{in} \pm .05\,\text{in}$. In such a context, the gauge capability ratio (2.30) could be estimated to be between

$$\frac{6(.0073)}{.55-.45}=.44 \text{ and } \frac{6(.0224)}{.55-.45}=1.34.$$

These values are not small. (See again the discussion on page 65.) This measurement "system" is not really adequate to check conformance to even these crude $\pm.05\,\text{in}$ product requirements.

Some observations regarding the planning of a gauge R&R study are in order at this point. The precisions with which one can estimate σ, $\sigma_{reproducibility}$, and $\sigma_{R\&R}$ obviously depend upon I, J, and m. Roughly speaking, precision of estimation of σ is governed by the product $(m-1)IJ$, so increasing any of the "dimensions" of the data array will improve estimation of repeatability. However, it is primarily J that governs the precision with which $\sigma_{reproducibility}$ and $\sigma_{R\&R}$ can be estimated. Only by increasing the number of operators in a gauge R&R study can one substantially improve the estimation of reproducibility variation.

While this fact about the estimation of reproducibility is perfectly plausible, its implications are not always fully appreciated (or at least not kept clearly in mind) by quality assurance practitioners. For example, many standard gauge R&R data collection forms allow for at most $J = 3$ operators. But 3 is a *very* small sample size when it comes to estimating a variance or standard deviation. So although the data in Table 2.1 are perhaps more or less typical of many R&R data sets, the small ($J = 3$) number of operators evident there should *not* be thought of as in any way ideal. To get a really good handle on the size of reproducibility variation, many more operators would be needed.

Section 2.4 Exercises

1. Consider again the situation of problem 3 of the Sect. 2.3 exercises and the data from the **Fiber Angle** case used there. (Operator 1 measured fiber angle for three different specimens four times each.) Recast that scenario into the two-way framework of this section.

    (a) Give the values of I, J, and m.

    (b) Find a range-based estimate of σ_{device}.

    (c) Find a range-based estimate of σ_x.

2. Based only on the data of problem 3 of the Sect. 2.3 exercises, can $\sigma_{\text{reproducibility}}$ be estimated? Why or why not?

3. Consider again the situation of problems 2 and 1 of the Sect. 2.3 exercises and the data from the **Fiber Angle** case used there. (Fiber angle for specimen 1 was measured four times by each of operators 1, 2, and 3.) Recast that scenario into the two-way framework of this section.

    (a) Give the values of I, J, and m.

    (b) Find a range-based estimate of $\sigma_{\text{repeatability}}$.

    (c) Find a range-based estimate of $\sigma_{\text{reproducibility}}$.

    (d) Based only on the data considered here, can σ_x be estimated? Why or why not?

4. **Washer Assembly.** Sudam, Heimer, and Mueller studied a clothes washer base assembly. Two operators measured the distance from one edge of a washer base assembly to an attachment. For a single base assembly, the same distance was measured four times by each operator. This was repeated on three different base assemblies. The target distance was 13.320 with an upper specification of $U = 13.42$ and a lower specification of $L = 13.22$. A standard gauge R&R study was conducted and data like those below were obtained. (Units are 10^{-1} in.)

Part	Operator 1	Operator 2
1	13.285, 13.284, 13.283, 13.282	13.284, 13.288, 13.287, 13.283
2	13.298, 13.293, 13.291, 13.291	13.297, 13.292, 13.292, 13.293
3	13.357, 13.356, 13.354, 13.356	13.355, 13.354, 13.352, 13.357

(a) What were the values of I, J, and m in this study?

(b) Based on the ANOVA table for the data given below, find the estimates for $\sigma_{\text{repeatability}}$, $\sigma_{\text{reproducibility}}$, and $\sigma_{\text{R\&R}}$.

(c) Give 95 % confidence limits for $\sigma_{\text{repeatability}}$, $\sigma_{\text{reproducibility}}$, and $\sigma_{\text{R\&R}}$.

(d) Find 95 % confidence limits for the GCR. (Hint: Use the last of your answers to (c).)

ANOVA table

Source	SS	df	MS
Part	.0236793	2	.0118396
Operator	.0000007	1	.0000007
Part×Operator	.0000106	2	.0000053
Error	.0000895	18	.0000050
Total	.0237800	23	

2.5 Simple Linear Regression and Calibration Studies

Calibration is an essential activity in the qualification and maintenance of measurement devices. In a calibration study, one uses a measurement device to produce measurements on "standard" specimens with (relatively well-) "known" values of measurands and sees how the measurements compare to the known values. If there are systematic discrepancies between what is known to be true and what the device reads, a conversion scheme is created to (in future use of the device) adjust what is read to something that is hopefully closer to the (future) truth. A slight extension of "regression" analysis (curve fitting) as presented in an elementary statistics course is the relevant statistical methodology in making this conversion. (See, e.g., Section 9.1 of Vardeman and Jobe's *Basic Engineering Data Collection and Analysis*.) This section discusses exactly how regression analysis is used in calibration.

Calibration studies employ true/gold-standard-measurement values of measurands x and "local" measurements y. (Strictly speaking, y need not even be in the same units as x.) Regression analysis can provide both "point conversions" and measures of uncertainty (the latter through inversion of "prediction limits"). The simplest version of this is where observed measurements are approximately linearly related to measurands, i.e.,

$$y \approx \beta_0 + \beta_1 x$$

This is "linear calibration." The standard statistical model for such a circumstance is

Simple Linear Regression Model

$$y = \beta_0 + \beta_1 x + \epsilon \qquad (2.43)$$

for a normal error ϵ with mean 0 and standard deviation σ. (σ describes how much ys vary for a fixed x and in the present context typically amounts to a repeatability standard deviation.) This model can be pictured as in Fig. 2.17.

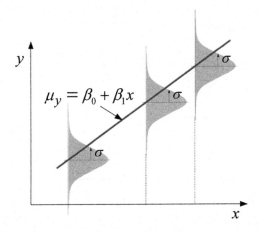

FIGURE 2.17. A schematic of the usual simple linear regression model (2.43)

For n data pairs (x_i, y_i), simple linear regression methodology allows one to make confidence intervals and tests associated with the model and *prediction limits for a new measurement* y_{new} associated with a new measurand, x_{new}. These are of the form

Prediction Limits for y_{new} in SLR

$$(b_0 + b_1 x_{\text{new}}) \pm t s_{\text{LF}} \sqrt{1 + \frac{1}{n} + \frac{(x_{\text{new}} - \bar{x})^2}{\sum_i (x_i - \bar{x})^2}} \qquad (2.44)$$

where the least squares line is $\hat{y} = b_0 + b_1 x$ and s_{LF} (a "line-fitting" sample standard deviation) is an estimate of σ derived from the fit of the line to the data. Any good statistical package will compute and plot these limits as functions of x_{new} along with a least squares line through the data set.

Example 17 *Measuring Cr^{6+} Concentration with a UV-Vis Spectrophotometer.* *The data below were taken from a web page of the School of Chemistry at the University of Witwatersrand developed and maintained by Dr. Dan Billing. They are measured absorbance values, y, for $n = 6$ solutions with "known" Cr^{6+} concentrations, x (in mg/l), from an analytical lab.*

x	0	1	2	4	6	8
y	.002	.078	.163	.297	.464	.600

Figure 2.18 is a plot of these data, the corresponding least squares line, and the prediction limits (2.44).

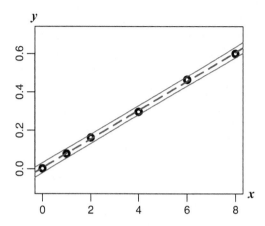

FIGURE 2.18. Scatterplot of the Cr^{6+} Concentration calibration data, least squares line, and prediction limits for y_{new}

What is here of most interest about simple linear regression technology is what it says about calibration and measurement in general. Some applications of inference methods based on the model (2.43) to metrology are the following.

1. From a simple linear regression output,

$$s_{\text{LF}} = \sqrt{MSE} = \sqrt{\frac{1}{n-2} \sum_{i=1}^{n} (y_i - \hat{y}_i)^2} = \text{``root mean square error''}$$

(2.45)

is an estimated repeatability standard deviation. One may make confidence intervals for $\sigma = \sigma_{\text{repeatability}}$ based on the estimate (2.45) using $\nu = n - 2$ degrees of freedom and limits

$$s_{\text{LF}} \sqrt{\frac{n-2}{\chi^2_{\text{upper}}}} \quad \text{and} \quad s_{\text{LF}} \sqrt{\frac{n-2}{\chi^2_{\text{lower}}}}.$$

(2.46)

Confidence Limits for σ in Model (2.43)

2. The least squares equation $\hat{y} = b_0 + b_1 x$ can be solved for x, giving

$$\hat{x}_{\text{new}} = \frac{y_{\text{new}} - b_0}{b_1}$$

(2.47)

Conversion Formula for a Future Measurement, y_{new}

78   Chapter 2. Statistics and Measurement

as a way of estimating a new "gold-standard" value (a new measurand, x_{new}) from a measured local value, y_{new}.

3. One can take the prediction limits (2.44) for y_{new} and "turn them around" to get confidence limits for the x_{new} corresponding to a measured local y_{new}. This provides a defensible way to set "error bounds" on what y_{new} indicates about x_{new}.

4. In cases (unlike Example 17) where y and x are in the same units, confidence limits for the slope β_1 of the simple linear regression model

Confidence Limits for β_1 in Model (2.43)

$$b_1 \pm t \frac{s_{\text{LF}}}{\sqrt{\sum (x_i - \bar{x})^2}} \quad (2.48)$$

provide a way of investigating the constancy of bias (linearity of the measurement device in the sense introduced on page 37). That is, when x and y are in the same units, $\beta_1 = 1.0$ is the case of constant bias. If confidence limits for β_1 fail to include 1.0, there is clear evidence of device nonlinearity.

Example 18 *(Example 17 continued.)* The use of the JMP statistical package with the data of Example 17 produces

$$y = .0048702 + .0749895x \text{ with } s_{\text{LF}} = .007855.$$

We might expect a local (y) repeatability standard deviation of around .008 (in the y absorbance units). In fact, 95% confidence limits for σ can be made (using $n - 2 = 4$ degrees of freedom and formula (2.46)) as

$$.007855\sqrt{\frac{4}{11.143}} \text{ and } .007855\sqrt{\frac{4}{.484}},$$

i.e.,

$$.0047 \text{ and } .0226.$$

Making use of the slope and intercept of the least squares line, a conversion formula for going from y_{new} to x_{new} is (as in display (2.47))

$$\hat{x}_{\text{new}} = \frac{y_{\text{new}} - .0048702}{.0749895},$$

So, for example, a future measured absorbance of $y_{\text{new}} = .20$ suggests a concentration of

$$\hat{x}_{\text{new}} = \frac{.20 - .0048702}{.0749895} = 2.60 \text{ mg}/1.$$

Finally, Fig. 2.19 on page 80 is a modification of Fig. 2.18 that illustrates how the plotted prediction limits (2.44) provide both 95% predictions for a new measurement on a fixed/known measurand and 95 % confidence limits on a new measurand, having observed a particular measurement. Reading from the figure, one is "95 % sure" that a future observed absorbance of .20 comes from a concentration between

$$2.28 \,\text{mg}/\text{l} \text{ and } 2.903 \,\text{mg}/\text{l}.$$

Example 19 *A Check on Device "Linearity." A calibration data set due to John Mandel compared $n = 14$ measured values y for a single laboratory to corresponding consensus values x for the same specimens derived from multiple labs. (The units are not available, but were the same for x and y values.) A simple linear regression analysis of the data pairs produced*

$$b_1 = .882 \text{ and } \frac{s_{\text{LF}}}{\sqrt{\sum(x_i - \bar{x})^2}} = .012$$

so that (using the upper 2.5 % point of the t_{12} distribution, 2.179, and formula (2.48)) 95 % confidence limits for β_1 are

$$.882 \pm 2.179\,(.012)$$

or

$$.882 \pm .026 \,.$$

A 95 % confidence interval for β_1 clearly does not include 1.0. So bias for the single laboratory was not constant. (The measurement "device" was not linear in the sense discussed on page 37.)

Section 2.5 Exercises

1. $n = 14$ polymer specimens of *known* weights, x, were weighed and the measured weights, y, recorded. The following table contains the data. (All units are gs.)

x	1	1	3	3	5	5	7
y	1.10	.95	2.98	3.01	5.02	4.99	6.97

x	7	10	10	12	12	14	14
y	7.10	10.03	9.99	12.00	11.98	14.10	14.00

   (a) Find the least squares line $\hat{y} = b_0 + b_1 x$ for these data.

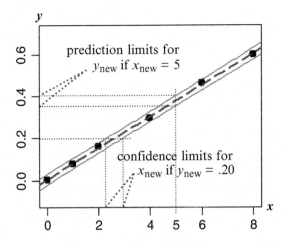

FIGURE 2.19. Confidence limits for x_{new} based on an observed y_{new} (and prediction limits (2.44))

 (b) Find the estimated repeatability standard deviation corresponding to your regression analysis.

 (c) Find 95 % confidence limits for the y repeatability standard deviation based on your answer to (b).

2. In the context of problem 1, suppose a new specimen is measured as having a weight of 6.10 g.

 (a) Find the "calibrated weight," $\hat{x}$, corresponding to this new specimen based on your regression analysis.

 (b) Find 95 % confidence limits for the slope of the relationship between measured and actual weight (β_1). Does the device used to produce the y measurements have constant bias (is it "linear")? Why or why not?

3. Based on your regression analysis in problem 1, find 95 % prediction limits for the next measured weight for a new specimen with standard known weight of 8 g.

4. Would it be wise to use the above regression analyses to adjust a measured specimen weight of $y_{new} = .2$ g? Why or why not?

2.6   R&R Considerations for Go/No-Go Inspection

Ideally, observation of a process results in quantitative measurements. But there are some contexts in which all that is determined is whether an item or process condition is of one of two types, which we will for the present call "conforming" and "nonconforming." It is, for example, common to check the conformance of machined metal parts to some engineering requirements via the use of a "go/no-go gauge." (A part is conforming if a critical dimension fits into the larger of two check fixtures and does not fit into the smaller of the two.) And it is common to task human beings with making visual inspections of manufactured items and producing an "ok/not-ok" call on each.

Engineers are sometimes then called upon to apply the qualitative "repeatability" and "reproducibility" concepts of metrology to such go/no-go or "0/1" contexts. One wants to separate some measure of overall inconsistency in 0/1 "calls" on items into pieces that can be mentally charged to inherent inconsistency in the equipment or method and the remainder that can be charged to differences between how operators use it. Exactly how to do this is presently not well established. The best available statistical methodology for this kind of problem is more complicated than can be presented here (involving so-called generalized linear models and random effects in these). What we *can* present is a rational way of making point estimates of what might be termed repeatability and reproducibility components of variation in 0/1 calls. (These are based on reasoning similar to that employed in Sect. 2.4.2 to find correct range-based estimates in usual measurement R&R contexts.) We then remind the reader of elementary methods of estimating differences in population proportions and in mean differences and point out their relevance in the present situation.

2.6.1 Some Simple Probability Modeling

To begin, think of coding a "nonconforming" call as "1" and a "conforming" call as "0" and having J operators each make m calls on a fixed part. Suppose that J operators have individual probabilities $p_1, p_2, \ldots, p_J$ of calling the part "nonconforming" on any single viewing and that across m viewings

$X_j =$ the number of nonconforming calls among the m made by operator j

is binomial (m, p_j). We'll assume that the p_j are random draws from some population with mean π and variance v.

The quantity

$$p_j (1 - p_j)$$

is a kind of "per-call variance" associated with the declarations of operator j and might serve as a kind of repeatability variance for that operator. (Given the value of p_j, elementary probability theory says that the variance of X_j is $mp_j(1-p_j)$.) The biggest problem here is that unlike what is true in the usual case of gauge R&R for measurements, this variance is not constant across operators. But its expected value, namely

$$\begin{aligned} \mathrm{E}\left(p_j(1-p_j)\right) &= \pi - \mathrm{E}p_j^2 \\ &= \pi - (v + \pi^2) \\ &= \pi(1-\pi) - v \end{aligned}$$

can be used as a sensible measure of variability in conforming/nonconforming classifications chargeable to repeatability sources. The variance v serves as a measure of reproducibility variance. This ultimately points to

$$\pi(1-\pi)$$

as the "total R&R variance" here. That is, we make definitions for 0/1 contexts

R&R Variance for One Part in a 0/1 Context

$$\sigma^2_{\text{R\&R}} = \pi(1-\pi) \qquad (2.49)$$

and

Repeatability Variance for One Part in a 0/1 Context

$$\sigma^2_{\text{repeatability}} = \pi(1-\pi) - v \qquad (2.50)$$

and

Reproducibility Variance for One Part in a 0/1 Context

$$\sigma^2_{\text{reproducibility}} = v \qquad (2.51)$$

2.6.2 Simple R&R Point Estimates for 0/1 Contexts

Still thinking of a single fixed part, let

$$\hat{p}_j = \frac{\text{the number of "nonconforming" calls made by operator } j}{m} = \frac{X_j}{m}$$

and define the (sample) average of these:

$$\bar{\hat{p}} = \frac{1}{J}\sum_{j=1}^{J} \hat{p}_j .$$

It is possible to argue that

$$\mathrm{E}\bar{\hat{p}} = \pi$$

so that a plausible estimate of $\sigma^2_{\text{R\&R}}$ is

Estimator of R&R Variance for a Single Part in a 0/1 Context

$$\hat{\sigma}^2_{\text{R\&R}} = \bar{\hat{p}}(1-\bar{\hat{p}}) \qquad (2.52)$$

Then, since $\hat{p}_j(1-\hat{p}_j)$ is a plausible estimate of the "per-call variance" associated with the declarations of operator j, $p_j(1-p_j)$, an estimate of $\sigma^2_{\text{repeatability}}$ is

Estimator of Repeatability Variance for a Single Part in a 0/1 Context

$$\hat{\sigma}^2_{\text{repeatability}} = \overline{\hat{p}(1-\hat{p})} \qquad (2.53)$$

(the sample average of the $\hat{p}_j(1-\hat{p}_j)$). Finally, a simple estimate of $\sigma^2_{\text{reproducibility}} = v$ is

$$\hat{\sigma}^2_{\text{reproducibility}} = \hat{\sigma}^2_{R\&R} - \hat{\sigma}^2_{\text{repeatability}}$$
$$= \bar{\hat{p}}(1-\bar{\hat{p}}) - \overline{\hat{p}(1-\hat{p})} \quad (2.54)$$

Estimator of Reproducibility Variance for a Single Part in a 0/1 Context

Again, the estimators (2.52), (2.53), and (2.54) are based on a single part. Exactly what to do based on multiple parts (say I of them) is not completely obvious. But in order to produce a simple methodology, *we will simply average estimates made one part at a time across multiple parts*, presuming that parts in hand are sensibly thought of as a random sample of parts to be checked and that this averaging is a reasonable way to combine information across parts.

TABLE 2.7. Hypothetical results of visual inspection of five parts by three operators

	Operator 1		Operator 2		Operator 3	
	$\hat{p}$	$\hat{p}(1-\hat{p})$	$\hat{p}$	$\hat{p}(1-\hat{p})$	$\hat{p}$	$\hat{p}(1-\hat{p})$
Part 1	.2	.16	.4	.24	.2	.16
Part 2	.6	.24	.6	.24	.7	.21
Part 3	1.0	0	.8	.16	.7	.21
Part 4	.1	.09	.1	.09	.1	.09
Part 5	.1	.09	.3	.21	.3	.21

In order for any of this to have a chance of working, m will need to be fairly large. The usual gauge R&R "$m = 2$ or 3" just isn't going to produce informative results in the present context. And in order for this to work in practice (so that an operator isn't just repeatedly looking at the same few parts over and over and remembering how he or she has called them in the past), a large value of I may also be needed.

TABLE 2.8. R&R calculations for the hypothetical visual inspection data

	$\overline{\hat{p}(1-\hat{p})} = \hat{\sigma}^2_{\text{repeatability}}$	$\bar{\hat{p}}$	$\bar{\hat{p}}(1-\bar{\hat{p}}) = \hat{\sigma}^2_{R\&R}$	$\bar{\hat{p}}(1-\bar{\hat{p}}) - \overline{\hat{p}(1-\hat{p})} = \hat{\sigma}^2_{\text{reproducibility}}$
Part 1	.187	.2667	.1956	.0090
Part 2	.230	.6333	.2322	.0022
Part 3	.123	.8333	.1389	.0156
Part 4	.090	.1	.0900	0
Part 5	.170	.2333	.1789	.0098
Average	.160		.1671	.0071

Example 20 *A Simple Numerical Example. For purposes of illustrating the formulas of this section, we will use a small numerical example due to Prof. Max Morris. Suppose that $I = 5$ parts are inspected by $J = 3$ operators, $m = 10$*

84   Chapter 2. Statistics and Measurement

times apiece, and that in Table 2.7 are sample fractions of "nonconforming" calls made by the operators and the corresponding estimates of per-call variance

Then the one-part-at-a-time and average-across-parts repeatability, R&R, and reproducibility estimates of variance are collected in Table 2.8 on page 83.

Then, for example, a fraction of only

$$\frac{.0071}{.1671} = 4.3\%$$

of the inconsistency in conforming/nonconforming calls seen in the original data seems to be attributable to clear differences in how the operators judge the parts (differences in the binomial "nonconforming call probabilities" p_j). Rather, the bulk of the variance seems to be attributable to unavoidable binomial variation. The ps are not close enough to either 0 or 1 to make the calls tend to be consistent. So the variation seen in the $\hat{p}$s in a given row is not clear evidence of large operator differences.

Of course, we need to remember that the computations above are on the variance (and not standard deviation) scale. On the (more natural) standard deviation scale, reproducibility variation

$$\sqrt{.0071} = .08$$

and repeatability variation

$$\sqrt{.160} = .40$$

are not quite so strikingly dissimilar.

2.6.3 Confidence Limits for Comparing Call Rates for Two Operators

It's possible to use elementary confidence interval methods to compare call rates for two particular operators. This can be done for a particular fixed part or for "all" parts (supposing that the ones included in a study are a random sample from the universe of parts of interest). The first possibility can be viewed as the problem of estimating the difference in two binomial parameters, say p_1 and p_2. The second can be approached as estimation of a mean difference in part-specific call rates, say μ_d.

A common elementary large-sample approximate confidence interval for $p_1 - p_2$ has end points

$$\hat{p}_1 - \hat{p}_2 \pm z\sqrt{\frac{\hat{p}_1(1-\hat{p}_1)}{n_1} + \frac{\hat{p}_2(1-\hat{p}_2)}{n_2}}.$$

But, as it turns out, this formula can fail badly if either p is extreme or n is small. So we will use a slight modification that is more reliable, namely

$$\hat{p}_1 - \hat{p}_2 \pm z \sqrt{\frac{\tilde{p}_1(1-\tilde{p}_1)}{n_1} + \frac{\tilde{p}_2(1-\tilde{p}_2)}{n_2}} \qquad (2.55)$$

Confidence Limits for $p_1 - p_2$

where

$$\tilde{p}_i = \frac{n_i \hat{p}_i + 2}{n_i + 4} \qquad (2.56)$$

Values to Use in Formula (2.55)

That is, under the square root of the usual formula, one essentially replaces the $\hat{p}$ values with $\tilde{p}$ values derived by adding two "successes" in four "additional trials" to the counts used to make up the $\hat{p}$ values. (This has the effect of making the standard large-sample interval a bit wider and correcting the problem that without this modification for small sample sizes and extreme values of p, it can fail to hold its nominal confidence level.)

Example 21 *(Example 20 continued.) Consider again part 1 from Example 20, and in particular consider the question of whether operator 1 and operator 2 have clearly different probabilities of calling that part nonconforming on a single call. With $\hat{p}_1 = .2$ and $\hat{p}_2 = .4$, formula (2.56) says that*

$$\tilde{p}_1 = \frac{2+2}{10+4} = .2857 \text{ and } \tilde{p}_2 = \frac{4+2}{10+4} = .4286$$

so that using formula (2.55) approximate 95% confidence limits for the difference $p_1 - p_2$ are

$$.2 - .4 \pm 1.96 \sqrt{\frac{.2857(1-.2857)}{10} + \frac{.4286(1-.4286)}{10}}$$

i.e.,

$$-.2 \pm .49$$

These limits cover 0 and there thus is no clear evidence in the $\hat{p}_1 = .2$ and $\hat{p}_2 = .4$ values (from the relatively small samples of sizes $m = 10$) that operators 1 and 2 have different probabilities of calling part 1 nonconforming.

The so-called "paired t" confidence limits (2.10) for the mean of a difference $d = x_1 - x_2$ (say μ_d) based on a random sample of normally distributed values $d_1, d_2, \ldots, d_n$ are presented in most elementary statistics courses. While a difference in observed call rates for operators 1 and 2 on a particular part ($d = \hat{p}_1 - \hat{p}_2$) will typically not be normally distributed, for rough purposes it is adequate to appeal to the "robustness" of this inference method (the fact that it is widely believed to be effective even when normality is not wholly appropriate as a modeling assumption) and employ it to compare operator 1 and operator 2 in terms of average part-specific call rates.

Example 22 *(Example 20 continued.)* Consider again Example 20, and in particular the question of whether operator 1 and operator 2 have clearly different average (across parts) probabilities of calling parts nonconforming on a single call. The $n = 5$ differences in $\hat{p}$s for the two operators are

$$.2 - .4 = -.2, .6 - .6 = 0, 1.0 - .8 = .2, .1 - .1 = 0, \text{ and } .1 - .3 = -.2.$$

These numbers have sample mean $\bar{d} = -.04$ and sample standard deviation $s_d = .17$. Then using the fact that the upper 5%, point of the t_4 distribution is 2.132, rough 90% two-sided confidence limits for the mean difference in call rates for the operators are

$$-.04 \pm 2.132 \frac{.17}{\sqrt{4}} \quad \text{that is,} \quad -.04 \pm .18,$$

and there is not definitive evidence in Table 2.7 of a consistent difference in how operators 1 and 2 call parts on average.

Section 2.6 Exercises

1. Suppose that ten parts are inspected by four operators 16 times apiece. Each inspection determines whether or not the item is conforming. The counts in the table below correspond to the numbers of "nonconforming" calls out of 16 inspections.

Part	Operator 1	Operator 2	Operator 3	Operator 4
1	10	11	11	10
2	11	9	12	10
3	8	8	9	7
4	15	14	14	16
5	12	14	11	12
6	15	15	16	15
7	14	11	14	12
8	16	16	15	15
9	13	15	14	15
10	16	15	16	16

   (a) Using the data above, fill in the table below.

Part	$\bar{\hat{p}}$	$\bar{\hat{p}}(1-\bar{\hat{p}})$	$\overline{\hat{p}(1-\hat{p})}$
1			
2			
⋮			
10			

(b) What is the fraction of inconsistency in conforming/nonconforming calls that can be attributed to clear differences in how the operators judged the parts (differences in the binomial "nonconforming call probabilities" p_j)? (Make your answer on the variance scale.)

(c) What is the estimated reproducibility variation (on the standard deviation scale)?

(d) What is the estimated repeatability variation (on the standard deviation scale)?

(e) For part 10, give a 90 % confidence interval for the difference (operator 1 minus operator 3) in probabilities of a nonconforming call. Does it appear the operators 1 and 3 have different probabilities of a nonconforming call on *any* one of the parts? Why?

(f) Compare operator 1 and operator 3 average "nonconforming" call rates using 90 % two-sided confidence limits for a mean difference.

2.7 Chapter Summary

This chapter has been concerned with how measurement error impacts what can be learned from empirical data. It presented some ideas from the probability modeling of measurement variation and considered how the interpretation of elementary statistical inferences is affected by measurement error. Then a variety of more advanced statistical tools were discussed, because of their usefulness in quantifying, partitioning, and (in some cases) removing the effects of measurement variation in quality assurance and improvement projects.

2.8 Chapter 2 Exercises

1. Does a perfectly calibrated device return measurements of a measurand that are completely free of error? Explain.

2. Is a standard (an item with corresponding "known" measurand) needed in *both* device calibration and estimation of σ_{device}? If not, which requires a standard? Explain.

3. A measurement device may have a bias as large as 1 unit (in absolute value) and a device standard deviation as large as 1 unit. You measure x and observe $y = 10$. If you believe in the simple (normal) measurement model and want to report an interval you are "at least 99 % sure" contains x, you should report what limits? (Hint: Before measurement, how far do you expect y to be from x with the indicated worst possible values of absolute

88   Chapter 2. Statistics and Measurement

bias and standard deviation? Interpret "99 % sure" in "plus or minus three standard deviations" terms.)

4. The same axel diameter is measured $n_1 = 25$ times with device 1 and $n_2 = 25$ times with device 2, with resulting means and standard deviations $\bar{y}_1 = 2.001$ in, $\bar{y}_2 = 2.004$ in, $s_1 = .003$ in, and $s_2 = .004$ in. The upper 2.5 % point of the $F_{24,24}$ distribution is about 2.27.

   (a) Give 95 % confidence limits for the difference in device biases.

   (b) Give 95 % confidence limits for the ratio of the two device standard deviations.

   (c) Is there a clear difference in device biases based on your interval in (a)? Why or why not?

   (d) Is there a clear difference in device standard deviations based on your interval in (b)? Why or why not?

5. Two different (physically stable) production lines produce plastic pop bottles. Suppose $n_1 = 25$ bottles from line 1 and $n_2 = 25$ bottles from line 2 are burst tested on a single tester, with resulting means and standard deviations $\bar{y}_1 = 201$ psi, $\bar{y}_2 = 202$ psi, $s_1 = 3$ psi, and $s_2 = 4$ psi.

   (a) Give a 95 % confidence interval for the difference between the mean burst strengths for lines 1 and 2 (line 1 minus line 2).

   (b) Give a 95 % confidence interval for the ratio of burst strength standard deviations (line 1 divided by line 2). The upper 2.5 % point of the $F_{24,24}$ distribution is about 2.27.

   (c) Is there a clear difference between mean burst strengths? Why or why not?

   (d) Is there a clear difference between the consistencies of burst strengths? Why or why not?

6. Using a single tester, a single metal specimen was tested for Brinell hardness 20 times with resulting sample standard deviation of hardness 10 HB. Subsequently, 40 different specimens cut from the same ingot of steel have sample standard deviation of measured hardness 20 HB (using the same tester):

   (a) Give 95 % confidence limits for a "test variability" standard deviation.

   (b) Give approximate 95 % confidence limits for a specimen-to-specimen standard deviation of actual Brinell hardness.

7. An ANOVA analysis of a gauge R&R data set produced $\hat{\sigma}_{R\&R} = 53$ (in appropriate units) and $\hat{\nu}_{R\&R} = 3$. In these units, engineering specifications on a critical dimension of a machined steel part are $nominal \pm 200$. Give approximate 95 % confidence limits for a GCR (gauge capability ratio) for checking conformance to these specifications.

8. 95 % confidence limits for a particular gauge capability ratio are 6 to 8. What does this indicate about the usability of the gauge for checking conformance to the specifications under consideration?

9. Below is an analysis of variance table from a calibration study. The data were light intensities, y (in unspecified analyzer units), for specimens of *known* riboflavin concentration x (in $\mu g/$ml).

    ANOVA table

Source	SS	df	MS
Model	10946.445	1	10946.445
Error	27.155	8	3.4
Total	10973.6	9	

    Parameter estimates for the simple linear regression model were $b_0 = 6.4634$ and $b_1 = 129.1768$.

    (a) Give a 95 % confidence interval for a repeatability standard deviation for this analyzer.

    (b) Suppose a new specimen with unknown concentration is analyzed and $y_{\text{new}} = 75$ is observed. Give a single-number estimate of the concentration in that specimen.

10. The final step in the production of some glass vials is a visual inspection presently carried out by human inspectors. A particular single vial (marked in an "invisible" ink that can be seen only under ultraviolet light) known to be defective is repeatedly run through the inspection process among a large number of newly produced vials. In fact, each of five company inspectors sees that vial ten times in a company study. Below are the rates at which that vial was identified as defective by the various operators ("1.0" means 100 %).

    $$.6, .9, .9, 1.0, 1.0$$

    (a) In general, what two values of $\hat{p}$ reflect perfect consistency of "defective/nondefective" calls made by a particular inspector?

    (b) What distribution models the number of correct "defective" calls (among ten calls) made by a particular inspector on the vial in question?

    (c) On the scale of (estimated) variances (not standard deviations), what is the fraction of overall variation seen in the "defective/nondefective" calls for this vial that should be attributed to operator-to-operator differences?

    (d) Give 95 % confidence limits for the long run difference in proportions of "defective" calls for the first operator (that made six out of ten "defective" calls) and the last operator (who made all "defective" calls).

11. **Laser Metal Cutting.** Davis, Martin and Popinga used a ytterbium argon gas laser to make some cuts in 316 stainless steel. Using 95- MJ/pulse and 20- Hz settings on the laser and a 15.5- mm distance to the steel specimens (set at a 45° angle to the laser beam), the students made cuts in specimens using 100, 500, and 1000 pulses. The measured depths of four different cuts (in machine units) at each pulse level are given below (assume the same operator made all measurements and that repeatability variation is negligible here).

100 Pulses	500 Pulses	1000 Pulses
7.4, 8.6, 5.6, 8.0	24.2, 29.5, 26.5, 23.8	33.4, 37.5, 35.9, 34.8

    (a) What is the response variable in this problem?

    (b) Give the sample average values for the 100, 500, and 1000 pulse levels. Calculate the sample range for the data at each pulse level. Give estimates of the standard deviation of cut depth for each level of pulse, first based on the sample range and then using the sample standard deviation. (You will have two estimates for each of the three population standard deviations.)

    (c) Assuming variability is the same for all three pulse levels, give an estimate of the common standard deviation based on the three sample ranges.

    (d) The concepts of measurement validity, precision, and accuracy are discussed in Sect. 2.1. The analysts decided to report the average cut depth for the different pulse levels. This averaging can be thought of in terms of improving which of (1) validity, (2) precision, or (3) accuracy (over the use of any single measurement)? The concept of calibration is most closely associated with which of the three?

12. **Fiber Angle.** Grunig, Hamdorf, Herman, and Potthoff studied a carpet-like product. They measured the angle at which fibers were glued to a sheet of base material. A piece of finished product was obtained and cut into five sections. Each of the four team members measured the fiber angle eight times for each section. The results of their measuring are given in Table 2.9 (in degrees above an undisclosed reference value). A corresponding ANOVA is also given in Table 2.10.

    (a) Say what each term in the equation $y_{ijk} = \mu + \alpha_i + \beta_j + \alpha\beta_{ij} + \epsilon_{ijk}$ means in this problem (including the subscripts i, j, and k).

    (b) Using ranges, estimate the repeatability and reproducibility standard deviations for angle measurement. Based on this analysis, what aspect of the measuring procedure seems to need the most attention? Explain.

    (c) Using ANOVA-based formulas, estimate the repeatability and reproducibility standard deviations for angle measurement. Is this analysis in essential agreement with that in part (b)? Explain.

TABLE 2.9. Data for problem 12

Angle	Analyst 1	Analyst 2	Analyst 3	Analyst 4
1	19, 20, 20, 23	20, 25, 17, 22	20, 19, 15, 16	10, 10, 10, 5
	20, 20, 20, 15	23, 15, 23, 20	20, 19, 12, 14	5, 5, 5, 5
2	15, 17, 20, 20	15, 13, 5, 10	15, 20, 14, 16	10, 10, 10, 10
	10, 15, 15, 15	8, 8, 10, 12	13, 20, 15, 15	10, 15, 15, 10
3	23, 20, 22, 20	20, 23, 20, 20	15, 20, 22, 18	10, 10, 10, 15
	25, 22, 20, 23	23, 23, 22, 20	15, 20, 16, 20	15, 10, 10, 10
4	15, 16, 22, 15	20, 22, 18, 23	13, 13, 15, 20	5, 10, 10, 10
	15, 15, 22, 17	23, 23, 24, 20	11, 20, 13, 15	10, 10, 10, 10
5	20, 20, 22, 20	18, 20, 18, 23	10, 14, 17, 12	5, 10, 10, 10
	27, 17, 20, 15	20, 20, 18, 15	11, 10, 15, 10	10, 10, 10, 10

TABLE 2.10. ANOVA for problem 12

Source	SS	df	MS
Angle	390.913	4	97.728
Analyst	2217.15	3	739.05
Angle×Analyst	797.788	12	66.482
Error	971.75	140	6.941
Total	4377.6	159	

(d) Using your answer to (c), give an estimate of the standard deviation that would be experienced by many analysts making a single measurement on the same angle (in the same section) assuming there is no repeatability component to the overall variation.

(e) Specifications on the fiber angle are *nominal* $\pm 5°$. Estimate the gauge capability ratio using first the ranges and then ANOVA-based estimates. Does it appear this measurement method is adequate to check conformance to the specifications? Why or why not?

13. Refer to the **Fiber Angle** case in problem 12.

(a) Is it preferable to have eight measurements on a given section by each analyst as opposed to, say, two measurements on a given section by each analyst? Why or why not?

(b) For a given number of angle measurements per analyst×section combination, is it preferable to have four analysts instead of two, six, or eight? Why or why not?

(c) When making angle measurements for a given section, does it matter if the angle at a fixed location on the piece is repeatedly measured, or is it acceptable (or even preferable?) for each analyst to measure at eight different locations on the section? Discuss.

(d) Continuing with (c), does it matter that the locations used on a given section varied analyst to analyst? Why or why not?

14. **Bolt Shanks.** A 1-in micrometer is used by an aircraft engine manufacturer to measure the diameter of a body-bound bolt shank. Specifications on this dimension have been set with a spread of .002 in. Three operators and ten body-bound bolt shanks were used in a gauge R&R study. Each bolt shank was measured twice by each operator (starting with part 1 and proceeding sequentially to part 10) to produce the data in Table 2.11 (in inches). A corresponding ANOVA is provided in Table 2.12 as well (SSs and MSs are in 10^{-6} in^2).

(a) Plot the bolt shank diameter measurements versus part number using a different plotting symbol for each operator. (You may wish to also plot part×operator means and connect consecutive ones for a given operator with line segments.) Discuss what your plot reveals about the measurement system.

TABLE 2.11. Data for problem 14

Part	Operator A	Operator B	Operator C	Part	Operator A	Operator B	Operator C
1	.3473	.3467	.3472	6	.3472	.3463	.3471
	.3473	.3465	.3471		.3472	.3464	.3471
2	.3471	.3465	.3471	7	.3473	.3465	.3472
	.3471	.3464	.3471		.3473	.3469	.3471
3	.3472	.3467	.3471	8	.3474	.3470	.3473
	.3472	.3464	.3471		.3473	.3470	.3473
4	.3474	.3470	.3473	9	.3472	.3465	.3472
	.3475	.3470	.3474		.3472	.3466	.3471
5	.3474	.3470	.3473	10	.3474	.3470	.3474
	.3474	.3470	.3473		.3474	.3470	.3473

(b) Find an ANOVA-based estimate of repeatability standard deviation.

(c) Find an ANOVA-based estimated standard deviation for reproducibility assuming there is no repeatability component of variation.

(d) Using your answers to (b) and (c), estimate the percent of total (R&R) measurement variance due to repeatability.

(e) Using your answers to (b) and (c), estimate the percent of total measurement (R&R) variance due to reproducibility.

TABLE 2.12. ANOVA for problem 14

Source	SS	df	MS
Part	1.3	9	.145
Operator	3.78	2	1.89
Part×Operator	.321	18	.0178
Error	.195	30	.0065
Total	5.601	59	

(f) Discuss the relationship of your plot in (a) to your answers to (b) through (e).

(g) Find an ANOVA-based estimate of the gauge capability ratio. Is the measurement process acceptable for checking conformance to the specifications? Why or why not?

15. Refer to the **Bolt Shanks** case in problem 14. The data in Table 2.13 are from three new operators with a different set of ten body-bound bolt shanks (numbered as part 11 through part 20). An appropriate ANOVA is also provided for these new data in Table 2.14 (units for the SS's and MS's are $10^{-6}\,\text{in}^2$).

TABLE 2.13. Data for problem 15

	Operator				Operator		
Part	D	E	F	Part	D	E	F
11	.3694	.3693	.3693	16	.3692	.3692	.3692
	.3694	.3693	.3693		.3693	.3692	.3691
12	.3693	.3693	.3692	17	.3696	.3695	.3695
	.3693	.3692	.3692		.3696	.3695	.3695
13	.3698	.3697	.3697	18	.3697	.3696	.3696
	.3697	.3697	.3697		.3696	.3696	.3696
14	.3697	.3698	.3697	19	.3697	.3696	.3695
	.3696	.3697	.3697		.3696	.3695	.3696
15	.3694	.3695	.3695	20	.3697	.3697	.3698
	.3693	.3695	.3694		.3697	.3698	.3697

(a) Answer (a) through (g) from problem 14 for these new data.

(b) Are your answers to (a) qualitatively different than those for problem 14? If your answer is yes, in what ways do the results differ, and what might be the sources of the differences?

(c) Do conclusions from this R&R study indicate a more consistent measurement process for body-bound bolt shanks than those in problem 14? Why or why not?

TABLE 2.14. ANOVA for problem 15

Source	SS	df	MS
Part	2.08	9	.231
Operator	.016	2	.008
Part×Operator	.0873	18	.00485
Error	.07	30	.00233
Total	2.254	59	

16. **Transmission Gear Measurement.** Cummins, Rosario, and Vanek studied two gauges used to measure ring gear height and bevel gear height in the production of transmission differentials. (Ring gear height and bevel gear height determine the milling points for the customized transmission housings, creating the horizontal location in the housing and the "tightness" of the casing against the differential.) A test stand (hydraulically) puts a 1000 pound force on the differential. This force is used to keep the differential from free spinning while allowing spin with some force applied. A 3-in Mitutoyo digital depth micrometer and a 6-in Mitutoyo digital depth micrometer were used to make the measurements. Vanek used the 3-in micrometer and took two ring gear height measurements on differential 8D4. Using the same 3-in Mitutoyo micrometer, Cummins made two ring gear height measurements on the same part. Vanek then took two bevel gear height measurements with the 6-in Mitutoyo micrometer on the same differential. Cummins followed with the same 6-in micrometer and took two bevel gear height measurements on differential 8D4. This protocol was repeated two more times for the differential 8D4. The whole procedure was then applied to differential 31D4. The data are given in Table 2.15. ANOVAs are given for both the ring gear data (SS and MS units are 10^{-4} in^2) and the bevel gear data (SS and MS units are 10^{-5} in^2) in Tables 2.16 and 2.17, respectively.

(a) Consider the ring gear heights measured with the 3-in Mitutoyo micrometer. Give the values of m, I, and J.

(b) In the context of the ring gear height measurements, what do m, I, and J represent?

(c) Give an ANOVA-based estimated repeatability standard deviation for ring gear height measuring. Find a range-based estimate of this quantity.

(d) Give an ANOVA-based estimated reproducibility standard deviation for ring gear height measuring.

(e) The upper and lower specifications for ring gear heights are, respectively, 1.92 in and 1.88 in. If the company requires the gauge capability ratio to be no larger than .05, does the 3-in Mitutoyo micrometer, as currently used, seem to meet this requirement? Why or why not?

TABLE 2.15. Data for problem 16

Ring gear heights (inches) (3-in Mitutoyo micrometer)			Bevel gear heights (inches) (6-in Mitutoyo micrometer)		
	Vanek	Cummins		Vanek	Cummins
8D4	1.88515	1.88470	8D4	5.49950	5.49850
	1.88515	1.88470		5.49985	5.49945
	1.88540	1.88380		5.49975	5.49945
	1.88530	1.88510		5.50000	5.50005
	1.88485	1.88435		5.49930	5.50070
	1.88490	1.88450		5.49945	5.49945
31D4	1.88365	1.88270	31D4	5.49785	5.49700
	1.88370	1.88295		5.49775	5.49710
	1.88330	1.88235		5.49765	5.49615
	1.88325	1.88235		5.49750	5.49615
	1.88270	1.88280		5.49670	5.49595
	1.88265	1.88260		5.49680	5.49620

TABLE 2.16. ANOVA for problem 16 ring gear data

Source	SS	df	MS
Differential	.219	1	.219
Operator	.021	1	.021
Differential×Operator	.0000042	1	.0000042
Error	.0249	20	.00124
Total	.2644	23	

TABLE 2.17. ANOVA for problem 16 bevel gear data

Source	SS	df	MS
Differential	4.44	1	4.44
Operator	.148	1	.148
Differential×Operator	.124	1	.124
Error	.550	20	.02752
Total	5.262	23	

(f) Repeat (a) through (e) for bevel gear heights measured with the 6-in Mitutoyo micrometer. Lower and upper specifications are, respectively, 5.50 in and 5.53 in for the bevel gear heights.

17. **Computer Locks.** Cheng, Lourits, Hugraha, and Sarief decided to study "tip diameter" for some computer safety locks produced by a campus machine shop. The team began its work with an evaluation of measurement precision for tip diameters. The data in Table 2.18 are in inches and represent two diameter measurements for each of two analysts made on all 25 locks machined on one day. An appropriate ANOVA is also given in Table 2.19. (The units for the SSs and MSs are 10^{-4} in^2.)

TABLE 2.18. Data for problem 17

Part	Lourits	Cheng	Part	Lourits	Cheng
1	.375, .375	.374, .374	14	.373, .373	.379, .374
2	.375, .375	.377, .376	15	.372, .373	.374, .373
3	.375, .373	.374, .375	16	.373, .373	.374, .374
4	.375, .373	.375, .374	17	.373, .373	.374, .373
5	.374, .374	.374, .374	18	.373, .373	.373, .373
6	.374, .374	.374, .375	19	.373, .373	.376, .373
7	.374, .375	.375, .376	20	.373, .373	.373, .373
8	.374, .375	.374, .373	21	.374, .374	.374, .375
9	.374, .374	.375, .375	22	.375, .375	.374, .377
10	.374, .374	.374, .374	23	.375, .375	.376, .377
11	.375, .373	.374, .374	24	.376, .375	.376, .374
12	.375, .374	.376, .374	25	.374, .374	.374, .375
13	.376, .373	.373, .374			

TABLE 2.19. ANOVA for problem 17

Source	SS	df	MS
Part	.58	24	.0242
Operator	.0625	1	.0625
Part×Operator	.22	24	.00917
Error	.445	50	.0089
Total	1.3075	99	

(a) Organizations typically establish their own guidelines for interpreting the results of gauge R&R studies. One set of guidelines is shown below. ($6\hat{\sigma}_{repeatability} \div (U - L)$ expressed as a percentage is sometimes called the "% gauge" for repeatability. $6\hat{\sigma}_{reproducibility} \div (U - L)$ expressed as a percentage is sometimes called the "% gauge" for reproducibility.)

% gauge	Rating
33 %	Unacceptable
20 %	Marginal
10 %	Acceptable
2 %	Good
1 %	Excellent

Suppose that specifications for the lock tip diameters are $.375 \pm .002$ in. According to the guidelines above and using ANOVA-based estimates, how does the diameter measuring process "rate" (based on "% gauge" for repeatability and "% gauge" for reproducibility)? Why?

(b) Find expressions for $\bar{y}_{\text{operator1}}$ and $\bar{y}_{\text{operator2}}$ as functions of the model terms used in the equation $y_{ijk} = \mu + \alpha_i + \beta_j + \alpha\beta_{ij} + \epsilon_{ijk}$.

(c) Continuing with (b) and applying logic consistent with that used to develop Eq. (2.31), what does $|\bar{y}_{\text{operator1}} - \bar{y}_{\text{operator2}}|/d_2(2)$ estimate in terms of $\sigma_\alpha^2, \sigma_\beta^2, \sigma_{\alpha\beta}^2$, and σ^2?

18. Refer to the **Computer Locks** case in problem 17. Consider the measurements made by Lourits. The sample average tip diameter for the ith randomly selected lock measured by Lourits can be written (holding only Lourits fixed) as

$$\bar{y}_{i\text{Lourits}} = \mu + \alpha_i + \beta_{\text{Lourits}} + \alpha\beta_{i\text{Lourits}} + \bar{\epsilon}_{i\text{Lourits}}.$$

(a) What is the random portion of $\bar{y}_{i\text{Lourits}}$?

(b) In terms of $\sigma^2, \sigma_\alpha^2, \sigma_\beta^2$, and $\sigma_{\alpha\beta}^2$, give the variance of your answer to part (a).

(c) Letting Γ be the range of the 25 variables $\bar{y}_{i\text{Lourits}}$, what does $\Gamma/d_2(25)$ estimate?

(d) Give the observed numerical value for $\Gamma/d_2(25)$ considered in part (c).

(e) In terms of $\sigma^2, \sigma_\alpha^2, \sigma_\beta^2$, and $\sigma_{\alpha\beta}^2$, what is the variance of (different) lock tip diameters as measured by a single operator (say Lourits) assuming there is no repeatability variation?

(f) In terms of $\sigma^2, \sigma_\alpha^2, \sigma_\beta^2$, and $\sigma_{\alpha\beta}^2$, what is the variance of (single-) diameter measurements made on (different) lock tips made by the same operator (say Lourits)? (Hint: This is your answer to (e) plus the repeatability variance, σ^2.)

(g) Using the Lourits data, find a range-based estimate of the repeatability variance.

(h) Using the Lourits data, find a range-based estimate of your answer to (e). (Hint: Use your answers for (d) and (g) appropriately.)

(i) Using the Lourits data, estimate your answer to (f). (Hint: Use your answers for (h) and (g) appropriately.)

19. **Implement Hardness.** Olsen, Hegstrom, and Casterton worked with a farm implement manufacturer on the hardness of a steel part. Before process monitoring and experimental design methodology were considered, the consistency of relevant hardness measurement was evaluated. Nine parts were obtained from a production line, and three operators agreed to participate in the measuring process evaluation. Each operator made two readings on each of nine parts. The data in Table 2.20 are in mm. An appropriate ANOVA is given in Table 2.21 (the units for the SSs and MSs are mm^2).

TABLE 2.20. Data for problem 19

Part	Operator A	B	C	Part	Operator A	B	C
1	3.30	3.25	3.30	6	3.30	3.30	3.25
	3.30	3.30	3.30		3.30	3.20	3.20
2	3.20	3.20	3.15	7	3.15	3.10	3.15
	3.25	3.30	3.30		3.20	3.20	3.20
3	3.20	3.20	3.25	8	3.25	3.20	3.20
	3.30	3.20	3.20		3.20	3.20	3.25
4	3.25	3.20	3.20	9	3.25	3.20	3.30
	3.30	3.25	3.20		3.30	3.30	3.40
5	3.25	3.10	3.20				
	3.30	3.10	3.15				

TABLE 2.21. ANOVA for problem 19

Source	SS	df	MS
Part	.08833	8	.01104
Operator	.01778	2	.00889
Part×Operator	.04139	16	.00259
Error	.0575	27	.002130
Total	.205	59	

(a) Say what each term in Eq. (2.26) means in the context of this problem.

(b) What are the values of I, J, and m in this study?

(c) Give an ANOVA-based estimate of the repeatability standard deviation, σ.

(d) Give an ANOVA-based estimate of the reproducibility standard deviation, $\sigma_{reproducibility}$.

(e) Estimate the gauge capability ratio using the ANOVA-based calculation if specifications on the hardness of this part are $nominal \pm 10$ mm.

(f) Using the corporate gauge rating table given in problem 17, rate the repeatability and the reproducibility of the hardness measurement method.

(g) Does it appear the current measuring process is adequate to check conformance to $nominal \pm 10$ mm hardness specifications? Why or why not?

20. Refer to the **Implement Hardness** case in problem 19.

    (a) Suppose each operator used a different gauge to measure hardness. How would this affect the interpretation of your calculations in problem 19?

    (b) If it were known that measuring alters the part hardness in the vicinity of the point tested, how should this be addressed in a gauge R&R study?

    (c) When an operator measures the same part two times in a row, it is likely the second measurement is "influenced" by the first in the sense that there is psychological pressure to produce a second measurement like the initial one. How might this affect results in a gauge R&R study? How could this problem be addressed/eliminated?

21. Is it important to include an evaluation of measuring processes early in a quality improvement effort? Why or why not?

22. Management tells engineers involved in a quality improvement project "We did a gauge R&R study last year and the estimated gauge capability ratio was .005. You don't need to redo the study." How should the engineers respond and why?

23. **Paper Weight.** Everingham, Hart, Hartong, Spears, and Jobe studied the top loading balance used by the Paper Science Department at Miami University, Oxford, Ohio. Two 20 cm × 20 cm (400 cm^2) pieces of 20 lb bond paper were cut from several hundred feet of paper made in a departmental laboratory. Weights of the pieces obtained using the balance are given below in grams. The numbers in parentheses specify the order in which the measurements were made. (Piece 1 was measured 15 times, three times by each operator. That is, piece 1 was measured first by Spears, second by Spears, third by Hart,..., 14th by Hartong, and lastly by Jobe.) Different orders were used for pieces 1 and 2, and both were determined using a random number generator. Usually, the upper specification minus the lower specification $(U - L)$ is about 4 g/m^2 for the density of this type of paper. An appropriate ANOVA is given below (units for the SSs and MSs are g^2).

Piece	Hartong	Hart	Spears	Everingham	Jobe
1	(14) 3.481	(3) 3.448	(1) 3.485	(13) 3.475	(10) 3.472
	(12) 3.477	(9) 3.472	(2) 3.464	(4) 3.472	(5) 3.470
	(7) 3.470	(6) 3.470	(11) 3.477	(8) 3.473	(15) 3.474
2	(1) 3.258	(13) 3.245	(7) 3.256	(6) 3.249	(11) 3.241
	(2) 3.254	(12) 3.247	(5) 3.257	(15) 3.238	(8) 3.250
	(3) 3.258	(9) 3.239	(10) 3.245	(14) 3.240	(4) 3.254

ANOVA table for weight

Source	SS	df	MS
Piece	.37386	1	.37386
Operator	.00061	4	.000152
Piece×Operator	.00013	4	.000032
Error	.00095	20	.000047
Total	.37555	29	

(a) What purpose is potentially served by randomizing the order of measurement as was done in this study?

(b) Give the table of operator×piece ranges, R_{ij}.

(c) Give the table of operator×piece averages, $\bar{y}_{ij}$.

(d) Give the ranges of the operator×piece means, Δ_i.

(e) Express the observed weight range determined by Spears for piece 2 in g/m². (Note: $10^4 \text{ cm}^2 = 1 \text{ m}^2$.)

(f) Find a gauge repeatability rating based on ranges. (See part (a) of problem 17.) Pay attention to units.

(g) Find a gauge reproducibility rating based on ranges. (Again see part (a) of problem 17 and pay attention to units.)

(h) Calculate an estimated gauge capability ratio. Pay attention to units.

(i) What minimum value for $(U-L)$ would guarantee an estimated gauge capability ratio of at most .1?

(j) Using ANOVA-based estimates, answer (f)–(h).

(k) Using ANOVA-based estimates, give an exact 95% confidence interval for $\sigma_{\text{repeatability}}$. Your units should be g/m².

(l) Using the ANOVA-based estimates, give 95% approximate confidence limits for $\sigma_{\text{reproducibility}}$. Your units should be g/m².

24. **Paper Thickness.** Everingham, Hart, Hartong, Spears, and Jobe continued their evaluation of the measuring equipment in the Paper Science Laboratory at Miami University by investigating the repeatability and reproducibility of the TMI automatic micrometer routinely used to measure

paper thickness. The same two 20 cm × 20 cm pieces of 20 lb bond paper referred to in problem 23 were used in this study. But unlike measuring weight, measuring thickness alters the properties of the portion of the paper tested (by compressing it and thus changing the thickness). So an 8 × 8 grid was marked on each piece of paper. The corresponding squares were labeled $1, 2, \ldots, 64$ left to right, top to bottom. Ten squares from a given piece were randomly allocated to each operator (50 squares from each piece were measured). Because so many measurements were to be made, only the "turn" for each analyst was determined randomly, and each operator made all ten of his measurements on a given piece consecutively. A second randomization and corresponding order of measurement was made for piece 2. Hartong measured third on piece 1 and fifth on piece 2, Hart was first on piece 1 and third on piece 2, Spears was fifth and fourth, Everingham was second and second, and Jobe was fourth and first. The data are in Table 2.22 (in mm). The numbers in parenthesis identify the squares (from a given piece) measured. (Thus, for piece 1, Hart began the measurement procedure by recording thicknesses for squares $51, 54, 18, 63, \ldots, 7$; then Everingham measured squares $33, 38, \ldots, 5$, etc. After the data for piece 1 were obtained, measurement on piece 2 began. Jobe measured squares $9, 3, \ldots, 22$; then Everingham measured squares $43, 21, \ldots, 57$, etc.) An appropriate ANOVA is also given in Table 2.23 (units for the SSs and MSs are mm^2).

(a) Say what each term in Eq. (2.26) means in the context of this problem.

(b) How is this study different from a "garden-variety" gauge R&R study?

(c) Will the nonstandard feature of this study tend to increase, decrease, or have no effect on the estimate of the repeatability standard deviation? Why?

(d) Will the nonstandard feature of this study tend to increase, decrease, or have no effect on the estimated standard deviation of measurements from a given piece across many operators? Why?

(e) Give the ANOVA-based estimated standard deviation of paper thickness measurements for a fixed piece × operator combination, i.e., approximate the repeatability standard deviation assuming that square-to-square variation is negligible.

(f) Give the ANOVA-based estimated standard deviation of thicknesses measured on a fixed piece across many operators. (The quantity being estimated should include but not be limited to variability for a fixed piece × operator combination.) That is, approximate the reproducibility standard deviation assuming square-to-square variation is negligible.

(g) What percent of the overall measurement variance is due to repeatability? What part is due to reproducibility?

TABLE 2.22. Data for problem 24

Piece	Hartong	Hart	Spears	Everingham	Jobe
1	(14) .201	(51) .195	(48) .192	(33) .183	(43) .185
	(25) .190	(54) .210	(58) .191	(38) .189	(40) .204
	(17) .190	(18) .200	(15) .198	(36) .196	(49) .194
	(21) .194	(63) .203	(55) .197	(3) .195	(12) .199
	(53) .212	(20) .196	(44) .207	(59) .192	(29) .192
	(16) .209	(50) .189	(23) .202	(45) .195	(13) .193
	(47) .208	(31) .205	(64) .196	(41) .185	(56) .190
	(42) .192	(37) .203	(57) .188	(9) .193	(2) .195
	(22) .198	(34) .195	(26) .201	(62) .194	(8) .199
	(35) .191	(7) .186	(1) .181	(5) .194	(6) .197
2	(5) .188	(14) .186	(55) .177	(43) .179	(9) .191
	(16) .173	(24) .171	(51) .174	(21) .194	(3) .180
	(11) .188	(62) .178	(36) .184	(18) .187	(42) .194
	(47) .180	(34) .175	(12) .180	(39) .175	(50) .183
	(25) .178	(29) .183	(38) .179	(6) .173	(53) .181
	(15) .188	(10) .185	(41) .186	(7) .179	(17) .188
	(56) .166	(30) .190	(63) .183	(64) .171	(33) .188
	(26) .173	(40) .177	(45) .172	(54) .184	(23) .173
	(8) .175	(58) .184	(31) .174	(59) .181	(60) .180
	(52) .183	(13) .186	(2) .178	(57) .187	(22) .176

TABLE 2.23. ANOVA for problem 24

Source	SS	df	MS
Piece	.00557	1	.00557
Operator	.00018	4	.000045
Piece×Operator	.00028	4	.00007
Error	.003986	90	.000044
Total	.010013	99	

25. **Paper Burst Strength.** An important property of finished paper is the force (lb/in^2) required to burst or break through it. Everingham, Hart, Hartong, Spears, and Jobe investigated the repeatability and reproducibility of existing measurement technology for this paper property. A Mullen tester in the Miami University Paper Science Department was studied. Since the same two 20 cm × 20 cm pieces of paper referred to in problems 23 and 24 were available, the team used them in its gauge R&R study for burst strength measurement. The burst test destroys the portion of paper tested, so repeat measurement of exactly the same paper specimen is not possible. Hence, a grid of 10 approximately equal-sized rectangles, 10 cm × 4 cm (each large

enough for the burst tester), was marked on each large paper piece. Each of the analysts was assigned to measure burst strength on two randomly selected rectangles from each piece. The measurement order was also randomized among the five operators for each paper piece. The data obtained are shown below. The ordered pairs specify the rectangle measured and the order of measurement. (For example, the ordered pair (2,9) in the top half of the table indicates that 8.8 lb/in^2 was obtained from rectangle number 2, the ninth rectangle measured from piece 1.) An ANOVA table for this study is also provided.

Piece	Hartong	Hart	Spears	Everingham	Jobe
1	(9,2) 13.5	(6,6) 10.5	(4,8) 12.9	(2,9) 8.8	(3,10) 12.4
	(7,5) 14.8	(5,1) 11.7	(1,4) 12.0	(8,3) 13.5	(10,7) 16.0
2	(3,9) 11.3	(1,8) 14.0	(5,6) 13.0	(6,7) 12.6	(2,1) 11.0
	(8,10) 12.0	(7,5) 12.5	(9,3) 13.1	(4,2) 12.7	(10,4) 10.6

ANOVA table for burst strength

Source	SS	df	MS
Piece	.5445	1	.5445
Operator	2.692	4	.6730
Piece×Operator	24.498	4	6.1245
Error	20.955	10	2.0955
Total	48.6895	19	

In the following, assume that specimen-to-specimen variation within a given piece of paper is negligible.

(a) To what set of operators can the conclusions of this study be applied?

(b) To what set of paper pieces can the conclusions of this study correctly be applied?

(c) What are the values of I, J, and m in this study?

(d) Give an ANOVA-based estimate of the repeatability standard deviation, σ.

(e) Give another estimate of the repeatability standard deviation, σ, this time based on ranges.

(f) Find an ANOVA-based estimate of $\sigma_{\text{reproducibility}}$.

(g) Find another estimate of $\sigma_{\text{reproducibility}}$, this one based on ranges

(h) Using the ANOVA, estimate the standard deviation of single burst measurements on a fixed piece of paper made by many operators, $\sigma_{\text{R\&R}}$.

26. **Paper Tensile Strength.** The final type of measurement method studied by Everingham, Hart, Hartong, Spears, and Jobe in the Paper Science Laboratory at Miami University was that for paper tensile strength. Since the burst tests discussed in problem 25 destroyed the 20 cm × 20 cm pieces of 20 lb bond paper referred to there, two new 20 cm × 20 cm pieces of paper were selected from the same run of paper. Ten 15 cm × 20 cm strips were cut from each 20 cm × 20 cm piece. Each set of ten strips was randomly allocated among the five operators (two strips per operator for each set of ten). The order of testing was randomized for the ten strips from each piece, and the same Thwing-Albert Intellect 500 tensile tester was used by each operator to measure the load required to pull apart the strips. The data appear below in kg. (Consider, e.g., the data given for piece 1, Hartong, (9,2) 4.95. A 4.95 - kg load was required to tear strip number 9 from piece 1 and the measurement was taken second in order among the ten strips measured for piece 1.) Since the testing destroyed the strip, the analysts had to assume strip-to-strip variation for a given piece to be negligible. An appropriate ANOVA is also given below (units for SSs and MSs are kg^2).

Piece	Everingham	Hart	Hartong	Spears	Jobe
1	(2,8) 4.34	(1,5) 4.34	(9,2) 4.95	(6,6) 4.03	(10,4) 4.51
	(8,10) 4.71	(4,3) 4.61	(7,7) 4.53	(3,9) 3.62	(5,1) 4.56
2	(4,7) 5.65	(6,6) 4.80	(1,1) 4.38	(2,2) 4.65	(9,5) 4.30
	(8,9) 4.51	(10,8) 4.75	(3,3) 3.89	(5,4) 5.06	(7,10) 3.87

ANOVA table for tensile strength

Source	SS	df	MS
Piece	.13778	1	.1378
Operator	.69077	4	.17269
Piece×Operator	1.88967	4	.47242
Error	1.226	10	.1226
Total	3.9442	19	

(a) Make a table of load averages, $\bar{y}_{ij}$, for the ten operator×piece combinations.

(b) Plot the load averages $\bar{y}_{ij}$ versus piece number for each of the operators (connect the two $\bar{y}_{ij}$s for each operator).

(c) Suppose the target tensile strength for strips of 20 lb bond paper is 4.8 kg. Typically, upper and lower specifications for paper properties are set 5% above and below a target. Estimate the gauge capability ratio under these conditions, using ANOVA-based calculations.

(d) If upper and lower specifications for tensile strength of 20 lb bond paper are equal distances above and below a target of 4.8 kg, find the upper and lower limits such that the estimated gauge capability ratio is .01.

(e) Redo part (d) for an estimated gauge capability ratio of .1.

(f) Is it easier to make a gauge capability ratio better (smaller) by increasing its denominator or decreasing its numerator? Will your answer lead to a more consistent final product? Why or why not?

CHAPTER 3

PROCESS MONITORING

This chapter discusses the important topic of process monitoring using so-called control charts. These are devices for the routine and organized plotting of process performance measures, with the goal of identifying process changes. When these are detected, those running the process can either intervene and set things aright (if the change is detrimental) or try to make permanent the source of an unexpected process improvement.

The discussion begins with some control charting philosophy in Sect. 3.1. Then the standard Shewhart control charts for both measurements/"variables data" and counts/"attributes data" are presented in consecutive Sects. 3.2 and 3.3. Section 3.4 discusses qualitative interpretation and practical implications of patterns sometimes seen on Shewhart charts and some sets of rules often applied to check for such patterns. Then there is a presentation of the so-called average run length concept that can be used to quantify what a particular process-monitoring scheme can be expected to provide in Sect. 3.5. Finally, in Sect. 3.6 the chapter closes with a discussion clarifying the relationship between "statistical process control" and "engineering control" and presenting some basic concepts of so-called proportional-integral-derivative (PID) engineering control schemes.

3.1 Generalities About Shewhart Control Charting

Section 1.2.1 introduced the notion of process "stability" as consistency over time in the pattern of process variation. Walter Shewhart, working at Bell Labs in the late 1920s and early 1930s, developed an extremely powerful and simple tool for investigating whether a process can be sensibly thought of as stable. He called it a "control chart." Some 90 plus years after the fact, your authors would prefer (for reasons laid out in Sect. 3.6) that Shewhart had chosen instead the name "monitoring chart." Nevertheless, this book will use Shewhart's terminology and the "monitoring chart" terminology interchangeably.

Shewhart's fundamental conceptualization was that while some variation is inevitable in any real process, overall variation seen in process data can be decomposed as

Shewhart's Grand Insight

$$\text{observed variation} = \text{baseline variation} + \text{variation that can be eliminated}. \quad (3.1)$$

Shewhart conceived of baseline variation as that variability in production and measurement which will remain even under the most careful process monitoring and appropriate physical intervention. It is an inherent property of a combination of system configuration and measurement methodology that cannot be reduced without basic changes in the physical process or how it is run or observed. This variation is sometimes called variation due to "system" or "common" (universal) causes. Other names for it that will be used in this book are "random" or "short-term" variation. It is the kind of variation expected under the best of circumstances, measuring item to consecutive item produced on a production line. It is a variation that comes from many small, unnameable, and unrecognized physical causes. When only this kind of variation is present, it is reasonable to call a process "stable" and model observations on it as independent random draws from a fixed population or universe.

The second component of overall variability portrayed in Eq. (3.1) is that which can potentially be eliminated by careful process monitoring and wise physical intervention (when such is warranted). This has variously been called "special cause" or "assignable" cause variation and "nonrandom" and "long-term" variation. It is the kind of systematic, persistent change that accompanies real (typically unintended) physical alteration of a process (or the measurement system used to observe it). It is a change that is large enough that one can potentially track down and eliminate its root cause, leaving behind a stable process.

If one accepts Shewhart's conceptualization (3.1), the problem then becomes one of detecting the presence of the second kind of variation so that appropriate steps can be taken to eliminate it. The Shewhart control chart is a tool for making such detection.

Shewhart's charting method is this. One periodically takes samples from the process of interest (more will be said later about the timing and nature of these samples) and computes a statistic meant to summarize process behavior at the period in question. Values of the statistic are plotted against time order of observation and compared to so-called **control limits** drawn on the chart. These separate values of the statistic that are plausible if the process is in fact stable, from those that are rare or implausible under this scenario. As long as the plotted points remain inside the control limits, one presumes that all is well (the process is stable) and does not intervene in its workings. (This is an oversimplification of how these charts are often used that will be corrected in Sect. 3.4. But for the time being, this simplified picture will suffice.) When a point plots outside control limits, there is an indication that a physical change has probably taken place and that intervention is appropriate. Figure 3.1 shows a generic Shewhart control chart where the plotted statistic is Q; upper and lower control limits are UCL_Q, and LCL_Q respectively; and there is one "out-of-control" point.

Control Limits

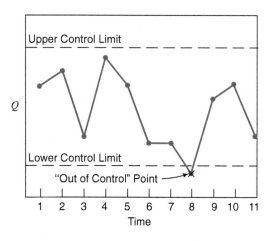

FIGURE 3.1. Generic Shewhart control chart (for a statistic Q)

There are many different kinds of Shewhart charts, corresponding to various choices of the plotted statistic, Q. Some of these chart types will be discussed in the next two sections. But before moving to discussion of specific charts, several generalities remain to be considered. First, there is the question of how one sets the control limits, UCL_Q and LCL_Q.

Shewhart's suggestion for setting control limits was essentially the following. If one can model the process output under stable conditions (i.e., if one can specify a sensible probability distribution for individual observations made on the process), then probability theory can often be invoked to produce a corresponding distribution for Q. Then small upper and lower percentage points for this distribution can provide the necessary control limits. The thinking is that only rarely will values outside these be seen under stable process conditions. Further, rather than working explicitly with probability tables or formulas for a distribution of Q, one often simply makes use of the fact that for many probability distributions, most of the probability is within three standard deviations of the mean. So, if μ_Q and σ_Q are, respectively, a stable process mean and standard deviation for Q, then common control limits are

Generic 3-Sigma Control Limits

$$UCL_Q = \mu_Q + 3\sigma_Q \quad \text{and} \quad LCL_Q = \mu_Q - 3\sigma_Q . \tag{3.2}$$

Further, it is common to draw in a "center line" on a Shewhart control chart at

Generic Center Line

$$CL_Q = \mu_Q . \tag{3.3}$$

To make this discussion slightly more concrete, consider briefly the situation where the plotted statistic is the sample mean of n individual measurements, $Q = \bar{x}$. If the process output can be modeled as independent selections from a distribution with mean μ and standard deviation σ, the statistic $\bar{x}$ has a distribution with mean $\mu_Q = \mu_{\bar{x}} = \mu$ and standard deviation $\sigma_Q = \sigma_{\bar{x}} = \sigma/\sqrt{n}$. Then applying relationships (3.2) and (3.3), it follows that typical control limits for $\bar{x}$ are

$$UCL_{\bar{x}} = \mu + 3\frac{\sigma}{\sqrt{n}} \quad \text{and} \quad LCL_{\bar{x}} = \mu - 3\frac{\sigma}{\sqrt{n}} , \tag{3.4}$$

with a center line drawn at μ.

Display (3.4) helps bring into focus another general issue regarding Shewhart control charting. As in limits (3.4), process parameters (like μ and σ) typically appear in formulas for control limits. Values for them must come from somewhere in order to apply a control chart, and there are two possibilities in this regard. Sometimes past experience with a process, engineering standards, or other considerations made prior to process monitoring specify what values should be used.

Standards Given Context

This kind of situation is commonly known as a **standards given** scenario. In other circumstances, one has no information on a process outside a series of samples that are presented along with the question "Is it plausible that the process was physically stable over the period represented by these data?" In such a case, all that one can do is tentatively assume that in fact the process was stable, make provisional estimates of process parameters and plug them into formulas for control limits, and apply those limits to the data in hand as a means of criticizing the tentative assumption of stability. This kind of situation is sometimes called an

Retrospective Context

as past data scenario and will often be referred to in this text as a **retrospective** scenario.

The difference between what is possible in standards given and retrospective contexts can be thought of in terms of two different questions addressed in the two situations. In a standards given context, with each new sample, one can face the question

Are process parameters currently at their standard values?

In a retrospective context, one can only wait until a number of samples have been collected (often, a minimum of 20–25 time periods is recommended) and then looking back over the data ask the question

Are these data consistent with *any* fixed set of process parameters?

Having introduced the notion of control limits, it is important to warn readers of a common pitfall. That is the confusion that students (and even practicing engineers) often have regarding the *much different* concepts of control limits and engineering specifications. Control limits have to do with assessing process stability. They refer to a statistic Q. They are usually derived from what a process has done in the past or is currently doing. On the other hand, engineering specifications have to do with assessing product acceptability or functionality. They almost always refer to individual measurements. They are usually derived from product performance requirements and may have little or nothing to do with the inherent capability of a process to produce a product meeting those requirements.

Control Limits vs. Engineering Specifications

Despite these real differences in meaning, people often confuse these concepts (e.g., applying specifications to sample means as if they were control limits or arguing that since a mean or individual is inside control limits for $\bar{x}$, the product being monitored is acceptable). But it is vital that these notions be kept separate and applied in their proper contexts. (Notice that a process that is stable and producing Qs inside appropriate control limits need *not* be producing mostly acceptable product. And conversely, a process may produce product that is acceptable by current engineering standards, but nevertheless be very unstable!)

Another issue needing some general discussion here is the matter of sampling. How should one go about gathering the data to be used in control charting? This matter includes the issue sometimes referred to as **rational subgrouping** or rational sampling. When one is collecting process-monitoring data, it is important that anything one intends to call a single "sample" be collected over a short enough time span that there is little question that the process was physically stable during the data collection period. It must be clear that an "independent draws from a single-population/single-universe" model is appropriate for describing data in a given sample. This is because the variation within such a sample essentially specifies the level of background noise against which one looks for process changes. If what one calls "samples" often contain data from genuinely different process conditions, the apparent level of background noise will be so large that it will be hard to see important process changes. In high-volume manufacturing applications of control charts, single samples (rational subgroups) typically consist of n consecutive items taken from a production line. On the other hand, in extremely low-volume operations, where one unit might take many hours to produce and there is significant opportunity for real process change between consecutive units, the only natural samples may be of size $n = 1$.

Once one has determined to group only observations close together in time into samples or subgroups, there is still the question of how often these samples should be taken. When monitoring a machine that turns out 1000 parts per hour, where samples are going to consist of $n = 5$ consecutive parts produced on the machine, does one sample once every minute, once every hour, once every day, or what? An answer to this kind of question depends upon what one expects in terms of process performance and the consequences of process changes. If the consequences of a process change are disastrous, one is pushed toward frequent samples. The same is true if significant process upsets are a frequent occurrence. On the other hand, if a process rarely experiences changes and even when those occur, only a moderate loss is incurred when it takes a while to discover them, long intervals between samples are sensible.

As a final matter in this introductory discussion of Shewhart charting philosophy, we should say what control charting can and cannot reasonably be expected to provide. It can signal the need for process intervention and can keep one from ill-advised and detrimental overadjustment of a process that is behaving in a stable fashion. But in doing so, what is achieved is simply reducing variation to the minimum possible for a given system configuration (in terms of equipment, methods of operation, methods of measurement, etc.). Once that minimum has been reached, what is accomplished is maintaining a *status quo* best possible process performance. (Remember, e.g., the use of the "control" step in the Six Sigma paradigm discussed on page 9.) In today's global economy, standing still is never good enough for very long. Achieving process stability provides a solid background against which to evaluate possible innovations and fundamental/order-of-magnitude improvements in production methods. But it does not itself guide their discovery. Of the tools discussed in this book, it is the methods of experimental design and analysis covered in Chaps. 5 and 6 that have the most to say about aiding fundamental innovations.

Section 3.1 Exercises

1. What can control charting contribute to a process-improvement effort?

2. What is the difference between "standards given" and "retrospective" control charting?

3. What is the difference between common cause and special cause variation? Which type of variation are control charts designed to detect?

4. What happens to the control limits (3.4) for an $\bar{x}$ chart as the subgroup size gets large?

5. How do you expect the behavior of a control charting scheme to change if a value smaller than 3 is used in limits (3.2)?

6. How do you expect the behavior of a control charting scheme to change if a value larger than 3 is used in limits (3.2)?

7. If the plotted statistic Q is inside appropriately constructed control limits (indicating that a process is stable), does that necessarily imply that the process is producing acceptable product? Briefly explain.

8. If the plotted statistic Q is regularly outside appropriately constructed control limits (indicating that a process is unstable), does that necessarily imply that the process is producing unacceptable product? Briefly explain.

9. The same item is being produced on two production lines. Every 15 min, 5 items are sampled from each line, and a feature of interest is measured on each item. Some statistic Q is calculated for each set of 5 measurements from each line and plotted versus time. Analyst 1 puts all 10 items together into a single group (5 from line 1 and 5 from line 2), calculates a value of the statistic Q, and plots it. (This person says, "After all, is not a larger sample size better?") Analyst 2 keeps the data from the two different lines separate and makes a different control chart for each production line.

   (a) What subgroup size is Analyst 1 using?

   (b) What subgroup size is Analyst 2 using?

   (c) Which analyst is making the most appropriate chart? Why? (Hint: Consider the concept of rational subgrouping. See also the discussion of stratification on page 141.)

3.2 Shewhart Charts for Measurements/"Variables Data"

This section considers the problem of process monitoring when the data available are measurements (as opposed to counts or the kind of 0/1 calls considered in Sect. 2.6). Sometimes the terminology "variables data" is used in this context. In such situations, it is common to make charts for both the process location and also for the process spread (size of the process short-term variability). So this section will consider the making of $\bar{x}$ and median ($\tilde{x}$) charts for location and R and s charts for spread.

3.2.1 Charts for Process Location

The most common of all Shewhart control charts is that for means of samples of n measurements, the case where $Q = \bar{x}$. As was discussed in the previous section (and portrayed in display (3.4)), the fact that sampling from a distribution with mean μ and standard deviation σ produces sample averages with expected value $\mu_{\bar{x}} = \mu$ and standard deviation $\sigma_{\bar{x}} = \sigma/\sqrt{n}$ suggests **standards given** Shewhart control limits for $\bar{x}$

$$UCL_{\bar{x}} = \mu + 3\frac{\sigma}{\sqrt{n}} \quad \text{and} \quad LCL_{\bar{x}} = \mu - 3\frac{\sigma}{\sqrt{n}}, \quad (3.5)$$

Standards Given $\bar{x}$ Chart Control Limits

Chapter 3. Process Monitoring

and center line at

Standards Given $\bar{x}$ Chart Center Line

$$CL_{\bar{x}} = \mu .$$

Example 23 *Monitoring the Surface Roughness of Reamed Holes.* Dohm, Hong, Hugget, and Knoot worked with a manufacturer on a project involving roughness measurement after the reaming of preformed holes in a metal part. Table 3.1 contains some summary statistics (the sample mean $\bar{x}$, the sample median $\tilde{x}$, the sample range R, and the sample standard deviation s) for 20 samples (taken over a period of 10 days) of $n = 5$ consecutive reamed holes.

Suppose for the time being that standards (established on the basis of previous experience with this reaming process) for surface roughness are $\mu = 30$ and $\sigma = 4$. Then, standards given control limits for the $\bar{x}$ values in Table 3.1 are

$$UCL_{\bar{x}} = 30 + 3\frac{4}{\sqrt{5}} = 35.37$$

and

$$LCL_{\bar{x}} = 30 - 3\frac{4}{\sqrt{5}} = 24.63 .$$

TABLE 3.1. Summary statistics for 20 samples of five surface roughness measurements on reamed holes (μ in)

Sample	$\bar{x}$	$\tilde{x}$	R	s
1	34.6	35	9	3.4
2	46.8	45	23	8.8
3	32.6	34	12	4.6
4	42.6	41	6	2.7
5	26.6	28	5	2.4
6	29.6	30	2	0.9
7	33.6	31	13	6.0
8	28.2	30	5	2.5
9	25.8	26	9	3.2
10	32.6	30	15	7.5
11	34.0	30	22	9.1
12	34.8	35	5	1.9
13	36.2	36	3	1.3
14	27.4	23	24	9.6
15	27.2	28	3	1.3
16	32.8	32	5	2.2
17	31.0	30	6	2.5
18	33.8	32	6	2.7
19	30.8	30	4	1.6
20	21.0	21	2	1.0

Figure 3.2 is a standards given $\bar{x}$ chart for the surface roughness measurements. Based on this chart, one would detect the fact that the reaming process is not stable at standard process parameters as early as the second sample. Several of the sample means fall outside control limits, and had the control limits been applied to the data as they were collected, the need for physical intervention would have been signaled.

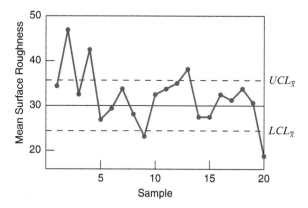

FIGURE 3.2. Standards given $\bar{x}$ chart for surface roughness

In order to make a **retrospective** $\bar{x}$ chart, one must derive estimates of the process parameters μ and σ from data in hand (temporarily assuming process stability) and plug them into the formulas (3.5). There are many possible ways of doing this, each leading to slightly different retrospective control limits. Here only the most common ones will be considered and we begin with the matter of estimating μ.

Let r stand for the number of samples available in a retrospective $\bar{x}$ chart analysis. One way of estimating a supposedly common process mean for the r periods is to simply average the r sample means. Standard control charting practice is to use

$$\bar{\bar{x}} = \frac{1}{r} \sum_{i=1}^{r} \bar{x}_i$$

Average Sample Mean

as an estimator of μ in making retrospective control limits for $\bar{x}$.

An answer to the question of how to estimate σ is not so obvious. The estimator of σ with the best theoretical properties is obtained by pooling the r sample variances to obtain (in the constant sample size case)

$$s^2_{\text{pooled}} = \frac{1}{r} \sum_{i=1}^{r} s_i^2,$$

and then taking the square root. However, this method is not common in practice (due to historical precedent). Instead, common practice is to use estimators based on the average sample range or the average sample standard deviation.

Consider first the estimation of σ based on

Average Sample Range

$$\overline{R} = \frac{1}{r}\sum_{i=1}^{r} R_i.$$

As in the discussion of range-based estimation in gauge R&R on page 66, if process output is normally distributed at time period i,

$$\mathrm{E}R_i = d_2 \sigma$$

and thus

$$\mathrm{E}\left(\frac{R_i}{d_2}\right) = \sigma.$$

(The dependence of d_2 on n is not being displayed here, since there is no chance of confusion regarding which "sample size" is under discussion.) So assuming the process is stable over all r periods, all sample sizes are n, and that a normal distribution governs the data generation process,

$$\frac{\overline{R}}{d_2}$$

is a sensible estimator of σ. Plugging this and $\overline{\overline{x}}$ into the standards given control limits for $\overline{x}$ provided in display (3.5), one obtains retrospective Shewhart control limits for $\overline{x}$:

$$UCL_{\overline{x}} = \overline{\overline{x}} + 3\frac{\overline{R}}{d_2\sqrt{n}} \quad \text{and} \quad LCL_{\overline{x}} = \overline{\overline{x}} - 3\frac{\overline{R}}{d_2\sqrt{n}}. \quad (3.6)$$

Further, one can define a constant A_2 (depending upon n) by

$$A_2 = \frac{3}{d_2\sqrt{n}},$$

and rewrite display (3.6) more compactly as

Retrospective Control Limits for $\overline{x}$ Based on the Average Range

$$UCL_{\overline{x}} = \overline{\overline{x}} + A_2\overline{R} \quad \text{and} \quad LCL_{\overline{x}} = \overline{\overline{x}} - A_2\overline{R}. \quad (3.7)$$

Values of A_2 can be found in the table of control chart constants, Table A.5.

As an alternative to estimating σ on the basis of sample ranges, next consider estimating σ based on the average sample standard deviation

Average Sample Standard Deviation

$$\overline{s} = \frac{1}{r}\sum_{i=1}^{r} s_i.$$

When sampling from a normal distribution with standard deviation σ, the sample standard deviation, s, has a mean that is not quite σ. The ratio of the mean of s to σ is commonly called c_4. (c_4 depends upon the sample size and again is tabled in Table A.5, but it will not be necessary to display the dependence of c_4 on n.) Thus, if one assumes that process output is normally distributed at period i,

$$\mathrm{E}\left(\frac{s_i}{c_4}\right) = \sigma.$$

So assuming the process is stable over all r periods, all sample sizes are n, and that a normal distribution governs the data generation process,

$$\frac{\bar{s}}{c_4}$$

is a sensible estimator of σ. Plugging this and $\bar{\bar{x}}$ into the standards given control limits for $\bar{x}$ provided in display (3.5) one obtains retrospective Shewhart control limits for $\bar{x}$

$$UCL_{\bar{x}} = \bar{\bar{x}} + 3\frac{\bar{s}}{c_4\sqrt{n}} \quad \text{and} \quad LCL_{\bar{x}} = \bar{\bar{x}} - 3\frac{\bar{s}}{c_4\sqrt{n}}. \quad (3.8)$$

Further, one can define another constant A_3 (depending upon n) by

$$A_3 = \frac{3}{c_4\sqrt{n}},$$

and rewrite display (3.8) more compactly as

$$UCL_{\bar{x}} = \bar{\bar{x}} + A_3\bar{s} \quad \text{and} \quad LCL_{\bar{x}} = \bar{\bar{x}} - A_3\bar{s}. \quad (3.9)$$

Values of A_3 can also be found in the table of control chart constants, Table A.5.

Retrospective Control Limits for $\bar{x}$ Based on the Average Standard Deviation

Example 24 *(Example 23 continued.) Returning to the reaming study, from Table 3.1*

$$\bar{\bar{x}} = 32.1, \bar{R} = 8.95, \text{ and } \bar{s} = 3.76.$$

Further, for $n = 5$ (which was the sample size used in the study), Table A.5 shows that $A_2 = .577$ and $A_3 = 1.427$. Thus, from formulas (3.7), retrospective control limits for $\bar{x}$ based on $\bar{R}$ are

$$UCL_{\bar{x}} = 32.1 + .577(8.95) = 37.26 \quad \text{and} \quad LCL_{\bar{x}} = 32.1 - .577(8.95) = 26.94.$$

And from formulas (3.9), retrospective control limits for $\bar{x}$ based on $\bar{s}$ are

$$UCL_{\bar{x}} = 32.1 + 1.427(3.76) = 37.47 \text{ and } LCL_{\bar{x}} = 32.1 - 1.427(3.76) = 26.73.$$

Figure 3.3 on page 118 shows the retrospective $\bar{x}$ control chart with control limits based on $\bar{R}$. It is clear from this chart (as it would be using the limits based on $\bar{s}$) that the reaming process was not stable over the period of the study. The mean measured roughness fluctuated far more than one would expect under any stable process model.

118   Chapter 3. Process Monitoring

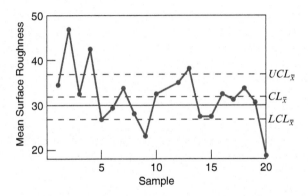

FIGURE 3.3. Retrospective $\bar{x}$ chart for surface roughness

$\bar{x}$ charts are by far the most common charts for monitoring process location, but there is an alternative worth mentioning. That is to use **sample medians** in place of sample means ($\tilde{x}$ in place of $\bar{x}$). This alternative has the advantage of requiring less in the way of computational skills from those who must compute the values to be plotted, but has the drawback of being somewhat less sensitive to changes in process location than the $\bar{x}$ chart.

The basic probability facts that lead to control limits for $\tilde{x}$ concern sampling from a normal distribution. For a sample of size n from a normal distribution with mean μ and standard deviation σ, the random variable $\tilde{x}$ has mean $\mu_{\tilde{x}} = \mu$ and standard deviation $\sigma_{\tilde{x}} = \kappa \sigma_{\bar{x}} = \kappa \sigma / \sqrt{n}$ for a constant κ (depending upon n). Table 3.2 gives a few values of κ.

TABLE 3.2. Ratios κ between $\sigma_{\tilde{x}}$ and $\sigma_{\bar{x}}$ when sampling from a normal distribution

n	3	5	7	9	11	∞
κ	1.160	1.197	1.214	1.223	1.229	$\sqrt{\pi/2}$

Applying these facts about the probability distribution of $\tilde{x}$ under a normal process model and the generic Shewhart control limits given in display (3.2) produces standards given control limits for $\tilde{x}$

Standards Given Control Limits for Medians

$$UCL_{\tilde{x}} = \mu + 3\kappa \frac{\sigma}{\sqrt{n}} \quad \text{and} \quad LCL_{\tilde{x}} = \mu - 3\kappa \frac{\sigma}{\sqrt{n}}. \quad (3.10)$$

Retrospective limits can be made by replacing μ and σ with any sensible estimates.

Example 25 *(Examples 23 and 24 continued.) Returning to the reaming study, suppose once more that process standards are $\mu = 30$ and $\sigma = 4$. Then for samples of size $n = 5$ (like those used in the students' project), control limits for sample medians are:*

$$UCL_{\tilde{x}} = 30 + 3(1.197)\frac{4}{\sqrt{5}} = 36.42$$

and
$$LCL_{\bar{x}} = 30 - 3(1.197)\frac{4}{\sqrt{5}} = 23.58 \ .$$

Had these limits been applied to the data of Table 3.1 as they were collected, the need for physical intervention would have been signaled as early as the second sample.

3.2.2 Charts for Process Spread

Our exposition of control charts for measurements began with the $\bar{x}$ chart for location because it is surely the single most commonly used process-monitoring tool and because facts from elementary probability can be invoked to quickly motivate the notion of control limits for $\bar{x}$. However, in practice it is often important to deal *first* with the issue of consistency of process spread before going on to consider consistency of process location. After all, such consistency of spread (constancy of σ) is already implicitly assumed when one sets about to compute control limits for $\bar{x}$. So it is important to now consider charts intended to monitor this aspect of process behavior. The discussion here will center on charts for ranges and standard deviations, beginning with the **range chart**.

In deriving $\overline{R}/d_2$ as an estimator of σ, we have employed the fact that when sampling from a normal universe with mean μ and standard deviation σ,

$$\mathrm{E}R = \mu_R = d_2 \sigma \ . \tag{3.11}$$

The same kind of mathematics that stands behind relationship (3.11) can be used to also derive a standard deviation to associate with R. (This is a measure of spread for the probability distribution of R, which is itself a measure of spread of the sample.) It turns out that the standard deviation of R is proportional to σ. The constant of proportionality is called d_3 and is tabled in Table A.5. (Again, d_3 depends on n, but it will not be useful to display that dependence here.) That is,

$$\sigma_R = d_3 \sigma \ . \tag{3.12}$$

Now the relationships (3.11) and (3.12) together with the generic formula for Shewhart control limits given in display (3.2) and center line given in display (3.3) imply that standards given control limits for R are

$$UCL_R = (d_2 + 3d_3)\sigma \quad \text{and} \quad LCL_R = (d_2 - 3d_3)\sigma \tag{3.13}$$

with a center line at

$$CL_R = d_2 \sigma \ . \tag{3.14}$$

Standards Given R Chart Center Line

Further, if one adopts the notations $D_2 = d_2 + 3d_3$ and $D_1 = d_2 - 3d_3$, the relationships (3.13) can be written somewhat more compactly as

$$UCL_R = D_2 \sigma \quad \text{and} \quad LCL_R = D_1 \sigma \ . \tag{3.15}$$

Standards Given R Chart Control Limits

Values of the constants D_1 and D_2 may again be found in the table of control chart constants, Table A.5.

It is instructive to look at the tabled values of D_1. There are no tabled values for sample sizes $n \leq 6$. For such sample sizes, the difference $d_2 - 3d_3$ turns out to be negative. Since ranges are nonnegative, a negative lower control limit would make no sense. So standard practice for $n \leq 6$ is to use no lower control limit for R.

Consider also the implications of the fact that for $n > 6$, one typically employs a positive lower control limit for R. This means that it is possible for an R chart to signal an "out-of-control" situation because R is *too small*. This fact sometimes causes students confusion. After all, is it not the goal to produce *small* variation? Then why signal an alarm when R is small? The answer to this conundrum lies in remembering precisely what a control chart is meant to detect, namely, *process instability/change*. It is possible for unintended causes to occasionally act on a process to reduce variability. A lower control limit on an R chart simply allows one to detect such happy events. If one can detect such a change and identify its physical source, there is the possibility of making that assignable cause part of standard practice and the accompanying decrease in σ permanent. So the practice of using positive lower control limits for R when n is sufficiently big is one that makes perfectly good practical sense.

Example 26 (*Examples 23 through 25 continued.*) *Consider once more the reaming example of Dohm, Hong, Hugget, and Knoot from a standards given perspective with $\sigma = 4$. For samples of size $n = 5$, Table A.5 provides the values $d_2 = 2.326$ and $D_2 = 4.918$. So using formulas (3.14) and (3.15), standards given control chart values for R are*

$$UCL_R = 4.918(4) = 19.7 \quad \text{and} \quad CL_R = 2.326(4) = 9.3 \; .$$

Figure 3.4 is the corresponding standards given control chart for the students' ranges. There are three out-of-control points on the chart, the first coming as early as the second sample. The reaming process did not behave in a manner consistent with the $\sigma = 4$ standard over the period of the study. Samples 2, 11, and 14 simply have too much internal variability to make consistency of σ at the value 4 believable. One wonders if perhaps the reamer was changed in the middle of these samples, with the effect that some holes were very rough, while others were very smooth.

Retrospective control limits for R come about by plugging an estimate for σ derived from samples in hand into the formulas (3.14) and (3.15). A particularly natural choice for an estimator of σ in this context is $\overline{R}/d_2$. Substituting this into relationship (3.14), one gets the perfectly obvious retrospective center line for an R chart,

Retrospective R Chart Center Line

$$CL_R = \overline{R} \; . \tag{3.16}$$

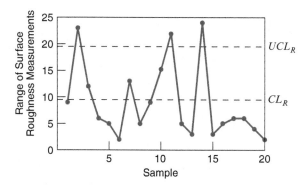

FIGURE 3.4. Standards given R chart for surface roughness

Further, substituting $\overline{R}/d_2$ into Eq. (3.15) for σ, one gets retrospective control limits for R

$$UCL_R = D_2 \left(\frac{\overline{R}}{d_2}\right) \quad \text{and} \quad LCL_R = D_1 \left(\frac{\overline{R}}{d_2}\right). \quad (3.17)$$

And adopting the notations $D_4 = D_2/d_2$ and $D_3 = D_1/d_2$, it is possible to write the relationships (3.17) more compactly as

$$UCL_R = D_4 \overline{R} \quad \text{and} \quad LCL_R = D_3 \overline{R}. \quad (3.18)$$

Retrospective R Chart Control Limits

As is by now to be expected, the constants D_3 and D_4 are tabled in Table A.5. And the table contains no values of D_3 for $n \leq 6$.

Example 27 *(Examples 23 through 26 continued.)* *Recall that the 20 samples in Table 3.1 have $\overline{R} = 8.95$ and note that for $n = 5$, $D_4 = 2.114$. So from displays (3.16) and (3.18), a retrospective control chart for the ranges (based on $\overline{R}/d_2$ as an estimator of σ) has center line at*

$$CL_R = \overline{R} = 8.95$$

and upper control limit

$$UCL_R = D_4 \overline{R} = 2.114(8.95) = 18.9.$$

A plot of this retrospective R chart would look very little different from Fig. 3.4. The same three ranges plot outside control limits. Not only is a "σ constant at 4" view of the students' data not plausible, but neither is a "σ constant at some value" view. There is solid evidence of reaming process instability in the ranges of Table 3.1. The short-term process variability changes over time.

The R chart is the most common Shewhart control chart for monitoring process spread. It requires very little from its user in the way of calculations, is based on a statistic that is very easy to understand, and is firmly entrenched in quality assurance practice dating from the days of Shewhart himself. There is, however, an alternative to the R chart that tends to detect changes in process spread more quickly, at the price of increased computational complexity. Where the quantitative sophistication of a user is high and calculations are not a problem, the **s chart** is a viable competitor for the R chart.

The fact that (when sampling from a normal distribution)

$$E s = c_4 \sigma , \qquad (3.19)$$

has already proved useful when making retrospective control limits for $\bar{x}$ based on $\bar{s}$. The same kind of mathematics that leads to relationship (3.19) can be used to find the standard deviation of s (based on a sample from a normal universe). (This is a measure of spread for the probability distribution of the random variable s that is itself a measure of spread of the sample.) It happens that this standard deviation is a multiple of σ. The multiplier is called c_5 and it turns out that $c_5 = \sqrt{1 - c_4^2}$. That is,

$$\sigma_s = \sigma \sqrt{1 - c_4^2} = c_5 \sigma . \qquad (3.20)$$

Now relationships (3.19) and (3.20) together with the generic Shewhart control limits and center line specified in displays (3.2) and (3.3) lead immediately to standards given control limits and center line for an s chart. That is,

$$UCL_s = (c_4 + 3c_5)\sigma \quad \text{and} \quad LCL_s = (c_4 - 3c_5)\sigma \qquad (3.21)$$

and

Standards Given s Chart Center Line

$$CL_s = c_4 \sigma . \qquad (3.22)$$

Further, if one adopts the notations $B_6 = c_4 + 3c_5$ and $B_5 = c_4 - 3c_5$, the relationships (3.21) can be written as

Standards Given s Chart Control Limits

$$UCL_s = B_6 \sigma \quad \text{and} \quad LCL_s = B_5 \sigma . \qquad (3.23)$$

Values of the constants B_5 and B_6 may again be found in the table of control chart constants, Table A.5. For $n \leq 5$, there are no values of B_5 given in Table A.5 because for such sample sizes, $c_4 - 3c_5$ is negative. For $n > 5$, B_5 is positive, allowing the s chart to provide for detection of a decrease in σ (just as is possible with an R chart and $n > 6$).

Retrospective control limits for s can be made by substituting any sensible estimate of σ into the standards given formulas (3.23) and (3.22). A particularly natural choice in this context is $\bar{s}/c_4$. Substituting this into relationship (3.23), one gets the obvious retrospective center line for an s chart

Retrospective s Chart Center Line

$$CL_s = \bar{s} . \qquad (3.24)$$

Further, substituting $\bar{s}/c_4$ into Eqs. (3.23) for σ produces retrospective control limits for s

$$UCL_s = B_6\left(\frac{\bar{s}}{c_4}\right) \quad \text{and} \quad LCL_s = B_5\left(\frac{\bar{s}}{c_4}\right). \quad (3.25)$$

And adopting the notations $B_4 = B_6/c_4$ and $B_3 = B_5/c_4$, it is possible to write the relationships (3.25) more compactly as

$$UCL_s = B_4\bar{s} \quad \text{and} \quad LCL_s = B_3\bar{s}. \quad (3.26)$$

Retrospective s Chart Control Limits

As usual, the constants B_3 and B_4 are tabled in Table A.5, and the table contains no values of B_3 for $n \leq 5$.

Example 28 (*Examples 23 through 27 continued.*) *The 20 samples in Table 3.1 have $\bar{s} = 3.76$. For $n = 5$, $B_4 = 2.089$. So from displays (3.24) and (3.26) a retrospective control chart for the standard deviations (based on $\bar{s}/c_4$ as an estimator of σ) has center line at*

$$CL_s = \bar{s} = 3.76$$

and upper control limit

$$UCL_s = B_4\bar{s} = 2.089(3.76) = 7.85.$$

Figure 3.5 is a retrospective s chart for the sample standard deviations of Table 3.1. It carries the same message as does a retrospective analysis of the sample ranges for this example. Not only is a "σ constant at 4" view of the students' data not plausible, neither is a "σ constant at some value" view. There is solid evidence of reaming process instability in the standard deviations of Table 3.1.

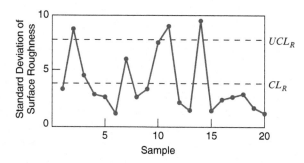

FIGURE 3.5. Retrospective s chart for surface roughness

3.2.3 What If $n = 1$?

To call n observations a "sample" or a "rational subgroup" is to implicitly guarantee that they were collected under essentially constant process conditions. The discussion in Sect. 3.1 has already raised the possibility (particularly in some low-volume production contexts) that a natural sample or subgroup size can be $n = 1$. Sometimes it is simply not safe to assume that even two successive process outcomes are necessarily generated under the same conditions.

There are two commonly raised questions about control charting for measurements when $n = 1$. These are:

1. Exactly what should one chart (in particular, should one chart so-called moving ranges)? and

2. How does one estimate a process standard deviation, σ?

We consider these questions before closing this section on Shewhart charting for measurement data.

Where rational subgroups are of size $n = 1$, there is really only one possible choice for a plotted statistic Q, namely, x. One can "transform" the most natural measurement to some other scale (e.g., by taking logarithms), but ultimately it is $Q = x$ that is available for plotting. However, people used to making $\bar{x}$ and R chart (or $\bar{x}$ and s chart) pairs in cases where $n > 1$ sometimes reason that it might be useful to supplement an x chart (or **individuals chart**) with a chart on which one plots **moving ranges**

Moving Range for an ith Observation, x_i

$$MR_i = |x_i - x_{i-1}|$$

The most commonly suggested version of this is where standards given control limits for x (the limits (3.5) for $\bar{x}$ when $n = 1$)

Standards Given Control Limits for Individuals, x

$$UCL_x = \mu + 3\sigma \quad \text{and} \quad LCL_x = \mu - 3\sigma \qquad (3.27)$$

are used together with

$$UCL_{MR} = D_2 \sigma \qquad (3.28)$$

(for D_2 based on the pseudo-sample size of $n = 2$). This practice turns out to produce a very large "false alarm rate" when in fact the process is stable. And attempts to remedy this by applying control limits looser than (3.27) and (3.28) are typically not repaid with improved ability to detect process changes over what is available using only an x chart with limits (3.27). Adding a moving range chart to an individuals chart just turns out to be a bad idea that should probably disappear from control charting practice. There is, to our knowledge, only one circumstance in which adding the moving range chart to an individuals chart makes sense. That is a case where the departure from stable process behavior that one fears and needs to detect is one of *oscillation* in consecutive individuals. There, a moving range chart is (not surprisingly) more effective than the individuals chart at "seeing" the nonstandard process behavior. To be absolutely explicit, in cases where $n = 1$, *the best thing to do about control charting is typically to use only an individuals chart with corresponding control limits (3.27).*

Chart Only x When $n = 1$

Consider then the second question above. When $n = 1$, there are no sample ranges or standard deviations to use in estimating σ. In fact, there is no really honest way to estimate a process standard deviation unless one has a sample or samples with $n \geq 2$. But some "dishonest" methods are less dishonest than others, and the best known method (the least dishonest method) is based on an average of moving ranges of successive observations. (Notice that this is *not* a matter of process monitoring based on moving ranges, but rather using moving ranges to estimate process standard deviation.)

The rationale for using moving ranges of successive observations in estimating σ is this. If process conditions can change observation to observation, observations will vary not only because $\sigma \neq 0$, but because the process mean changes. However, it is not unreasonable to expect the variability in pairs of successive observations to be less affected by mean changes than the variability of any other type of group of observations that could be made up. It is thus reasonable to use moving ranges to make an estimate of process standard deviation. While such an estimate is potentially inflated by variation in the process mean, it can be expected to be *less so* than any other estimate one might make.

The exact form of estimator of σ we will use (based on samples of size $n = 1$) is

$$\hat{\sigma} = \frac{\overline{MR}}{d_2} \quad (3.29)$$

Moving Range-Based Estimate of σ

where d_2 is for "sample size" 2 (as there are two observations represented in each moving range). This is a *conservative* estimator, as it will tend to overestimate σ when μ is not constant. But it is the best one available.

Example 29 *A Numerical Example. Consider the eight successive observations in the table below and the corresponding 7 moving ranges.*

Sample	1	2	3	4	5	6	7	8
x	5	3	9	10	17	4	6	2
MR		2	6	1	7	13	2	4

The values $5, 3, 9, 10, 17, 4, 6,$ *and* 2 *certainly vary because* $\sigma \neq 0$. *They may vary beyond what can be attributed to inherent process short-term variability if* μ *is not constant. That is, the 7 moving ranges should not be thought of as honest sample ranges, but as potentially overrepresenting* σ. *Nevertheless, the best available estimate of* σ *in this* $n = 1$ *context is from formula (3.29):*

$$\hat{\sigma} = \frac{\overline{MR}}{d_2}$$

$$= \frac{(2+6+1+7+13+2+4)/7}{1.128}$$
$$= 4.43.$$

Section 3.2 Exercises

1. Some specialized containers are produced by a process that runs 8 hours per day. Nine containers are sampled hourly, each day for five days. The distance from the bottom of the container to the container's handle is of interest. The target value for this dimension is 4 cm, and the process standard deviation for this quality dimension is .1 cm. (This is known from extensive experience with the process.)

    (a) What is a subgroup in this context? What is the subgroup size? How many subgroups make up the entire study?

    (b) Give control limits for process monitoring when subgroup averages are plotted versus time.

    (c) In addition to the chart in (b), a control chart for short-term process variation is to be made. Suppose that only subgroup averages and the smallest and largest values in a subgroup are available for analysis. What subgroup statistic can be used to do the charting? Give appropriate control limits and center line for the chart.

    (d) Are the limits in (b) and (c) standards given or retrospective limits? Why?

    (e) Suppose both the charts in (b) and (c) indicate that the process is stable. Is it then possible that any plotted subgroup mean is outside the limits from (b)? Is it possible that there are plotted values on the second chart outside control limits from (c)? Explain.

2. Continue in the context of problem 1, except now assume that no target values for the critical dimension or process standard deviation have previously been established. The average of the $r = 40$ subgroup averages was 3.9 cm, the average of the subgroup ranges was .56 cm, and the average of the 40 subgroup standard deviations was .48 cm.

    (a) Find control limits and center line to assess the consistency of "hourly variation" quantified as subgroup ranges.

    (b) Find control limits and center line to assess the consistency of process aim hour to hour based on subgroup averages. (Estimate the "within-hour" standard deviation based on subgroup ranges.)

    (c) Repeat (a), except now use the subgroup standard deviations instead of ranges.

(d) Repeat (b), except now use the subgroup standard deviations to estimate the "within-hour" standard deviation.

(e) Suppose that none of the charts in (a) to (d) suggest lack of process stability (so that it makes sense to talk about a single process mean and single process standard deviation). Give a valid estimate of the process average distance from the container bottom to the handle. Give two valid estimates of the standard deviation of the distance from the container bottom to the handle. (Provide both the formulas you use and numerical answers.)

3. Below are sample means and standard deviations from ten samples, each of size $n = 4$.

Sample	1	2	3	4	5	6	7	8	9	10	Sum
$\bar{x}$	7.0	7.9	7.1	7.7	5.2	5.4	6.4	6.5	5.8	6.8	65.8
s	1.5	3.1	3.4	1.1	1.4	1.0	2.5	.7	1.4	1.1	17.2

(a) Suppose process standards $\mu = 6.0$ and $\sigma = 1.5$ are provided. Find the standards given center line and control limits for an $\bar{x}$ chart. If these limits had been applied to the values in the table as they were collected, would there have been out-of-control signals?

(b) Using the standards in (a), find the standards given center line and control limits for an s chart. If these limits had been applied to the values in the table as they were collected, would there have been out-of-control signals?

(c) Suppose that the standards in (a) were not available. Make retrospective $\bar{x}$ and s charts, and assess whether there is evidence of process instability in the values in the table.

(d) What is an estimate of σ based on the average sample standard deviation? Use this estimate and estimate the mean of a *range* for an additional sample of size $n = 6$.

4. **Transmission Housings.** Apple, Hammerand, Nelson, and Seow analyzed data taken from a set of "series XX" transmission housings. One critical dimension they examined was the diameter for a particular hole on the side cover of the housing. A total of 35 consecutively produced housings were examined and the corresponding x = hole diameter measured and recorded (in inches). Specifications for the diameter were $3.7814 \pm .002$ in. Below are the first 10 recorded diameters. Summary statistics for all 35 housings are $\sum x = 132.319$ in and $\sum MR = .02472$ in.

Housing	1	2	3	4	5	6	7	8	9	10
x	3.7804	3.7803	3.7806	3.7811	3.7812	3.7809	3.7816	3.7814	3.7809	3.7814

(a) What is a subgroup here and what is the subgroup size?

(b) The 35 consecutive hole diameters produce how many moving ranges?

(c) Compute the first two moving ranges.

(d) Make an estimate of σ. Use your estimate and the sample mean diameter to replace process parameters in the limits (3.27). Are the resulting limits for individuals standards given or retrospective limits? Why? Apply your limits to the first ten hole diameters. Do these values provide evidence of process instability?

3.3 Shewhart Charts for Counts/"Attributes Data"

The control charts for measurements introduced in Sect. 3.2 are the most important of the Shewhart control charts. Where it is at all possible to make measurements, they will almost always provide more information on process behavior than will a corresponding number of qualitative observations. However, there are occasions where only attributes data can be collected. So this section presents Shewhart control charting methods for such cases. The section considers charting *counts* and corresponding *rates* of occurrence for nonconforming items (or defectives) and for nonconformities (or defects). The case of so-called np charts and p charts for "percent nonconforming" (or percent defective) contexts is treated first. Then follows a discussion of c and u charts for "nonconformities per unit" (or defects per unit) situations.

3.3.1 Charts for Fraction Nonconforming

Consider now a situation where one periodically samples n items or outcomes from a process and (making careful use of operational definitions) classifies each one as "nonconforming" or "conforming." (The old terminology for these possibilities is "defective" and "nondefective." The newer terminology is used in recognition of the fact that some kinds of failures to meet inspection criteria do not render a product functionally deficient. There is also reluctance in today's litigious society to ever admit that anything produced by an organization could possibly be "defective.")

Then let

$$X = \text{the number nonconforming in a sample of } n \text{ items or outcomes} \quad (3.30)$$

and

$$\hat{p} = \frac{X}{n} = \text{the fraction nonconforming in a sample of } n \text{ items or outcomes}. \quad (3.31)$$

Shewhart **np charts** are for the plotting of $Q = X$, and **p charts** are for the monitoring of $Q = \hat{p}$. Both are based on the same probability model for the variable X. (The fact that $\hat{p}$ is simply X divided by n implies that control limits for $\hat{p}$ should simply be those for X, divided by n.) Under stable process conditions for the creation of the n items or outcomes in a sample (under the assumption that the sample in question is a rational subgroup), it is reasonable to model the variable X with a binomial distribution for n "trials" and "success probability," p, equal to the process propensity for producing nonconforming outcomes.

Elementary properties of the binomial distribution can be invoked to conclude that

$$\mu_X = \mathrm{E}X = np \quad \text{and} \quad \sigma_X = \sqrt{\mathrm{Var}\,X} = \sqrt{np(1-p)}. \quad (3.32)$$

Then the mean and standard deviation in display (3.32) and the generic Shewhart control limits and center line specified in displays (3.2) and (3.3) lead to standards given control limits for both X and $\hat{p}$. That is,

$$CL_X = np \quad (3.33)$$

Standards Given np Chart Center Line

while

$$UCL_X = np + 3\sqrt{np(1-p)} \quad \text{and} \quad LCL_X = np - 3\sqrt{np(1-p)}. \quad (3.34)$$

Standards Given np Chart Control Limits

And dividing the expressions (3.33) and (3.34) through by n, one arrives at standards given values for $\hat{p}$,

$$CL_{\hat{p}} = p, \quad (3.35)$$

Standards Given p Chart Center Line

$$UCL_{\hat{p}} = p + 3\sqrt{\frac{p(1-p)}{n}} \quad \text{and} \quad LCL_{\hat{p}} = p - 3\sqrt{\frac{p(1-p)}{n}}. \quad (3.36)$$

Standards Given p Chart Control Limits

Example 30 *Monitoring the Fraction Nonconforming in a Pelletizing Process. Kaminiski, Rasavaghn, Smith, and Weitekamper worked with a manufacturer of hexamine pellets. Their work covered a time period of several days of production. Early efforts with the pelletizing machine (using shop standard operating procedures) produced a standard fraction nonconforming of approximately $p = .60$. On the final day of the study, after adjusting the "mix" of the powder being fed into the machine, the counts and proportions of nonconforming pellets in samples of size $n = 30$ portrayed in Table 3.3 were collected.*

From Eqs. (3.34), standards given control limits for the numbers of nonconforming pellets in the samples represented by Table 3.3 are

$$UCL_X = 30(.6) + 3\sqrt{30(.6)(.4)} = 26.05$$

130   Chapter 3. Process Monitoring

TABLE 3.3. Counts and fractions of nonconforming pellets in samples of size 30

Sample	X	$\hat{p}$	Sample	X	$\hat{p}$
1	14	.47	14	9	.30
2	20	.67	15	16	.53
3	17	.57	16	16	.53
4	13	.43	17	15	.50
5	12	.40	18	11	.37
6	12	.40	19	17	.57
7	14	.47	20	8	.27
8	15	.50	21	16	.53
9	19	.63	22	13	.43
10	21	.70	23	16	.53
11	18	.60	24	15	.50
12	14	.47	25	13	.43
13	13	.43			

and

$$LCL_X = 30(.6) - 3\sqrt{30(.6)(.4)} = 9.95,$$

and from display (3.33) a center line at

$$CL_X = 30(.6) = 18$$

is in order. Figure 3.6 is the standards given np control chart for the data of Table 3.3.

It is evident from Fig. 3.6 that the pelletizing process was not stable at the standard value of $p = .60$ on the final day of the students' study. Notice that there are two out-of-control points on the chart (and most of the plotted points run below the center line established on the basis of the standard value of p). The message that was delivered at samples 14 and 20 (if not even before, on the basis of the plotted values running consistently below 18) was one of clear process improvement, presumably traceable to the change in powder mix.

Example 30 nicely illustrates the fact that a positive lower control limit on an np chart or on a p chart makes perfectly good sense in terms of allowing identification of unexpectedly good process output. Remember that the objective of Shewhart charting is to detect process instability/change. On occasion, that change can be for the good.

Retrospective control limits for X or $\hat{p}$ require that one take the data in hand and produce a provisional estimate of (a supposedly constant) p for plugging into formulas (3.33) through (3.36) in place of p. If samples (of possibly different sizes) are available from r different periods, then a most natural estimator of a

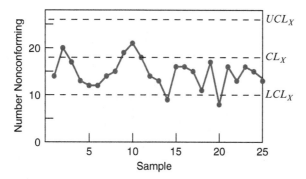

FIGURE 3.6. Standards given np chart for counts of nonconforming pellets

common p is the pooled sample fraction nonconforming

$$\hat{p}_{\text{pooled}} = \frac{\sum_{i=1}^{r} n_i \hat{p}_i}{\sum_{i=1}^{r} n_i} = \frac{\sum_{i=1}^{r} X_i}{\sum_{i=1}^{r} n_i} = \frac{\text{total nonconforming}}{\text{total of the sample sizes}} . \quad (3.37)$$

Pooled Fraction Nonconforming

Example 31 *(Example 30 continued.)* *Returning again to the pelletizing example, the counts of nonconforming pellets in Table 3.3 total to 367. There were* $30(25) = 750$ *pellets inspected, so from relationship (3.37),* $\hat{p}_{\text{pooled}} = 367/750 = .4893$. *Substituting this into Eqs. (3.33) and (3.34) in place of* p, *one arrives at retrospective values*

$$CL_X = 30(.4893) = 14.68 ,$$

$$UCL_X = 30(.4893) + 3\sqrt{30(.4893)(.5107)} = 22.89 ,$$

and

$$LCL_X = 30(.4893) - 3\sqrt{30(.4893)(.5107)} = 6.47 .$$

Figure 3.7 is a retrospective np chart made using these values and the data of Table 3.3. The figure shows that although it is not plausible that the pelletizing process was stable at the standard value of p (.60) on the final day of the students' study, it is *plausible that the process was stable at* some *value of p, and .4893 is a reasonable guess at that value.*

A few words need to be said about cases where sample sizes vary in a fraction nonconforming context. In such situations, it makes much more sense to plot $\hat{p}$ values than it does to plot Xs based on differing sample sizes. Then at least, one has a constant center line (given by expression (3.35)). Of course, the control limits represented in display (3.36) will vary with the sample size. Equations (3.36) show that the larger the sample size, the tighter will be the control limits about the central value p. This is perfectly sensible. The larger the sample, the more

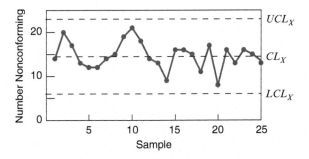

FIGURE 3.7. Retrospective np chart for counts of nonconforming pellets

information about the current process propensity for producing nonconforming outcomes, and the *less* variation one should allow from the central value before declaring that there is evidence of process instability.

3.3.2 Charts for Mean Nonconformities per Unit

A second kind of situation leading to count and rate data (that is fundamentally different from the fraction nonconforming scenario) is the so-called "mean nonconformances/nonconformities per unit" ("or mean defects per unit") situation. In such a context, one periodically selects k inspection units from a process output and counts

$$X = \text{the total number of nonconformities on the } k \text{ units} \quad (3.38)$$

(older terminology for nonconformities is "defects" or "flaws"). In cases where k is always equal to 1, the count X itself is plotted, and the resulting chart is called a **c chart**. Where k varies and/or is not equal to 1, it is common to plot instead

$$\hat{u} = \frac{X}{k} = \text{the sample mean nonconformities per unit} \quad (3.39)$$

and the resulting chart is called a **u chart**.

Control limits for c and u charts are based on the Poisson process model. If one assumes that under stable process conditions, the generation of nonconformities can be described by a Poisson process with (constant) rate parameter λ, the number of defects on one inspection unit has a Poisson distribution with mean λ. And X, the number of defects on k inspection units, is a Poisson random variable with mean $k\lambda$. Thus, under stable process conditions,

$$\mu_X = \mathrm{E}X = k\lambda \quad \text{and} \quad \sigma_X = \sqrt{\mathrm{Var}\,X} = \sqrt{k\lambda}. \quad (3.40)$$

So using facts (3.40) and the generic Shewhart control limits and center line specified in displays (3.2) and (3.3), in the c chart situation ($k \equiv 1$), standards given values are

Standards Given c Chart Center Line

$$\boxed{CL_X = \lambda,} \quad (3.41)$$

and

$$UCL_X = \lambda + 3\sqrt{\lambda} \quad \text{and} \quad LCL_X = \lambda - 3\sqrt{\lambda}. \quad (3.42)$$

Standards Given c Chart Control Limits

It follows from the definition of $\hat{u}$ in display (3.39) and relationships (3.40) that

$$\mu_{\hat{u}} = E\hat{u} = \lambda \quad \text{and} \quad \sigma_{\hat{u}} = \sqrt{\text{Var}\,\hat{u}} = \sqrt{\frac{\lambda}{k}}. \quad (3.43)$$

Then using again the generic Shewhart control limits and center line and applying the facts (3.43), standards given values for a u chart are

$$CL_{\hat{u}} = \lambda, \quad (3.44)$$

Standards Given u Chart Center Line

and

$$UCL_{\hat{u}} = \lambda + 3\sqrt{\frac{\lambda}{k}} \quad \text{and} \quad LCL_{\hat{u}} = \lambda - 3\sqrt{\frac{\lambda}{k}}. \quad (3.45)$$

Standards Given u Chart Control Limits

Notice that in the case $k = 1$, the u chart control limits reduce (as they should) to the c chart limits.

Retrospective control limits for X or $\hat{u}$ require that one take the data in hand and produce a provisional estimate of (a supposedly constant) λ for plugging into formulas (3.41), (3.42), (3.44), and (3.45) in place of λ. If data from r different periods are available, then a most natural estimator of a common λ is the pooled mean nonconformities per unit

$$\hat{\lambda}_{\text{pooled}} = \frac{\sum_{i=1}^{r} k_i \hat{u}_i}{\sum_{i=1}^{r} k_i} = \frac{\sum_{i=1}^{r} X_i}{\sum_{i=1}^{r} k_i} = \frac{\text{total nonconformities}}{\text{total units inspected}}. \quad (3.46)$$

Pooled Mean Nonconformities Per Unit

Example 32 *Monitoring the Number of Leaks in Assembled Radiators. The article "Quality Control Proves Itself in Assembly," by Wilbur Burns (reprinted from* Industrial Quality Control*) in Volume 2, Number 1 of* Quality Engineering, *contains a classic set of data on the numbers of leaks found in samples of auto radiators at final assembly. These are reproduced in Table 3.4.*

This is a nonconformities per unit situation. Each unit (each radiator) presents the opportunity for the occurrence of any number of leaks, several units are being inspected, and the total number of leaks on those units is being counted. The leaks per radiator are calculated as in display (3.39), and if one wishes to investigate the statistical evidence for process instability, a u chart is in order.

The article gives no shop standard value for λ, so consider a retrospective analysis of the data in Table 3.4. There are 116 total leaks represented in Table 3.4, and 841 radiators were tested. So from relationship (3.46),

$$\hat{\lambda}_{\text{pooled}} = \frac{116}{841} = .138,$$

TABLE 3.4. Counts and occurrence rates of outlet leaks found in 18 daily samples of radiators

Day	X (leaks)	k (radiators)	$\hat{u}$ (leaks/radiator)
1	14	39	.36
2	4	45	.09
3	5	46	.11
4	13	48	.27
5	6	40	.15
6	2	58	.03
7	4	50	.08
8	11	50	.22
9	8	50	.16
10	10	50	.20
11	3	32	.09
12	11	50	.22
13	1	33	.03
14	3	50	.06
15	6	50	.12
16	8	50	.16
17	5	50	.10
18	2	50	.04

and a center line for a retrospective u chart for these data can be drawn at this value. From Eqs. (3.45) (using .138 for λ), the control limits change with k, larger k leading to tighter limits about the center line. As an example of using Eq. (3.45), note that for those $\hat{u}$ values based on tests of $k = 50$ radiators,

$$UCL_{\hat{u}} = .138 + 3\sqrt{\frac{.138}{50}} = .296 \,.$$

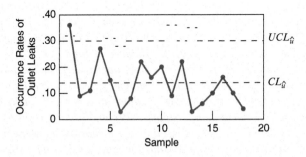

FIGURE 3.8. Retrospective u chart for rates of radiator outlet leaks

On the other hand, since the formula (3.45) for $LCL_{\hat{u}}$ produces a negative value for the intrinsically nonnegative $\hat{u}$, no lower control limit would be used for $\hat{u}$

based on 50 radiators. (As a matter of fact, no k in Table 3.4 is large enough to lead to the use of a lower control limit.)

Figure 3.8 is a retrospective u chart for the radiator leak data. It shows that the initial day's experience does not "fit" with the subsequent 17 days. There is evidence of process change/instability, and appearances are that things improved in the radiator assembly process after the first day.

This section opened with the disclaimer that where possible, the charts for measurements introduced in the previous section should be used in preference to the ones presented here. That advice bears repeating. The two examples in this section are reasonably convincing, but they are so in part because the relevant fraction nonconforming and mean nonconformities per unit are fairly large. Modern business pressures make standard defect rates in the "parts per million" range common. And there is really no way to effectively monitor processes that are supposed to have such performance with attributes control charts (sample sizes in the millions would be required for effective detection of even doubling of defect rates).

Section 3.3 Exercises

1. In a packaging department of a food processor, types of packaging "imperfections" are carefully defined and include creases, holes, printing smudges, and broken seals. 30 packages each hour are sampled, and X = the total number of imperfections identified on the 30 packages is recorded. On average about .05 imperfections per package have been seen in the past. Below are data from 7 hours one day in this department.

Hour	1	2	3	4	5	6	7
X	1	0	2	0	1	1	3

   (a) Are the data above variables or attributes data? Why?

   (b) What distribution (fully specify it, giving the value of any parameter(s)) can be used to model the number of imperfections observed on a single package?

   (c) What is the expected total number of imperfections observed on a set of 30 boxes? What probability distribution can be used to model this variable?

   (d) What is the standard deviation of the total number of imperfections on 30 boxes?

   (e) Find the standards given control limits and center line for a chart based on the data above, where the plotted statistic will be $X/30$. Do any values of $X/30$ plot outside your control limits?

   (f) What is the name of the type of chart you made in (e)?

(g) Suppose no standard is given for the rate of imperfections. Using values above, find appropriate retrospective control limits and center line for plotting $X/30$.

2. Consider a variant of problem 1 where any package with at least one imperfection (a crease, a hole, a smudge, or a broken seal) is considered to be nonconforming. Reinterpret the values X in the table of problem 1 as counts of nonconforming packages in samples of size 30. Suppose that in the past .05 (5 %) of packages have been nonconforming.

   (a) Does this variant of problem 1 involve variables data or attributes data? Why?

   (b) What probability distribution (fully specify it, giving the value of any parameter(s)) can be used to model the number of nonconforming packages in a sample of 30?

   (c) What is the mean number of nonconforming packages in a sample of 30?

   (d) What is the standard deviation of the number of nonconforming packages in a sample of 30?

   (e) Find the standards given control limits and center line for monitoring the proportion of nonconforming packages in samples of size 30.

   (f) Repeat (e) for monitoring the number of nonconforming packages in samples of size 30.

   (g) Suppose no standard is given for the fraction of nonconforming packages. Based on the data in the table above, find appropriate retrospective control limits and center line for an np chart.

3.4 Patterns on Shewhart Charts and Special Alarm Rules

To this point all that we have discussed doing with values Q plotted on a Shewhart chart is to compare them to control limits one at a time. If that were the whole story, there would be little reason to actually make the plots. Simple numerical comparisons would suffice. But the plots offer the possibility of *seeing* other important things in process-monitoring data besides only where points plot outside control limits. And it is standard control charting practice to examine Shewhart control charts for these other kinds of indications of process change. The purpose of this section is to discuss some types of revealing patterns that occasionally show up on control charts (providing both jargon for naming them and discussion of the kinds of physical phenomena that can stand behind them) and to provide some sets of rules that can be applied to identify them.

Under stable process conditions (leading to Qs that can be modeled as independent and identically distributed), one expects to see a sequence of plotted values that

1. are without obvious pattern or trend,
2. only on rare occasions fall outside control limits,
3. tend to cluster about the center line, about equally often above and below it,
4. on occasion approach the control limits.

(The tacit assumption in most applications is that the stable process distribution of Q is reasonably "mound shaped" and centered at the chart's center line.) When something other than this kind of "random scatter" picture shows up on a control chart, it can be possible to get clues to what kinds of physical causes are acting on the process that can in turn be used in process-improvement efforts.

On occasion one notices **systematic variation/cycles**, regular "up then back down again" patterns on a Shewhart chart like those pictured in Fig. 3.9. This suggests that there are important variables acting on the process whose effects are periodic. Identification of the period of variation can give strong hints where to start looking for physical causes. Examples of factors that can produce cycles on a Shewhart chart are seasonal and diurnal variables like ambient temperature. And sometimes regular rotation of fixtures or gages or shift changes in operators running equipment or making measurements can stand behind systematic variation.

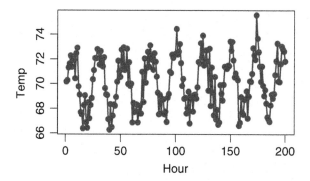

FIGURE 3.9. A plot of factory ambient temperature vs. time exhibiting systematic variation or cycles

While systematic variation is a variation of the "second kind" on the right side of Eq. (3.1), it may not always be economically feasible to eliminate it. For example, in some applications it may be preferable to live with effects of ambient temperature rather than try to control the environment in which a process operates. But recognition of its presence at least allows one to intelligently consider options regarding remedial measures and to mentally remove that kind of variation from the baseline against which one looks for the effects of other special causes.

138   Chapter 3. Process Monitoring

Instability is a word that has traditionally been applied to patterns on control charts where many points plot near or beyond control limits. This text has used (and will continue to use) the word to refer to physical changes in a process that lead to individual points plotting outside of control limits. But this traditional usage refers more to a pattern on the chart and specifically to one where points outside of control limits are very frequent. Figure 3.10 contrasts variation on a Shewhart chart that one expects to see, to a pattern of instability. Standing behind such a pattern can be more or less erratic and unexpected causes, like different lots of raw material with different physical properties mixed as process input.

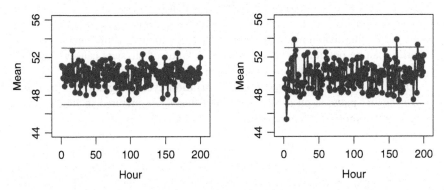

FIGURE 3.10. Two $\bar{x}$ charts, the second of which shows "instability"

Another important possible cause of many points at or beyond control limits is that of unwise operator overadjustment of equipment. Control charting is useful both because it signals the existence of conditions that deserve physical intervention *and* because it tells one to leave equipment untouched when it seems to be operating as consistently as possible. When that "hands-off" advice is not followed and humans tinker with physically stable processes, reacting to every small appearance of variation, the end result is not to decrease process variation, but rather to increase it. And such fiddling can turn a process that would otherwise be generating plotted values inside control limits into one that is regularly producing Qs near or beyond control limits.

Changes in level are sometimes seen on control charts, where the average plotted value seems to move decisively up or down. The change can be sudden as pictured on Fig. 3.11 and traceable to some basic change at the time of the shift. The introduction of new equipment or a clear change in the quality of a raw material can produce such a sudden change in level.

A change in level can also be like that pictured in Fig. 3.12, more gradual and attributable to an important cause starting to act at the beginning of the change in level, but so to speak "gathering steam" as time goes on until its full effect is felt. For example, effective worker training in machine operation and measuring techniques could well begin a gradual decrease in level on an R chart that over time and with practice will reach its full potential for reducing observed variation.

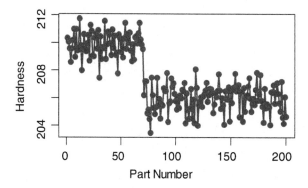

FIGURE 3.11. Run chart with sudden change in part hardness level

Where a gradual change in level does not end with stabilization around a new mean, but would go on unabated in the absence of physical intervention, it is traditional to say that there is a **trend** on a control chart. Figure 3.13 on page 140 pictures such a trend on a run chart. Many physical causes acting on manufacturing processes will produce trends if they remain unaddressed. An example is tool wear in machining processes. As a cutting tool wears, the parts being machined will tend to grow larger. If adjustments are not made and the tool is not periodically changed, machined dimensions of parts will eventually be so far from ideal as to make the parts practically unusable.

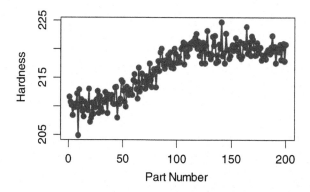

FIGURE 3.12. Run chart with gradual change in part hardness level

There is another phenomenon that occasionally produces strange-looking patterns on Shewhart control charts. This is something the early users of control charts called the occurrence of **mixtures.** These are the combination of two or more distinct patterns of variation (in either a plotted statistic Q or in an underlying distribution of individual observations leading to Q) that get put together on a single control chart. In "stable" mixtures, the proportions of the component patterns remain relatively constant over time, while in "unstable" versions the proportions vary with time.

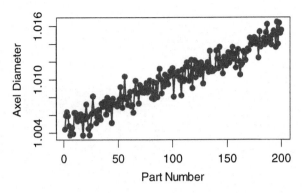

FIGURE 3.13. A run chart with an unabated trend

Where an underlying distribution of observations has two or more radically different components, a plotted statistic Q can be either unexpectedly variable or surprisingly consistent. Consider first the phenomenon of unexpectedly large variation in Q traceable to a mixture phenomenon. Where blunders like incomplete or omitted manufacturing operations or equipment malfunctions lead to occasional wild individual observations and correspondingly wild values of Q, the terminology **freaks** is often used. The resulting impact of mixing normal and aberrant observations can be as pictured in Fig. 3.14. Where individual observations or values of Q of a given magnitude tend to occur together in time as pictured in Fig. 3.15, the terminology **grouping** or **bunching** is common. Different work methods employed by different operators or changes in the calibration of a measurement instrument can be responsible for grouping or bunching. So, how mixture phenomena sometimes lead to unexpectedly large variation on a control chart is fairly obvious.

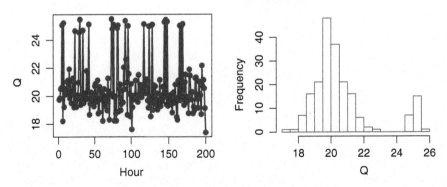

FIGURE 3.14. An example of a pattern that could be described as exhibiting "freaks" (and the corresponding histogram)

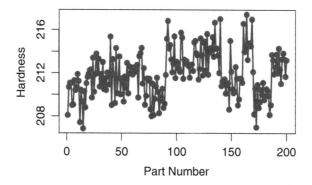

FIGURE 3.15. A run chart showing grouping or bunching

How a mixture can lead to unexpectedly small variation in a plotted statistic is more subtle, but very important. It involves a phenomenon sometimes known in quality assurance circles as **stratification.** If an underlying distribution of observations has radically different components, each with small associated variation, and these components are (wittingly or unwittingly) sampled in a systematic fashion, a series of plotted values Q with unbelievably *small* variation can result. One might, for example, be sampling different raw material streams or the output of different machines and unthinkingly calling the resulting values a single "sample" (in violation, by the way, of the notion of rational subgrouping). The result can be a Shewhart control chart like the one in Fig. 3.16.

Stratification

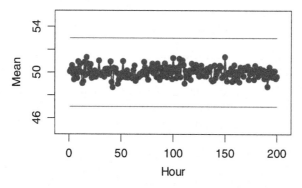

FIGURE 3.16. Unexpectedly small variation on an $\bar{x}$ chart, potentially due to stratification

To see how stratification can lead to surprisingly small variation in Q, consider the case of a p chart and a hypothetical situation where a 10-head machine has one completely bad head and 9 perfect ones. If the items from this machine are taken off the heads in sequence and placed into a production stream, "samples" of 10 consecutive items will have fractions defective that are *absolutely constant* at $\hat{p} = .10$. A p chart for the process will look unbelievably stable about a center line at .10. (A similar hypothetical example involving $\bar{x}$ and R charts can be invented by thinking of 9 of the 10 heads as turning out widget diameters of essentially exactly 5.000, while the tenth turns out widget diameters of essentially exactly 7.000. Ranges of "samples" of 10 consecutive parts will be unbelievably stable at 2.000, and means will be unbelievably stable at 5.200.)

So, too much consistency on a control chart is not a cause for rejoicing and relaxation. When plotted points hug a center line and never approach control limits, something is not as it should be. There may be a simple blunder in the computation of the control limits, or the intrinsic variation in the process may be grossly overestimated. (For example, an excessive standard value for σ produces $\bar{x}$ and R chart control limits that are too wide and plotted points that never approach them under stable conditions.) And on occasion stratification may be present. When it is and it goes unrecognized, one will never be in a position to discover and eliminate the cause(s) of the differences between the components of the underlying distribution of observations. In the 10-head machine example, someone naively happy with the "$\hat{p}$ constant at .10" phenomenon will never be in a position to discover that the one head is defective and remedy it. So, a chart that looks too good to be true is as much a cause for physical investigation as is one producing points outside control limits.

Once one recognizes the possibility of looking for patterns on a Shewhart control chart, the question becomes exactly what to consider to be an occurrence of a pattern. This is important for two reasons. In the first place, there is the matter of consistency within an organization. If control charts are going to be used by more than one person, those people need a common set of ground rules for interpreting the charts that they together use. Second, without a fair amount of theoretical experience in probability and/or practical experience in using control charts, people tend to want to "see" patterns that are in actuality very easily produced by a stable process.

Since the simple "one point outside control limits" rule is blind to the interesting kinds of patterns discussed here and there is a need for some standardization of the criteria used to judge whether a pattern is present, organizations often develop sets of "special checks for unnatural patterns" for application to Shewhart control charts. These are usually based on segmenting the set of possible Qs into various zones defined in terms of multiples of σ_Q above and below the central value μ_Q. Figure 3.17 shows a generic Shewhart chart with typical zones marked on it.

The most famous set of special checks is the set of "Western Electric alarm rules" given in Table 3.5. They are discussed extensively in the *Statistical Quality Control Handbook* published originally by Western Electric and later by AT&T. Two other possible sets of rules, one taken from A.J. Duncan's excellent *Quality Control and Industrial Statistics* and the other published by Lloyd Nelson in the *Journal of Quality Technology* in 1984, are given in Tables 3.6 and 3.7, respectively. The reader should be able to see in these sets of rules attempts to provide operational definitions for the kinds of patterns discussed in this section. It is not at all obvious which set should be considered best or even what are rational criteria for comparing them and the many other sets that have been suggested. But the motivation behind them should be clear.

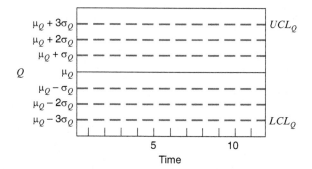

FIGURE 3.17. Generic Shewhart control chart with "1-sigma," "2-sigma," and "3-sigma" zones marked on it

TABLE 3.5. Western electric alarm rules

A single point outside 3-sigma control limits
2 out of any 3 consecutive points outside 2-sigma limits on one side of the center
4 out of any 5 consecutive points outside 1-sigma limits on one side of the center
8 consecutive points on one side of the center

TABLE 3.6. Alarm rules from Duncan's *Quality Control and Engineering Statistics*

A single point outside 3-sigma control limits
A run of 7 consecutive points up and down or on one side of the center
2 consecutive points outside 2-sigma limits
4 consecutive points outside 1-sigma limits
"Obvious" cycles up and down

TABLE 3.7. Nelson's alarm rules from the *Journal of Quality Technology*

A single point outside 3-sigma control limits
9 consecutive points on one side of the center
6 consecutive points increasing or decreasing
14 consecutive points alternating up and down
2 out of any 3 consecutive points outside 2-sigma limits on one side of the center
4 out of any 5 consecutive points outside 1-sigma limits on one side of the center
15 consecutive points inside 1-sigma limits
8 consecutive points with none inside 1-sigma limits

Section 3.4 Exercises

1. When a process is stable, what do you expect to see on a control chart for a statistic Q?

2. What motivates the use of multiple rules for identifying out-of-control situations?

3. When "extra alarm rules" (beyond the "single point outside 3-sigma control limits" rule) are used in process monitoring, do you expect the frequency of false alarms to decrease, stay the same, or increase? (A "false alarm" occurs when the chart signals, but no physical special cause can be identified.)

3.5 The Average Run Length Concept

Realizing that alternative schemes for issuing out-of-control signals based on process-monitoring data are possible, the need arises to quantify what a given scheme can be expected to do. For example, to choose intelligently between the sets of alarm rules in Tables 3.5 through 3.7, one needs some way of predicting behavior of the alternative monitoring schemes. The most effective tool available for making this kind of prediction is the "average run length" (ARL) notion. This section introduces the concept and illustrates its use in some very simple situations.

Consider a context where based on values of Q plotted at periods $1, 2, 3, \ldots$, one will monitor a process until an out-of-control signal is issued. Let

$$T = \text{the period at which the process-monitoring scheme first signals.} \quad (3.47)$$

T is a random variable and is called the **run length** for the scheme. The probability distribution of T is called the **run length distribution**, and the mean or average value of this distribution is called the **average run length** (ARL) for the process-monitoring scheme. That is,

$$ARL = \mathrm{E}T = \mu_T. \quad (3.48)$$

It is desirable that a process-monitoring scheme has a large ARL when the process is stable at standard values for process parameters and small ARLs under other conditions.

Finding formulas and numerical values for ARLs is usually not elementary. Some advanced probability and numerical analysis are often required. But there is one kind of circumstance where an explicit formula for ARLs is possible and we can illustrate the meaning and usefulness of the ARL concept in elementary terms. That is the situation where

1. the process-monitoring scheme employs only the single alarm rule "signal The first time that a point Q plots outside control limits," and

2. it is sensible to think of the process as physically stable (though perhaps not at standard values for process parameters).

Under condition 2, the values $Q_1, Q_2, Q_3, \ldots$ can be modeled as independent random variables with the same individual distribution, and the notation

$$q = P[Q_1 \text{ plots outside control limits}] \qquad (3.49)$$

Probability of an Immediate Alarm

will prove useful.

In this simple case, the random variable T has a geometric distribution with probability function

$$f(t) = \begin{cases} q(1-q)^{t-1} & \text{for } t = 1, 2, 3, \ldots \\ 0 & \text{otherwise} \end{cases}$$

It then follows from the properties of the geometric distribution and relationship (3.48) that

$$ARL = \mathrm{E}T = \frac{1}{q}. \qquad (3.50)$$

ARL for a "One Point Outside Control Limits" Scheme

Example 33 *Some ARLs for Shewhart $\bar{x}$ Charts.* *To illustrate the meaning of relationship (3.50), consider finding ARLs for a standards given Shewhart $\bar{x}$ chart based on samples of size $n = 5$. Note that if standard values for the process mean and standard deviation are, respectively, μ and σ, the relevant control limits are*

$$UCL_{\bar{x}} = \mu + 3\frac{\sigma}{\sqrt{5}} \quad \text{and} \quad LCL_{\bar{x}} = \mu - 3\frac{\sigma}{\sqrt{5}}.$$

Thus, from Eq. (3.49)

$$q = P\left[\bar{x} < \mu - 3\frac{\sigma}{\sqrt{5}} \text{ or } \bar{x} > \mu + 3\frac{\sigma}{\sqrt{5}}\right].$$

First suppose that "all is well," and the process is stable at standard values of the process parameters. Then $\mu_{\bar{x}} = \mu$ and $\sigma_{\bar{x}} = \sigma/\sqrt{5}$ and if the process output is normal, so also is the random variable $\bar{x}$. Thus

$$q = 1 - P\left[\mu - 3\frac{\sigma}{\sqrt{5}} < \bar{x} < \mu + 3\frac{\sigma}{\sqrt{5}}\right] = 1 - P\left[-3 < \frac{\bar{x} - \mu}{\sigma/\sqrt{5}} < 3\right]$$

can be evaluated using the fact that

$$Z = \frac{\bar{x} - \mu}{\sigma/\sqrt{5}}$$

is a standard normal random variable. Using a normal table with an additional significant digit beyond the one in this text, it is possible to establish that

$$q = 1 - P[-3 < Z < 3] = .0027$$

to four digits. Therefore, from relationship (3.50) it follows that

$$ARL = \frac{1}{.0027} = 370.$$

The interpretation of this is that when all is OK (i.e., the process is stable and parameters are at their standard values), the $\bar{x}$ chart will issue (false alarm) signals on average only once every 370 plotted points.

In contrast to the situation where process parameters are at their standard values, consider next the possibility that the process standard deviation is at its standard value, but the process mean is one standard deviation above its standard value. In these circumstances one still has $\sigma_{\bar{x}} = \sigma/\sqrt{5}$, but now $\mu_{\bar{x}} = \mu + \sigma$ (μ and σ are still the standard values of, respectively, the process mean and standard deviation). Then,

$$\begin{aligned} q &= 1 - P\left[\mu - 3\frac{\sigma}{\sqrt{5}} < \bar{x} < \mu + 3\frac{\sigma}{\sqrt{5}}\right], \\ &= 1 - P\left[\frac{\mu - 3\sigma/\sqrt{5} - (\mu+\sigma)}{\sigma/\sqrt{5}} < \frac{\bar{x} - (\mu+\sigma)}{\sigma/\sqrt{5}} < \frac{\mu + 3\sigma/\sqrt{5} - (\mu+\sigma)}{\sigma/\sqrt{5}}\right], \\ &= 1 - P[-5.24 < Z < .76], \\ &= .2236. \end{aligned}$$

Figure 3.18 illustrates the calculation being done here and shows the roughly 22% chance that under these circumstances the sample mean will plot outside $\bar{x}$ chart control limits. Finally, using relationship (3.50),

$$ARL = \frac{1}{.2236} = 4.5.$$

That is, if the process mean is off target by as much as one process standard deviation, then it will take on average only 4.5 samples of size $n = 5$ to detect this kind of misadjustment.

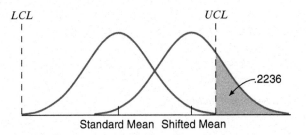

FIGURE 3.18. Two distributions for $\bar{x}$ and standards given control limits

Example 33 should agree completely with the reader's intuition about "how things should be." It says that when a process is on target, one can expect long periods between signals from an $\bar{x}$ chart. On the other hand, should the process mean shift off target by a substantial amount, there will typically be quick detection of that change.

Example 34 *Some ARLs for Shewhart c Charts. As a second example of the meaning of Eq. (3.50), consider finding some ARLs for two different versions of a Shewhart c chart when the standard rate of nonconformities is 1.5 nonconformities per unit. To begin, suppose that only one unit is inspected each period. Using relationships (3.42) with* $\lambda = 1.5$, *it follows that since* $1.5 - 3\sqrt{1.5} < 0$, *no lower control limit is used for the number of nonconformities found on an inspection unit, and*

$$UCL_X = 1.5 + 3\sqrt{1.5} = 5.2 \ .$$

So for this situation

$$q = P[X > 5.2] = 1 - P[X \leq 5] \ .$$

Consider evaluating q both when the nonconformity rate is at its standard value (of $\lambda = 1.5$ *nonconformities per unit) and when it is at three times its standard value (i.e., is 4.5 nonconformities per unit). When the rate is standard, one uses a Poisson distribution with mean 1.5 for X and finds*

$$q = 1 - P[X \leq 5] = .005 \quad \text{and} \quad ARL = \frac{1}{.005} = 200 \ .$$

When the rate is three times standard, one uses a Poisson distribution with mean 4.5 for X and finds

$$q = 1 - P[X \leq 5] = .298 \quad \text{and} \quad ARL = \frac{1}{.298} = 3.4 \ .$$

That is, completely in accord with intuition, the mean waiting time until an alarm is much smaller when quality deteriorates than when the process defect rate is standard.

Now suppose that two units will be inspected each period. One can then either use a u chart or equivalently simply apply a c chart where the standard value of λ *is 3.0 nonconformities per two units. Applying this second way of thinking and relationships (3.42) with* $\lambda = 3.0$, *it follows that since* $3.0 - 3\sqrt{3.0} < 0$, *no lower control limit is used for the number of nonconformities found on two inspection units, and*

$$UCL_X = 3.0 + 3\sqrt{3.0} = 8.2 \ .$$

So, for this situation

$$q = P[X > 8.2] = 1 - P[X \leq 8] \ .$$

Consider again the ARLs both where the nonconformity rate is at its standard value (of $\lambda = 3.0$ *nonconformities per two units) and where it is at three times* its standard value (i.e., is 9.0 nonconformities per two units). When the rate is standard, one uses a Poisson distribution with mean 3.0 for X and finds

$$q = 1 - P[X \leq 8] = .004 \quad \text{and} \quad ARL = \frac{1}{.004} = 250.$$

When the rate is three times standard, one uses a Poisson distribution with mean 9.0 for X and finds

$$q = 1 - P[X \leq 8] = .545 \quad \text{and} \quad ARL = \frac{1}{.545} = 1.8.$$

TABLE 3.8. ARLs for two c chart monitoring schemes for a standard nonconformity rate of 1.5 defects per unit

	Standard Defect rate	3×Standard Defect rate
One unit inspected	200	3.4
Two units inspected	250	1.8

Table 3.8 summarizes the calculations of this example. It shows the superiority of the monitoring scheme based on two units rather than one unit per period. The two-unit-per-period monitoring scheme has both a larger ARL when quality is standard and a smaller ARL when the nonconformity rate degrades by a factor of 3.0 than the one-unit-per-period scheme. This, of course, does not come without a price. One must do twice as much inspection for the second plan as for the first.

Examples 33 and 34 illustrate the ARL concept in very simple contexts that are covered by an elementary formula. Where the rules used to convert observed values $Q_1, Q_2, Q_3, \ldots$ into out-of-control signals or the probability model for these variables are at all complicated, explicit formulas and elementary computations are impossible. But it is not necessary to understand the nature of the numerical analysis needed to compute ARLs for more complicated cases to appreciate what an ARL tells one about a monitoring scheme.

For example, a paper by Champ and Woodall appearing in *Technometrics* in 1987 considered ARL computations for monitoring schemes using various combinations of the four Western Electric alarm rules. Example 33 showed the "all OK" ARL for an $\bar{x}$ chart scheme using only the "one point outside $3\sigma_{\bar{x}}$ control limits" rule to be about 370. When all four Western Electric alarm rules are employed simultaneously, Champ and Woodall found that the $\bar{x}$ chart "all OK" ARL is far less than 370 (or what naive users of the rules might expect), namely, approximately 92. The reduction from 370 to 92 shows the effects (in terms of increased frequency of false alarms) of allowing for other signals of process change in addition to individual points outside control limits.

Section 3.5 Exercises

1. Interpret the terms "ARL" and "all OK ARL."

2. What kind of ARL does one want under a "stable at standard parameter values" process model? What kind of ARL does one hope to have under any other circumstance?

3. $n = 4$ values are sampled every hour from a process that under "all OK stable process" conditions produces observations x that are normal with mean 20 and standard deviation 4. A typical Shewhart $\bar{x}$ chart is set up.

   (a) What is the all OK ARL of the monitoring scheme?

   (b) An upward shift in the process mean of at least 1 unit occurs, while the process standard deviation variation does not change. At worst, how many hours on average will pass before this change produces a subgroup average outside the control limits?

4. Consider a production process where one item (the subgroup size is 1) is periodically sampled and the number of nonconformities is observed. Suppose standard nonconformity rate per item is $\lambda = 4$.

   (a) Find the all OK ARL.

   (b) Find the ARL if an increase to a rate of $\lambda = 8$ occurs.

   (c) Answer (a) and (b) if two items make up each subgroup.

5. Control charting Method A is preferred to Method B relative to an "all OK" and some "not all OK" process conditions. Which of the following is true?

   (a) $ARL_A > ARL_B$ when "all is OK" and $ARL_A > ARL_B$ when "all is not OK."

   (b) $ARL_A > ARL_B$ when "all is OK" and $ARL_A < ARL_B$ when "all is not OK."

   (c) $ARL_A < ARL_B$ when "all is OK" and $ARL_A > ARL_B$ when "all is not OK."

   (d) $ARL_A < ARL_B$ when "all is OK" and $ARL_A < ARL_B$ when "all is not OK."

6. Process standards are $\mu = 100$ and $\sigma = 7$, and observations from the process are normally distributed. A Shewhart $\bar{x}$ chart is being considered for use in monitoring the process.

   (a) The charts with $n = 5$ and $n = 10$ will have different control limits. Why?

(b) The charts with $n = 5$ and $n = 10$ will have the same ARL if process parameters remain at standard values. Why?

3.6 Statistical Process Monitoring and Engineering Control

We have said that "statistical process *control*" is really better called "statistical process *monitoring*." "Engineering *control*" is a very important subject that is largely distinct from the considerations laid out thus far in this chapter. Unfortunately, there has been a fair amount of confusion about what the two methodologies offer, how they differ, and what are their proper roles in the running of industrial processes. This section is intended to help readers better understand the relationship between them. It begins with an elementary introduction to one simple kind of engineering control, called PID control. It then proceeds to a number of general comments comparing and contrasting statistical process monitoring and engineering control.

3.6.1 Discrete Time PID Control

Engineering control has to do with guiding processes by the deliberate manipulation of appropriate process parameters. For example, in a chemical process, a temperature in a reaction vessel might be kept constant by appropriate manipulation of the position of an inlet steam valve. A very common version of engineering control in industry can be represented in terms of a feedback control diagram like that in Fig. 3.19.

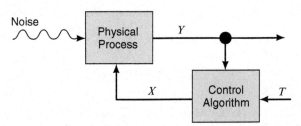

FIGURE 3.19. Schematic of an engineering feedback control system

In Fig. 3.19, a process outputs a value of a variable Y, which is fed into a control algorithm along with a value of a target T for the next output, resulting in a value for some manipulated process variable X, which together with (unavoidable) noise (somehow) produces a subsequent value of Y and so on. Depending upon what is known about the various elements in Fig. 3.19, different means of choosing a control algorithm can be applied. A method that requires very little in the way of detailed knowledge about how X or the noise impacts Y is that of **proportional-integral-derivative (PID) control**.

The discussion here will treat the discrete time version of PID control. So consider discrete integer times $t = 1, 2, 3, \ldots$ (typically evenly spaced in real time), and as in Fig. 3.19, suppose that

$Y(t)$ = the value of the controlled or output variable at time t,

$T(t)$ = the value of a target for Y at time t, and

$X(t)$ = the value of a (manipulated) process variable that is chosen after observing $Y(t)$.

A control algorithm converts knowledge of $Y(1), Y(2), \ldots, Y(t)$ and $T(s)$ for all s into a choice of $X(t)$. For example, in machining $Y(t)$ could be a measured widget diameter, $T(t)$ a target diameter, and $X(t)$ a cutting tool position. A control algorithm orders a tool position in light of all past and present diameters and all targets for past, present, and future diameters.

The practice of PID control does not typically invest much effort in modeling exactly how changes in X get reflected in Y. (If the goal of a study *was* to understand that relationship, tools of regression analysis might well be helpful.) Nevertheless, in understanding the goals of engineering control, it is useful to consider two kinds of process behavior with which engineering control algorithms must sometimes deal.

For one thing, some physical processes react to changes in manipulated variables only gradually. One behavior predicted by many models of physical science is that when initially at "steady state" at time t_0, a change of ΔX in a manipulated variable introduces a change in the output at time $t > t_0$ of the form

$$\Delta Y(t) = Y(t) - Y(t_0) = G \Delta X \left(1 - \exp\left(\frac{-(t - t_0)}{\tau} \right) \right), \quad (3.51)$$

for process-dependent constants G and τ. Figure 3.20 on page 152 shows a plot of ΔY in display (3.51) as a function of time. In cases where relationship (3.51) holds, G is the limit of the ratio $\Delta Y / \Delta X$ and is called the **control gain**. τ governs how quickly the limiting change in Y is reached (τ is the time required to reach a fraction $1 - e^{-1} \approx .63$ of the limiting change in Y). It is called the **time constant** for a system obeying relationship (3.51).

Another phenomenon that is sometimes part of the environment in which engineering control systems must operate is that of **dead time** or **delay** between when a change is made in X and when any effect of the change begins to be seen in Y. If there are δ units of dead time and thereafter a relationship similar to that in Eq. (3.51) holds, one might see a pattern like that shown in Fig. 3.21 on page 152 following an adjustment ΔX made at time t_0 on a process at steady state at that time.

Of course, not all physical systems involve the kind of gradual impact of process changes illustrated in Fig. 3.20, nor do they necessarily involve dead time. (For example, real-time feedback control of machine tools will typically involve

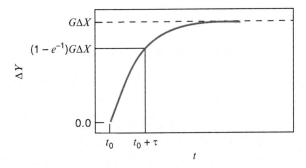

FIGURE 3.20. Change in the output Y (initially at steady state) in response to a ΔX change in the manipulated variable X

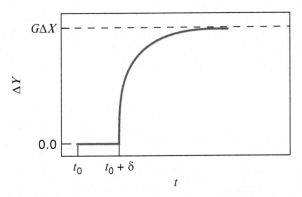

FIGURE 3.21. Change in the output Y (initially at steady state) in response to a ΔX change in the manipulated variable at time t_0 if there are δ units of dead time

changes in tool positions that take their full effect "immediately" after being ordered.) But where these phenomena are present, they increase the difficulty of finding effective control algorithms, the dead time problem being particularly troublesome where δ is large.

To get to the point of introducing the general PID control algorithm, consider a situation where it is sensible to expect that increasing X will tend to increase Y. Define the observed "error" at time t

Error at Time t

$$E(t) = T(t) - Y(t),$$

and the first and second differences of errors,

First Difference in Errors at Time t

$$\Delta E(t) = E(t) - E(t-1)$$

and

Second Difference in Errors at Time t

$$\Delta^2 E(t) = \Delta(\Delta E(t)) = \Delta E(t) - \Delta E(t-1).$$

With some additional algebra

$$\Delta^2 E(t) = (E(t) - E(t-1)) - (E(t-1) - E(t-2))$$
$$= E(t) - 2E(t-1) + E(t-2).$$

Then, for constants κ_1, κ_2, and κ_3, a PID control algorithm sets

$$\Delta X(t) = \kappa_1 \Delta E(t) + \kappa_2 E(t) + \kappa_3 \Delta^2 E(t). \quad (3.52)$$

PID Control Algorithm

(In cases where Y tends to increase with X, the constants κ_1, κ_2, and κ_3 are typically nonnegative.) The three terms summed on the right of Eq. (3.52) are, respectively, the **proportional, integral,** and **derivative** parts of the control algorithm.

Example 35 *PID Control of Final Dry Weight of 20-lb Bond Paper. Through the kind cooperation of the Miami University Paper Science Laboratory and Mr. Doug Hart, Research Associate at the lab, one of your authors was able to help implement a PID controller on a 13-in Fourdrinier paper-making machine. This machine produces paper in a long continuous sheet beginning with vats of pulp mix. The final dry weight of paper is measured as the paper leaves the machine and can be controlled by the rate at which a Masterflex peristaltic pump delivers pulp mix to the machine. A manual knob is used to vary the pump speed and can be adjusted in "ticks." (Each 1-tick change corresponds approximately to a change of pump speed equal to .2% of its maximum capacity.) Past experience with the machine indicated that for 20-lb bond pulp mixture, a 1-tick increase in pump speed produces approximately a $.3\,\mathrm{g/m^2}$ increase in paper dry weight. But unavoidable variations in the process (including the "thickness" of the mix available to the pump) produce variation in the paper dry weight and need to be compensated for by varying the pump speed.*

Since there is over a 4-min lag between when a pump speed change is made and when paper affected by the speed change reaches the scanner that measures dry weight at the end of the machine, measurements and corresponding adjustments to pump speed were made only once every 5 min. (This choice eliminates the effect of dead time on the control algorithm, which would be a concern if measurements and adjustments were made closer together.) Some experimentation with the machine led to the conclusion that a sensible PID control algorithm for the machine (using the 5-min intervals and measuring the control variable changes in terms of ticks) has

$$\kappa_1 = .83, \quad \kappa_2 = 1.66, \quad \text{and} \quad \kappa_3 = .83$$

in formula (3.52). Table 3.9 on page 154 shows an actual series of dry weight measurements and PID controller calculations made using these constants. (Since it was impossible to move the pump speed knob in fractions of a tick, the actual adjustments applied were those in the table rounded off to the nearest tick.) The production run was begun with the knob (X) in the standard or default position for the production of 20-lb bond paper.

For example, for $t = 3$,

$$E(3) = T(3) - Y(3) = 70.0 - 68.6 = 1.4,$$

TABLE 3.9. PID control calculations for the control of paper dry weight ($T, Y, E, \Delta E$, and $\Delta^2 E$ in g/m^2 and ΔX in ticks)

Period, t	$T(t)$	$Y(t)$	$E(t)$	$\Delta E(t)$	$\Delta^2 E(t)$	$\Delta X(t) = .83\Delta E(t)$ $+1.66 E(t) + .83\Delta^2 E(t)$
1	70.0	65.0	5.0			
2	70.0	67.0	3.0	−2.0		
3	70.0	68.6	1.4	−1.6	.4	1.328
4	70.0	68.0	2.0	.6	2.2	5.644
5	70.0	67.8	2.2	.2	−.4	3.486
6	70.0	69.2	.8	−1.4	−1.6	−1.162
7	70.0	70.6	−.6	−1.4	0	−2.158
8	70.0	69.5	.5	1.1	2.5	3.818
9	70.0	70.3	−.3	−.8	−1.9	−2.739
10	70.0	70.7	−.7	−.4	.4	−1.162
11	70.0	70.1	−.1	.6	1.0	1.162

$$\Delta E(3) = E(3) - E(2) = 1.4 - 3.0 = -1.6,$$
$$\Delta^2 E(3) = \Delta E(3) - \Delta E(2) = -1.6 - (-2.0) = .4,$$

and so the indicated adjustment (increment) on the pump speed knob is $\Delta X(3) = .83\Delta E(3) + 1.66 E(3) + .83\Delta^2 E(3) = .83(-1.6) + 1.66(1.4) + .83(.4) = 1.328$ ticks. *(As actually implemented, this led to a 1-tick increase in the knob position after measurement 3.)*

It is useful to separately consider the proportional, integral, and derivative parts of algorithm (3.52), beginning with the integral part. With $\kappa_2 > 0$, this part of the algorithm increases X when E is positive and thus $T > Y$. It is this part of a PID control algorithm that reacts to (attempts to cancel) **deviations from target**. Its function is to try to move Y in the direction of T.

To grasp why $\kappa_2 E(t)$ might be called the "integral" part of the control algorithm, consider a case where both $\kappa_1 = 0$ and $\kappa_3 = 0$ so that one has an "integral-only" controller. In this case (supposing that $Y(t)$s and $T(t)$s with $t < 1$ are available so that one can begin using relationship (3.52) at time $t = 1$), note that

$$\sum_{s=1}^{t} \Delta X(s) = \kappa_2 \sum_{s=1}^{t} E(s). \tag{3.53}$$

But the sum on the left of Eq. (3.53) telescopes to $X(t) - X(0)$ so that one has

$$X(t) = X(0) + \kappa_2 \sum_{s=1}^{t} E(s).$$

That is, the value of the manipulated variable is $X(0)$ plus a sum or "integral" of the error.

"Integral-only" control (especially in the presence of a large time constant and/or large dead time) often tends to overshoot target values and set up oscillations in the variable Y. The proportional and derivative parts of a PID algorithm are meant to reduce overshoot and damp oscillations. Consider next the proportional term from Eq. (3.52), namely, $\kappa_1 \Delta E(t)$.

The proportional part of a PID control algorithm reacts to **changes in the error**. In graphical terms, it reacts to a nonzero slope on a plot of $E(t)$ versus t. Where $\kappa_1 > 0$, this part of the algorithm increases X if the error increases and decreases X if E decreases. In some sense, this part of the algorithm works to hold the error constant (whether at 0 or otherwise).

When κ_1 and κ_2 have the same sign, the proportional part of a PID control algorithm augments the integral part when E is moving away from 0 and "brakes" or cancels part of the integral part when E is moving toward 0. Figure 3.22 pictures two plots of $Y(t)$ versus t for cases where the target T is constant. In the first plot, Y is approaching T from below. $E(t) > 0$, while $\Delta E(t) < 0$. This is a case where the proportional part of the algorithm brakes the integral part. In the second plot, Y is above T and diverging from it. There, both $E(t) < 0$ and $\Delta E(t) < 0$, and the proportional part of the algorithm augments the integral part. The braking behavior of the proportional part of a PID algorithm helps to resist the kind of oscillation/overshoot problem produced by "integral-only" control.

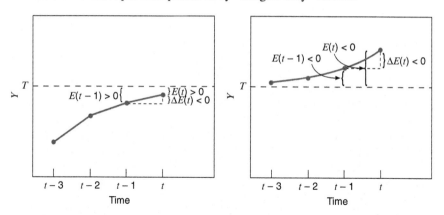

FIGURE 3.22. Two plots of Y against time and a constant target, T

To see why $\kappa_1 \Delta E(t)$ might be called the "proportional" part of the control algorithm, consider a case where both $\kappa_2 = 0$ and $\kappa_3 = 0$ so that one has a "proportional-only" controller. In this case (supposing that $Y(t)$s and $T(t)$s with $t < 1$ are available so that one can begin using relationship (3.52) at time $t = 1$),

$$\sum_{s=1}^{t} \Delta X(s) = \kappa_1 \sum_{s=1}^{t} \Delta E(s). \tag{3.54}$$

But the sums on both sides of Eq. (3.54) telescope, so that one has

$$X(t) = X(0) - \kappa_1 E(0) + \kappa_1 E(t).$$

That is, the value of the manipulated variable is $X(0) - \kappa_1 E(0)$ plus a term "proportional" to the error.

Finally, consider the derivative part of the algorithm (3.52), namely, $\kappa_3 \Delta^2 E(t)$. This part of the algorithm reacts to curvature or **changes in slope** on a plot of $E(t)$ versus t. That is, it reacts to changes in $\Delta E(t)$. If a plot of errors versus t is linear ($\Delta E(t)$ is constant), this part of the algorithm does nothing to change X. If $\kappa_3 > 0$ and a plot of errors versus t is concave up, the derivative part of algorithm (3.52) will increase X (and thus Y, decreasing E), while if the plot is concave down, it will decrease X. For constant target T, this will tend to "straighten out" a plot of $E(t)$ or $Y(t)$ versus t (presumably then allowing the proportional part of the algorithm to reduce the slope to 0 and the integral part to put the process on target). Once again, since "integral-only" control often produces unwanted oscillations of Y about a target and it is impossible to oscillate without local curvature in a plot of Y versus t, the derivative part of the algorithm can be considered as corrective to a deficiency in the naive "integral-only" idea.

The rationale for calling $\kappa_3 \Delta^2 E(t)$ the "derivative" part of the PID algorithm (3.52) is similar to the arguments made about the other two parts. Namely, if κ_1 and κ_2 are both 0 (so that one has "derivative-only" control),

$$\sum_{s=1}^{t} \Delta X(s) = \kappa_3 \sum_{s=1}^{t} \Delta^2 E(s). \tag{3.55}$$

Telescoping both sides of Eq. (3.55), one then has

$$X(t) = X(0) - \kappa_3 \Delta E(0) + \kappa_3 \Delta E(t),$$

and the value of the manipulated variable is $X(0) - \kappa_3 \Delta E(0)$ plus a term proportional to the change in (or "derivative" of) the error.

The primary practical problem associated with the use of PID controllers is the matter of choosing the constants κ_1, κ_2, and κ_3, sometimes called, respectively, the **proportional, integral, and derivative gains** for the control algorithm. In simple situations where engineers have good mathematical models for the physical system involved, those can sometimes provide at least starting values for searches to find good values of these constants. Where such models are lacking, various rules of thumb aid searches for workable values of κ_1, κ_2, and κ_3. For instance, one such rule is to initially set κ_1 and κ_3 to zero, increase κ_2 until oscillations occur, then halve that value of κ_2, and begin searching over κ_1 and κ_3. And it is pretty clear that in systems where a relationship like (3.51) holds, the gains κ_1, κ_2, and κ_3 should be inversely proportional to G. Further, conventional wisdom also says that in systems where there is dead time $\delta > 0$, the control gains should decrease (exponentially?) in δ. (One should not be changing a manipulated variable wildly if there is to be a long delay before one gets to measure the impact of those changes and to begin to correct any unfortunate effects one sees.)

Ultimately, the matter of finding good values for the gains κ_1, κ_2, and κ_3 is typically a problem of empirical optimization. Section 6.2 of this book discusses some experimental strategies in process optimization. These can be applied to the problem of finding good constants κ_1, κ_2, and κ_3 in the following way. For given choices of the constants, one may run the process using the PID controller (3.52) for some number of periods, say m. Then a sensible figure of merit for that particular set of constants is the random variable

$$S = \frac{1}{m}\sum_{t=1}^{m}(E(t))^2,$$

the average squared error. The empirical optimization strategies of Sect. 6.2 may then be applied in an attempt to find a set of values for κ_1, κ_2, and κ_3 with minimum associated mean for S, μ_s. Chapter problems 38 through 44 describe how the average squared error idea was used to arrive at the control algorithm of Example 35.

3.6.2 Comparisons and Contrasts

The PID ideas just discussed are not the only ones used to produce engineering control algorithms. For example, where good models are available for both uncontrolled process behavior and for the impact of control actions on process outputs, mathematically optimal control algorithms (that need not be of the PID type) can sometimes be derived. And the introduction just given completely ignores real issues like the multivariate nature of most industrial applications. (The Y and X just considered are one-dimensional, while real process outputs and possible manipulated variables are often multidimensional.) But the foregoing brief discussion is intended only to give the reader enough of an idea of how engineering control operates to allow the following comments on the proper roles of engineering control and statistical process monitoring to make sense.

The relative merits of the two methodologies when applied in production contexts have been at times been hotly debated by their proponents. On some occasions, zealots on one side or the other of the debate have essentially claimed that their methods are universally applicable, and those of the other side are either without merit or are simply a weak version of their own. The truth is that the methods of statistical process monitoring and engineering control are not competitors. They are, in fact, completely complementary, each having its own purposes and appropriate areas of application. When applied to the running of industrial processes, both are aimed at the reduction of unwanted variability. In many applications, they can and should be used *together* in an effort to reduce process variation and improve quality, engineering control helping to create stable conditions that are monitored using statistical process-monitoring methods.

In cases where a process is already physically stable about a target value, statistical process-monitoring tools should only infrequently (and wrongly) signal the need for intervention, and engineering control *is of no help in reducing variation*. That is, in the classical stable process situation, tweaking process parameters can only make variation worse, not better. On the other hand, if successive observa-

158   Chapter 3. Process Monitoring

tions on a process look as if they are dependent or if they have means (either constant or moving) different from a target, engineering control may be able to improve process performance (uniformity of output) essentially by canceling predictable misadjustments of the process. Statistical process monitoring will then protect one from unexpected process changes.

Table 3.10 puts side by side a number of pairs of statements that should help the reader keep clear the basic differences between engineering control and statistical process monitoring as they are applied to industrial processes. The late Dr. Bill Tucker was fond of saying "You can't steer a car with statistical process control and you can't fix a car with engineering control." His apt and easily remembered analogy brings into focus the differences in intent of the two methodologies.

Section 3.6 Exercises

1. The target value for a process output variable, Y, is four units, and a controllable process parameter X is thought to impact Y in a direct fashion. In three successive periods, $Y(1) = 2, Y(2) = 1$, and $Y(3) = 0$. You may finish filling in a table like that below to help you answer this question.

Period, t	$T(t)$	$Y(t)$	$E(t)$	$\Delta E(t)$	$\Delta^2 E(t)$	$\Delta X(t)$
1		2				
2		1				
3		0				

    (a) What are your values of $T(t)$ here? What is the practical meaning of this variable?

    (b) What values do you get for $E(t), \Delta E(t)$, and $\Delta^2 E(t)$ here? Describe what these are measuring.

    (c) Use control gains $\kappa_1 = .8, \kappa_2 = 1.6$, and $\kappa_3 = 1.9$, and compute a PID control action $\Delta X(3)$ to be taken after observing $Y(3)$.

    (d) How will this table and future values of Y be used to extend the PID control of part (c) beyond period 3?

2. In the context of problem 1, suppose that no one is really sure whether Y is affected by changes in X and, if it is, whether the relationship is "direct" or "inverse".

    (a) Speculate on what might happen if the PID controller of part (c) above is implemented where Y is completely unrelated to X. What might happen if in fact Y is inversely related to X?

    (b) How would you propose to figure out what, if any, PID control based on X might be fruitful?

3. In what sense are control charts tools for "controlling" a process? In what meaning of the word "control" are they *not* tools for controlling a process?

TABLE 3.10. Contrasts between engineering control and statistical process control for industrial processes

Engineering control	Statistical process control
• In a typical application, there is a sensor on a process and an electromechanical adjustment mechanism that responds to orders (for change of some process parameter) sent by a computer "brain" based on signals from the sensor	• This is either manual or automatic plotting of process performance statistics to warn of process changes
• This is an adjustment/compensation methodology. Formulas prescribe explicit reactions to deviations from target	• This is a detection methodology. Corrective measures for process changes that are detected are not specified
• This is a methodology for ongoing small process adjustments	• There is a tacit assumption here that wise intervention following detection of a process change will set things perfectly aright (for an extended period)
• There is an explicit expectation of process instability/drift in this methodology	• There is a tacit assumption here of process stability over long periods
• This is typically computer (or at least mechanically) controlled	• There is typically a human agent involved in monitoring and interventions
• The ultimate effect is to keep a process optimally adjusted	• The ultimate effect is to warn of the presence of sources of special cause variation, to help identify them, and to lead to their permanent removal
• This is often "tactical" and applied to process parameters	• This is often "strategic" and applied to final quality variables
• In its "optimal stochastic control" version, this is what one does within a particular probability model (for process behavior) to best exploit the probabilistic predictability of a process	• This is what one does to monitor for "the unexpected" (departures from a stable process model of expected behavior)

3.7 Chapter Summary

Shewhart control charts are an engineer's most widely applicable and easily understood process-monitoring tools. The first four sections of this chapter have introduced these charts for both variables data and attributes data, considered their use in both standards given and retrospective contexts, and discussed their qualitative interpretation and supplementation with sets of "extra alarm rules." Table 3.11 summarizes many of the standard formulas used in the making of elementary Shewhart charts.

TABLE 3.11. Formulas for Shewhart control charting

Chart	Q	μ_Q	σ_Q	Standards given UCL_Q	LCL_Q	Retrospective UCL_Q	LCL_Q
$\bar{x}$	$\bar{x}$	μ	$\sigma/\sqrt{n}$	$\mu + 3\sigma/\sqrt{n}$	$\mu - 3\sigma/\sqrt{n}$	$\bar{\bar{x}}+A_2\bar{R}$	$\bar{\bar{x}}-A_2\bar{R}$
						$\bar{\bar{x}}+A_3\bar{s}$	$\bar{\bar{x}}-A_3\bar{s}$
Median	$\tilde{x}$	μ	$\kappa\sigma/\sqrt{n}$	$\mu + 3\kappa\sigma/\sqrt{n}$	$\mu - 3\kappa\sigma/\sqrt{n}$		
R	R	$d_2\sigma$	$d_3\sigma$	$D_2\sigma$	$D_1\sigma$	$D_4\bar{R}$	$D_3\bar{R}$
s	s	$c_4\sigma$	$c_5\sigma$	$B_6\sigma$	$B_5\sigma$	$B_4\bar{s}$	$B_3\bar{s}$
np	X	np	$\sqrt{np(1-p)}$	$np + 3\sqrt{np(1-p)}$	$np - 3\sqrt{np(1-p)}$	(use $\hat{p}_{\text{pooled}}$ for p)	
p	$\hat{p}$	p	$\sqrt{\frac{p(1-p)}{n}}$	$p + 3\sqrt{\frac{p(1-p)}{n}}$	$p - 3\sqrt{\frac{p(1-p)}{n}}$	(use $\hat{p}_{\text{pooled}}$ for p)	
c	X	λ	$\sqrt{\lambda}$	$\lambda + 3\sqrt{\lambda}$	$\lambda - 3\sqrt{\lambda}$	(use $\hat{\lambda}_{\text{pooled}}$ for λ)	
u	$\hat{u}$	λ	$\sqrt{\frac{\lambda}{k}}$	$\lambda + 3\sqrt{\frac{\lambda}{k}}$	$\lambda - 3\sqrt{\frac{\lambda}{k}}$	(use $\hat{\lambda}_{\text{pooled}}$ for λ)	

The final two sections of the chapter have provided context and perspective for the study of Shewhart charts and other process-monitoring tools. Section 3.5 introduced the ARL concept as a means of quantifying the likely performance of a monitoring scheme. Section 3.6 contrasted methods and goals of "engineering control" with those of process monitoring when they are both applied in production.

3.8 Chapter 3 Exercises

1. What is the purpose of control charting? What is suggested by out-of-control signals?

2. What makes step 3 in the quality assurance cycle presented in Chap. 1 difficult in service contexts? Explain.

3. Why is it essential to have a clear understanding of what constitutes a nonconformance if a Shewhart c or u chart is to be made?

4. Is control charting most directly concerned with "quality of design" or with "quality of conformance"?

5. Distinguish between "control limits" and "specification limits" for variables data.

6. Explain the difference between "control limits" and "specification limits" in an attributes data setting.

7. Explain the ARL concept in terms that a person with no statistical training could understand.

8. When designing a control chart, what kinds of ARL values are desirable for an on-target process? For an off-target process? Explain why your answers are correct from an economic point of view.

9. State why statistical methodology is an unavoidable part of quality assurance practice. (Review Chap. 1.)

10. Sometimes the plotted statistics appearing on a Shewhart control chart hug (or have little scatter around) a center line. Explain why this is not necessarily a good sign.

11. Uninformed engineers sometimes draw in lines on Shewhart $\bar{x}$ charts at engineering specifications for individual measurements. Why is that a bad practice?

12. It is common to hear people imply that the job of control charts is to warn of degradation in product quality. Do you agree with that? Why or why not?

13. What is the purpose of sets of "extra alarm rules" like the Western Electric alarm rules presented in Sect. 3.4?

14. What (relevant to quality-improvement efforts) does a multimodal shape of a histogram for a part dimension suggest? (Review Chap. 2.)

15. In colloquial terms, the language "control" chart perhaps suggests a plot associated with continuous regulatory efforts. Is this understanding correct? Why or why not? Suggest a better term than "control chart."

16. **Journal Diameters**. Table 3.12 provides some summary statistics (means and standard deviations) for journal diameters of tractor axles as the axles come off an automatic grinding machine. The statistics are based on subgroups of size $n = 4$ pieces taken once per hour. The values listed are in millimeters. Specifications on the journal diameter are from 44.975 to 44.990 mm. Note that $\sum \bar{x} = 899.5876$ and $\sum s = .0442$.

   (a) Are the above attributes data or variables data? Why?
   (b) Make a retrospective s chart for these values.
   (c) Make a retrospective $\bar{x}$ chart for these values.

TABLE 3.12. Summary statistics for problem 16

Subgroup	$\bar{x}$	s	Subgroup	$\bar{x}$	s
1	44.9875	.0029	11	44.9815	.0017
2	44.9813	.0025	12	44.9815	.0017
3	44.9808	.0030	13	44.9810	.0024
4	44.9750	.0000	14	44.9778	.0021
5	44.9783	.0039	15	44.9748	.0024
6	44.9795	.0033	16	44.9725	.0029
7	44.9828	.0021	17	44.9778	.0021
8	44.9820	.0024	18	44.9790	.0034
9	44.9770	.0024	19	44.9785	.0010
10	44.9795	.0010	20	44.9795	.0010

(d) What do these charts indicate (in retrospect) about the stability of the grinding process?

(e) Based on your conclusion in (d), can the fraction of journal diameters that currently meet specifications be reliably estimated? Why or why not?

(f) Independent of your conclusion in (d), if one judged the process to be stable based on the 20 subgroups summarized above, what could be used as an estimate of the fraction of journal diameters that currently meet specifications? (Give a number based on a normal distribution assumption for diameter measurements.)

(g) Suppose that henceforth (into the future) this process is to be monitored using subgroups of size $n = 5$. Give control limits for a (standards given) median chart based on the mid-specification (giving the center line) and your estimated process standard deviation from (b).

(h) Give control limits for future monitoring of sample ranges (use your estimated process standard deviation from (b) as a future standard value and assume $n = 5$).

17. Refer to the **Journal Diameter** case introduced in problem 16. Sometimes subgroup size is not constant. When using standard deviations from subgroup of varying sizes $n_1, n_2, \ldots, n_r$ to estimate σ, there are several possibilities. Of commonly used ones, the one with the best theoretical properties is

$$s_{\text{pooled}} = \sqrt{\frac{(n_1 - 1)s_1^2 + (n_2 - 1)s_2^2 + \cdots + (n_r - 1)s_r^2}{(n_1 - 1) + (n_2 - 1) + \cdots + (n_r - 1)}}.$$

Another possibility is

$$\hat{\sigma} = \frac{\frac{(n_1-1)s_1}{c_4(n_1)} + \frac{(n_2-1)s_2}{c_4(n_2)} + \cdots + \frac{(n_r-1)s_r}{c_4(n_r)}}{n_1 + n_2 + \cdots + n_r - r}.$$

(Since sample sizes vary in this development, we are displaying the dependence of c_4 on sample size here.) The most appropriate estimator of a common mean, μ, when sample sizes vary is

$$\bar{x}_{\text{pooled}} = \frac{n_1 \bar{x}_1 + n_2 \bar{x}_2 + \cdots + n_r \bar{x}_r}{n_1 + n_2 + \cdots + n_r}.$$

Consider the subgroup means and standard deviations given in problem 16. Suppose subgroups were of size $n = 4$ except for the ones indicated in the following table.

Subgroup	Sample size
1	2
8	2
10	8
15	5
18	3
19	3
20	9

(a) Find values for s_{pooled}, $\hat{\sigma}$, and $\bar{x}_{\text{pooled}}$.

(b) Give two estimates of (1) the standard deviation of a subgroup mean when $n = 2$ and (2) the standard deviation of a subgroup standard deviation when $n = 2$. (Hint: $\text{Var}\,\bar{x}_i = \sigma^2/n_i$ and $\text{Var}\,s_i = \sigma^2(1 - c_4^2(n_i))$.)

(c) With the new subgroup sizes, consider two retrospective control charts, one chart appropriate for assessing the constancy of variability of axle journal diameters and the other for monitoring average axle journal diameter. Would the control limits be constant across time for the two charts? (There is no need to actually make them here.) Why or why not? (See (a) and (b).)

(d) Do the center lines for the two charts in (c) change depending on subgroup size? (Again, there is no need to make the charts.) Why or why not?

18. **Rolled Paper.** Shervheim and Snider did a project with a company that cuts rolled paper into sheets. The students periodically sampled $n = 5$ consecutive sheets as they were cut and recorded their actual lengths, y. Data from 20 subgroups are summarized in Table 3.13. (Measurements corresponding to the values in the table were in 64ths of an inch above nominal, i.e., $x = y - nominal$.)

(a) Make a retrospective s chart.

(b) Make a retrospective $\bar{x}$ chart.

(c) What do these charts indicate (in retrospect) about the stability of the cutting process?

(d) Give an estimate of the process standard deviation based on $\bar{s}$.

(e) If one judges the process to be stable and sheet length to be normally distributed, estimate the fraction of sheets below nominal in length. (Hint: Find $P(x < 0)$ by transforming to a standard normal random variable Z.)

TABLE 3.13. Summary statistics for problem 18

Subgroup	$\bar{x}$	s	Subgroup	$\bar{x}$	s
1	12.2	.84	11	10.4	1.95
2	11.2	1.64	12	10.6	1.67
3	10.6	2.07	13	10.4	1.67
4	12.2	2.49	14	12.0	2.91
5	11.2	.84	15	11.2	.84
6	12.6	1.82	16	10.6	1.82
7	12.2	2.95	17	10.4	1.14
8	13.6	1.67	18	9.8	2.17
9	12.2	1.30	19	9.6	2.07
10	10.4	1.52	20	10.6	1.95
				224.0	35.33

(f) Each .25 in that the cutting process mean is above nominal represents a $100,000/year loss to the company from product "given away." On the other hand, the company wants to be sure that essentially no sheets are produced with below-nominal lengths (so it wants $\mu_x > 3\sigma$). With this in mind, what adjustment in mean length do you suggest, and what yearly savings or additional cost do you project if this adjustment is made?

(g) Suppose that the adjustment you recommend in (f) is made and henceforth the cutting process is to be monitored based on samples of size $n = 3$. What are standards given control limits for future monitoring of $\bar{x}$ and s?

(h) Suppose that while using your $\bar{x}$ chart from (g), the process mean suddenly drops to the point where 1 % of the sheets produced are below nominal in length. On average, how many samples will be required to detect this? (Hint: Find the "new μ_x" that will make $P(x < 0) = .01$; then using it find $P(\bar{x} < LCL) + P(\bar{x} > UCL)$.) How does this compare in terms of quickness of detection to a scheme (essentially a p chart) that signals the first time a sample of $n = 3$ contains at least one sheet with below-nominal length?

19. Refer to the **Rolled Paper** case in problem 18. Again use the means and standard deviations given there, but suppose that the number of sheets per subgroup was not constant. Instead, suppose subgroups contained five sheets except for the ones indicated in the following table.

Subgroup	Subgroup size
3	7
6	7
10	2
14	4
17	3
19	2
20	6

(a) Compute $\bar{x}_{\text{pooled}}$ and two different estimates of σ. (See problem 17.)

(b) For a subgroup size of $n = 7$, give two estimates of 1) the standard deviation of a subgroup mean and 2) the standard deviation of a subgroup standard deviation. (Hint: $\text{Var}\,\bar{x}_i = \sigma^2/n_i$ and $\text{Var}\,s_i = \sigma^2(1 - c_4^2(n_i))$.)

(c) With the variable subgroup sizes, consider two retrospective control charts, one s chart and one $\bar{x}$ chart. Would the control limits be constant across time for either chart? (There is no need to make the charts.) Why or why not? (See (a) and (b).)

(d) Do the center lines for the two charts in (c) remain constant across subgroup sizes? (Again, there is no need to make the charts.) Why or why not?

20. **U-bolt Threads.** A manufacturer of U-bolts for the auto industry measures and records thread lengths on bolts that it produces. Eighteen subgroups, each of $n = 5$ consecutive bolts were obtained, and actual thread lengths y were measured. These can be expressed as deviations from nominal by transforming as $x = y - nominal$. Some summary statistics are indicated in Table 3.14 (the units are .001 in above nominal).

(a) Estimate the supposedly common subgroup standard deviation, σ, using (1) the subgroup ranges (R_i) and (2) the subgroup standard deviations (s_i).

(b) Find control limits for the subgroup ranges. (Use the estimate of σ based on the s_i.)

(c) Find control limits for the subgroup standard deviations. (Use the estimate of σ based on the s_i.)

(d) Plot the ranges and standard deviations on Shewhart charts using the retrospective limits from (b) and (c). Is it plausible that variability of thread length was constant from sampling period to sampling period? Why or why not?

TABLE 3.14. Data and summary statistics for problem 20

Subgroup	Thread length	$\tilde{x}$	s	$\bar{x}$	R
1	11, 14, 14, 10, 8	11	2.61	11.4	6
2	14, 10, 11, 10, 11	11	1.64	11.2	4
3	8, 13, 14, 13, 10	13	2.51	11.6	6
4	11, 8, 13, 11, 13	11	2.05	11.2	5
5	13, 10, 11, 11, 11	11	1.10	11.2	3
6	11, 10, 10, 11, 13	11	1.22	11.0	3
7	8, 6, 11, 11, 11	11	2.30	9.4	5
8	10, 11, 10, 14, 10	10	1.73	11.0	4
9	11, 8, 11, 8, 10	10	1.52	9.6	3
10	6, 6, 11, 13, 11	11	3.21	9.4	7
11	11, 14, 13, 8, 11	11	2.30	11.4	6
12	8, 11, 10, 11, 14	11	2.17	10.8	6
13	11, 11, 13, 8, 13	11	2.05	11.2	5
14	11, 8, 11, 11, 11	11	1.34	10.4	3
15	11, 11, 13, 11, 11	11	.89	11.4	2
16	14, 13, 13, 13, 14	13	.55	13.4	1
17	14, 13, 14, 13, 11	13	1.22	13.0	3
18	13, 11, 11, 11, 13	11	1.10	11.8	2
		202	31.51	200.4	74

(e) Find retrospective control limits for the subgroup means. (Use your estimate of σ based on the s_i.) Plot the means on a Shewhart chart with these limits.

(f) Setting the center line at $\bar{\bar{x}}$, find upper and lower control limits for the subgroup medians. (Use your estimate of σ based on the s_i.) Plot the medians on a Shewhart chart with these limits.

(g) What do the charts in (e) and (f) suggest about the threading process?

(h) A U-bolt customer requires that essentially all U-bolt thread lengths are within .011 in of nominal. Assuming bolt manufacturing continues as represented by the values in the table, will the customer be satisfied with current production? Why or why not? Give a quantitative defense of your answer assuming normality of thread length. (Hint: Find $P(-11 < x < 11)$.)

21. Refer to the **U-bolt Threads** case in problem 20. Problem 17 presented ways of estimating σ when r subgroups are of varying size n_i. The formulas there are based on subgroup sample standard deviations s_i. Another expression sometimes used to estimate the process standard deviation is based on ranges, namely,

$$\frac{\frac{(n_1-1)R_1}{d_2(n_1)} + \frac{(n_2-1)R_2}{d_2(n_2)} + \cdots + \frac{(n_r-1)R_r}{d_2(n_r)}}{n_1 + n_2 + \cdots + n_r - r}.$$

Consider the subgroup means and ranges given in problem 20, and suppose that subgroups consisted of $n = 5$ bolts except for the subgroups indicated in the following table.

Subgroup	Subgroup size
2	8
5	4
6	6
7	2
11	3
14	7
15	2
18	2

(a) Give $\bar{x}_{\text{pooled}}$ and three estimates of σ. Base two of the estimates of σ on the subgroup standard deviations and the other on the ranges.

(b) Find three estimates of the standard deviation of a subgroup mean when $n = 8$. Base two of the estimates on subgroup standard deviations and one on the ranges. (Hint: $\text{Var } \bar{x}_i = \sigma^2/n_i$.)

(c) Find three estimates of the standard deviation of each subgroup sample standard deviation when $n = 8$. Base two of the estimates on subgroup standard deviations and one on the ranges.
(Hint: $\text{Var } s_i = \sigma^2(1 - c_4^2(n_i))$.)

(d) Find an estimate of the standard deviation of each subgroup range when $n = 8$. Base the estimate on the subgroup ranges. (Hint: $\text{Var } R_i = d_3^2(n_i)\sigma^2$.)

(e) Consider retrospective $\bar{x}$ and R charts using the new configuration of subgroup sizes. (There is no need to make the charts here.) Would control limits for either chart be constant across time? Why or why not?

(f) Are the center lines for the charts referred to in (e) constant across subgroups? Why or why not?

22. **Turning.** Allan, Robbins, and Wycoff worked with a machine shop that employs a CNC (computer numerically controlled) lathe in the machining of a part for a heavy equipment manufacturer. Some summary statistics for a particular part diameter (x) obtained from 25 subgroups of $n = 4$ parts turned on the lathe are given in Table 3.15. The units are inches.

(a) Find retrospective control limits for the values (both means and ranges). What do the $\bar{x}$ and R values indicate about the stability of the turning process?

(b) Suppose that one wishes to apply the four Western Electric alarm rules to the $\bar{x}$ values. Specify the zones to be used for the mean diameters.

TABLE 3.15. Summary statistics for problem 22

Subgroup	$\bar{x}$	R	Subgroup	$\bar{x}$	R
1	1.18093	.0001	14	1.18128	.0002
2	1.18085	.0002	15	1.18145	.0007
3	1.18095	.0002	16	1.18080	.0003
4	1.18063	.0008	17	1.18100	.0000
5	1.18053	.0007	18	1.18103	.0001
6	1.18053	.0005	19	1.18088	.0003
7	1.18058	.0005	20	1.18100	.0000
8	1.18195	.0001	21	1.18108	.0002
9	1.18100	.0003	22	1.18120	.0004
10	1.18095	.0001	23	1.18088	.0002
11	1.18095	.0006	24	1.18055	.0022
12	1.18098	.0001	25	1.18100	.0004
13	1.18123	.0009		29.52421	.0101

Are any of the rules violated in the first 10 samples? (If you find any violations, say which rule is violated for the first time where.)

(c) Give an estimate of the process short-term standard deviation derived from the ranges (use all 25 subgroups and the assumption that σ is constant over the study period).

(d) Engineering specifications on the diameter in question were in fact $1.1809 \pm .005$ in. Suppose that over short production runs, diameters can be described as normally distributed and that your estimate of σ from (c), is an appropriate description of the variation seen in short runs. Give an estimate of the best possible fraction of diameters meeting specifications available using this particular lathe.

(e) Make further use of your estimate of σ from (c), and set up control limits that could be used in the future monitoring of the process standard deviation via Shewhart charting of s based on samples of size $n = 5$.

(f) Again use your estimate of σ from (c) and consider future monitoring of $\bar{x}$ based on samples of size $n = 4$ using "3-sigma" limits and a center line at the target diameter, 1.1809. Assuming diameters are normally distributed, on average how many subgroups would be required to detect a change in mean diameter, μ, from 1.1809 to 1.1810?

23. Refer to the **Turning** case in problem 22. Problem 21 presented a method for estimating σ based on ranges of subgroups of varying size. Use that method in this problem. Use again the subgroup means and ranges given in problem 22, and suppose all subgroups were of size $n = 4$ parts except for the ones indicated in the following table.

Subgroup	Subgroup size
1	2
4	3
5	6
9	2
11	7
13	5
15	3
16	2
17	8
18	2

(a) Give $\bar{x}_{\text{pooled}}$ and an estimate of σ.

(b) Find an estimate of the standard deviation for a subgroup mean, $\bar{x}_i$, when $n = 7$.

(c) Find an estimate of the standard deviation for a subgroup range, R_i, when $n = 7$.

(d) Consider retrospective $\bar{x}$ and R charts using the new configuration of subgroup sizes. (There is no need to make the chart here.) Are the control limits for the two charts constant across subgroups? Why or why not?

(e) Are the center lines for the charts considered in (d) constant across subgroups? Why or why not?

24. **Package Sorting**. Budworth, Heimbuch, and Kennedy analyzed a company's package sorting system. As packages arrive at the sorting system, they are placed onto trays, and the bar codes affixed to the packages are scanned (in an operation much like the scanning process at a grocery store checkout). Bar code identification numbers begin with the zip code of the package destination. This permits packages to be sorted into 40 bins, each of which represents a different bulk mail center (BMC) or auxiliary service facility (ASF). All packages in a given bin are shipped by truck to the same mail center. The bulk transportation of these packages is much cheaper than if they were mailed directly by the nearest US Post Office. The large number of BMC packages handled daily by the company produces tremendous cost savings.

Initially, the team tackled the so-called "no chute open" problem. When one of the BMC bins is full, packages destined for that bin cannot be dropped into it. They end up in a "no chute open" bin. This eventuality produced many inefficiencies and even shutdowns of the entire system. In fact, the system was shut down about $10 \, \text{min/day}$ on average because of this problem. This lost time cost the company the ability to process about 400 packages/day, and accumulated over a year, this represents a serious loss.

TABLE 3.16. Data for problem 24

Date	Shift	Number in bin	Date	Shift	Number in bin
10/16	1	1510	10/20	2	2061
10/17	3	622	10/20	3	981
10/18	1	2132	10/21	1	1636
10/18	2	1549	10/21	2	2559
10/19	1	1203	10/21	3	1212
10/19	2	2752	10/22	1	2016
10/19	3	1531	10/22	2	2765
10/20	1	1314	10/22	3	574

The team decided to document the number of packages per shift dumped into the "no chute open" bin. The data they collected are in Table 3.16.

(a) Is this an attributes data problem or a variables data problem? Why?

(b) What constitutes a "subgroup" in the context of this problem?

(c) What probability model is a possible description of the number of packages routed to the "no chute open" bin during a given shift?

(d) Assuming the sorting process is stable, estimate the average number of packages routed to the "no chute open" bin during a particular shift. Estimate the standard deviation of the number of packages routed to the "no chute open" bin. These estimates should be consistent with your answer to (c).

(e) Was the number of packages in the "no chute open" bin apparently constant except for random fluctuation? Why or why not? Defend your answer using a control chart.

25. Refer to the **Package Sorting** case in problem 24. Budworth, Heimbuch, and Kennedy were told that the sorting system was set up to let a package circle on the conveyor belt for 10 cycles (once each cycle the package would fall into the correct chute if that chute was not occupied). If after ten cycles the correct chute was always occupied, a package would be consigned to the inefficient "no chute open" bin. Upon observing the system in operation, the team immediately recognized packages dropping into the "no chute open" bin after only 2 or 3 cycles. Management was notified and the programming of the system was corrected. The team gathered the data below after the correction.

Date	Shift	Number in bin
10/23	1	124
10/24	3	550
10/25	1	0
10/25	2	68
10/25	3	543
10/26	1	383
10/26	2	82
10/26	3	118

(a) Extend the control limits from your chart in part (e) of problem 24. Plot the data above on the same chart. Does it appear the system change was effective? Why or why not?

(b) Make a chart to assess stability of the number of packages in the "no chute open" bin using only the data above. Does it appear the system was stable? Why or why not?

(c) Has the team solved the problem of a large number of packages in the "no chute open" bin? Defend your answer.

26. Refer to the **Package Sorting** case of problems 24 and 25. Budworth, Heimbuch, and Kennedy also investigated the performance of the package scanning equipment. Just as items at a cashier's scanner often are not read on the first scan, so too were bar codes on packages not necessarily read on the first or second scan. Label damage and incorrect orientation, erroneous codes, and some simply unexplained failures all produced "no-read" packages. If a package was not read on the first pass, it continued on the carousel until reaching a second scanner at the end of the carousel. Failure to read at this second scanner resulted in the package being dropped into a "no-read" bin and scanned manually with a substantial loss in efficiency. The team took data over 30 consecutive 1-min periods on the variables

$n = $ the number of packages entering the system during the 1-min period,

$X_1 = $ the number of those packages failing the first scan, and

$X_2 = $ the number of those packages failing both scans.

The values they recorded are in Table 3.17.

(a) What constitutes a "subgroup" in this problem?

(b) Is this an attributes data or a variables data problem? Why?

(c) Make a retrospective control chart to assess consistency of the proportion of packages failing both scans, and comment on what it indicates.

TABLE 3.17. Data for problem 26

Minute	n	X_1	X_2	Minute	n	X_1	X_2
1	54	10	2	16	66	17	0
2	10	3	2	17	56	11	3
3	55	22	3	18	26	6	1
4	60	18	5	19	30	6	0
5	60	12	1	20	69	14	1
6	60	14	1	21	58	23	5
7	37	14	0	22	51	18	5
8	42	17	1	23	32	15	1
9	38	20	10	24	44	23	4
10	33	6	2	25	39	13	2
11	24	6	3	26	26	3	1
12	26	7	5	27	41	17	1
13	36	12	0	28	51	25	5
14	32	10	3	29	46	18	1
15	83	25	2	30	59	23	6

(d) Make a retrospective control chart to assess consistency of the proportion of packages that are not read on the first scan, and comment on what it indicates.

(e) Make a retrospective control chart to assess consistency of the proportion of all packages in a given minute that are not read on the first scan and are read on the second scan. Comment on what it indicates.

(f) Calculate the proportions of those packages failing the first scan that also fail the second scan.

(g) Make a retrospective control chart to assess consistency of the proportions in (f). Comment on what it indicates.

27. **Jet Engine Visual Inspection.** The data in Table 3.18 are representative of counts of nonconformances observed at final assembly at an aircraft engine company. Suppose that one final assembly is inspected per day.

   (a) Is this a variables data problem or is it an attributes data problem? Explain.

   (b) In the context of the problem, what is a "subgroup"?

   (c) What probability distribution is a likely model for counts of nonconformances on these engines? Defend your answer.

   (d) Find an estimated mean number of visually identified nonconformances and the corresponding estimated standard deviation.

   (e) Find appropriate upper and lower control limits and center line to apply to the counts. Make the corresponding control chart for these data. Does it appear that the process was stable over the period of

TABLE 3.18. Data for problem 27

Day	Number of Nonconformances	Day	Number of Nonconformances	Day	Number of Nonconformances
7/5	15	7/15	18	7/29	16
7/6	19	7/16	4	8/1	30
7/7	12	7/19	16	8/2	34
7/8	24	7/20	24	8/3	30
7/9	18	7/21	16	8/4	40
7/10	10	7/22	12	8/5	30
7/11	16	7/25	0	8/6	36
7/12	26	7/26	16	8/8	32
7/13	16	7/27	26	8/9	42
7/14	12	7/28	12	8/10	34

the study? Why or why not? Identify any out-of-control points. Apply Nelson's rules.

(f) Suppose two inspectors were involved in the data collection. Briefly discuss what must be true (in terms of data collection protocol) to assure that the chart and analysis in (e) are credible.

28. Refer to the **Jet Engine Visual Inspection** case in problem 27.

    (a) When possible causes for out-of-control points on a control chart are addressed and physically eliminated, it is common practice to discard the data associated with those out-of-control points and recalculate control limits. Apply this thinking to part (e) of problem 27, assuming causes of the out-of-control points have been addressed (you should "throw out" July 16 and 25 and August 1 through 10—a total of 11 out of 30 points).

    (b) Suppose the following data are obtained in visual inspection of final engine assemblies over the next three days:

Day	Assemblies inspected	Number of nonconformances
1	.5	8
2	2.0	31
3	1.5	26

    (Partial inspection of final engine assemblies could possibly occur because of unforeseen labor problems. More than one engine assembly might be inspected on days 2 and 3 to, in some sense, make up for the partial inspection on day 1.) Using the information from (a) above, find control limits for nonconformance rates on these three days (do not use the number of nonconformances during these 3 new days to find the limits). Also give the center line and three plotted values (nonconformances per engine assembly inspected).

(c) Do your values from part (b) suggest process instability? Explain.

(d) Your center line should be constant across the three days represented in (b). Why is this?

29. In a u charting context, the number of standard units inspected may vary from period to period. Let

$$X_i = \text{the number of nonconformances observed at period } i,$$
$$k_i = \text{the number of standard units inspected at period } i, \text{ and}$$
$$\hat{u}_i = X_i/k_i.$$

The following values were obtained over nine periods.

i	1	2	3	4	5	6	7	8	9
k_i	1	2	1	3	2	1	1	3	1
$\hat{u}_i$	0	3.00	0	1.33	4.00	0	0	.67	1.00

(a) From these values, what conclusions can you make about stability of the process being monitored? Make the appropriate control chart.

(b) Suppose that in the future k_i will be held constant at 1 and that 2.4 nonconformances per inspection unit will be considered to be "standard quality." Find the probability of an out-of-control signal on a 3-sigma Shewhart control chart, if the true nonconformance rate is at the standard quality level ($\lambda = 2.4$). Find the probability of an out-of-control signal if the true nonconformance rate changes to $\lambda = 4.8$. (Remember that the Poisson(μ) probability function is $P(X = x) = (\exp(-\mu)\mu^x)/x!$.)

(c) Suppose that in the future k_i will be held constant at 2. Find the probability of an out-of-control signal if the true nonconformance rate is at the standard quality level ($\lambda = 2.4$). Find the probability of an out-of-control signal if the true nonconformance rate changes to $\lambda = 4.8$

(d) Compare your answers to (b) and (c). Which subgroup size ($k = 1$ or $k = 2$) is more appealing? Why?

30. **Electrical Switches.** The following scenario is taken from an aircraft engine company's training material. One hundred electrical switches are sampled from each of 25 consecutive lots. Each sampled switch is tested and the sample numbers failing are recorded in Table 3.19.

(a) Find the sample fractions of switches failing the test.

(b) What is a plausible probability model for describing the count of switches in a particular sample failing the test? Explain.

(c) Plot the number failing versus the sample period. Plot an appropriate center line and control limits on the same graph.

TABLE 3.19. Data for problem 30

Sample	Number failing	Sample	Number failing
1	11	14	18
2	9	15	7
3	15	16	10
4	11	17	8
5	22	18	11
6	14	19	14
7	7	20	21
8	10	21	16
9	6	22	4
10	2	23	11
11	11	24	8
12	6	25	9
13	9		

(d) What does your plot in (c) monitor here?

(e) Interpret your plot in (c). Identify any out-of-control points.

(f) What is the usual name of the chart you prepared in part (c)?

(g) Suppose causes for out-of-control points identified in (e) are identified and physically removed. It would then make sense to delete the out-of-control points and recalculate limits. Do this recalculation and redo (c). (You should have identified and eliminated two out-of-control points.)

(h) Suppose the number of switches sampled and the number failing for the next three consecutive lots are as follows.

Number sampled	Number failing
75	8
144	12
90	11

Using your estimated fraction failing from (g) as a standard for judging these samples, find control limits and center lines appropriate for the three new "number failing" data points. Are the three sets of control limits and center lines the same? Why is this to be expected?

31. A data set in the book *Elementary Statistical Quality Control* by Burr indicates that in the Magnaflux inspection for cracks in a type of malleable casting, about $p \approx .11$ of the castings will have detectable cracks. Consider the examination of 12 such castings. Let X be the number of castings from the set of 12 identified as being cracked.

(a) Find $P[X = 5]$.

(b) Find $P[X > 5]$.

(c) Find EX.

(d) Find Var X.

(e) Ten sets of 12 castings are to be inspected. What is the probability that at least one set of 12 will have one or more cracked castings?

32. **Plastic Packaging.** This concerns a plastic packaging case investigated by Hsiao, Linse, and McKay. Plastic bags were supposed to hold three bagels each. An ideal bag is 6.75 in wide, has a 1.5- in lip, and has a total length of 12.5 in (including the lip). The ideal hole positions are on the lip. The hole position on selected bags was measured as the distance from the bottom of the bag to the hole. Five bags were obtained at six times on each of three days. Hole position, bag width, bag length, and lip width were measured and recorded for each bag. The data for hole position (in inches) are in Table 3.20.

   (a) What is a natural subgroup in this situation?

   (b) How many items are in each subgroup described in (a)? How many subgroups are there here in total?

   (c) Calculate the subgroup means and subgroup ranges.

   (d) Make a retrospective control chart for mean hole position. Give the center line, control limits, and zone limits using your answers from part (c).

   (e) Make a retrospective control chart for variability in position using your values from (c). Give the control limits and zone limits.

   (f) What is the usual name of the chart in (d)? What is the usual name of the chart in (e)?

   (g) Is it important which of the charts developed in (d) and (e) is analyzed first? Why or why not?

   (h) Find the estimated standard deviation of hole position based on the ranges.

33. Refer to the **Plastic Packaging** case in problem 32.

   (a) Calculate the 18 subgroup means and 18 subgroup ranges.

   (b) Use your answers from part (a), and for each day separately, make retrospective control charts for mean hole position. Give center lines, control limits, and zone limits. What do these charts suggest about process performance?

   (c) Use your answers from part (a), and for each day separately, make retrospective control charts for variability of hole position.

   (d) Based on your answer to (c), is variability of hole location constant within any one of the days? Why or why not?

TABLE 3.20. Data for problem 32

Day	Time	Hole position
1	10:10 am	1.87500, 1.84375, 1.87500, 1.84375, 1.84375
1	10:25 am	1.90625, 1.90625, 1.90625, 1.87500, 1.90625
1	10:55 am	1.87500, 1.93750, 1.93750, 1.93750, 1.96875
1	11:12 am	2.09375, 2.12500, 2.21875, 2.15625, 2.12500
1	11:35 am	2.00000, 2.00000, 2.00000, 2.00000, 2.03125
1	11:41 am	1.87500, 1.90625, 1.90625, 1.87500, 1.93750
2	8:15 am	1.62500, 1.62500, 1.59375, 1.65625, 1.59375
2	8:54 am	1.62500, 1.62500, 1.59375, 1.68750, 1.65625
2	9:21 am	1.62500, 1.59375, 1.62500, 1.59375, 1.62500
2	9:27 am	1.62500, 1.59375, 1.62500, 1.65625, 1.65625
2	9:51 am	1.56250, 1.59375, 1.56250, 1.56250, 1.56250
2	9:58 am	1.56250, 1.56250, 1.56250, 1.53125, 1.56250
3	10:18 am	1.50000, 1.56250, 1.53125, 1.53125, 1.50000
3	10:33 am	1.53125, 1.53125, 1.53125, 1.53125, 1.50000
3	10:45 am	1.50000, 1.53125, 1.50000, 1.53125, 1.46875
3	11:16 am	1.50000, 1.50000, 1.50000, 1.53125, 1.50000
3	11:24 am	1.53125, 1.53125, 1.50000, 1.50000, 1.50000
3	11:39 am	1.50000, 1.50000, 1.53125, 1.53125, 1.53125

(e) According to your charts in (c), is there a day in which a single standard deviation of hole position is plausible? Why or why not?

(f) Suppose your answer in (e) is "yes" for each day. Find estimated σs for the three different days treated separately. (Base your estimates on sample ranges.)

(g) Comment on how your estimates in (f) compare to the estimate in part (h) of problem 32.

34. Refer to the **Plastic Packaging** case in problems 32 and 33. The ideal lip width is 1.5 in. The lip width data in Table 3.21 (in inches) were taken on the same bags represented in problem 32.

   (a) Is this a variables data or an attributes data scenario? Why?

   (b) Find the subgroup means, ranges, and standard deviations.

   (c) Make retrospective control charts for lip width variability and lip width mean based on the sample ranges.

   (d) In view of the appearance of your chart for variability of lip width, does it make sense to seriously examine the chart for mean lip width? Why or why not?

   (e) Instead of making the completely retrospective charts asked for in (c), is it possible to incorporate some "standards" information and make a different chart for mean lip width? Explain.

TABLE 3.21. Data for problem 34

Day	Time	Lip width
1	10:10 am	1.75000, 1.62500, 1.62500, 1.65625, 1.62500
1	10:25 am	1.62500, 1.62500, 1.62500, 1.65625, 1.65625
1	10:55 am	1.53125, 1.53125, 1.50000, 1.50000, 1.50000
1	11:12 am	1.40625, 1.43750, 1.43750, 1.46875, 1.46875
1	11:35 am	1.46875, 1.46875, 1.46875, 1.46875, 1.40625
1	11:41 am	1.43750, 1.43750, 1.46875, 1.50000, 1.46875
2	8:15 am	1.37500, 1.40625, 1.37500, 1.40625, 1.37500
2	8:54 am	1.37500, 1.43750, 1.43750, 1.40625, 1.40625
2	9:21 am	1.40625, 1.37500, 1.43750, 1.40625, 1.40625
2	9:27 am	1.50000, 1.46875, 1.43750, 1.46875, 1.43750
2	9:51 am	1.43750, 1.43750, 1.43750, 1.43750, 1.43750
2	9:58 am	1.53125, 1.46875, 1.53125, 1.50000, 1.53125
3	10:18 am	1.53125, 1.56250, 1.50000, 1.50000, 1.53125
3	10:33 am	1.50000, 1.53125, 1.53125, 1.50000, 1.50000
3	10:45 am	1.34375, 1.34375, 1.34375, 1.37500, 1.37500
3	11:16 am	1.46875, 1.46875, 1.46875, 1.43750, 1.43750
3	11:24 am	1.37500, 1.40625, 1.40625, 1.40625, 1.40625
3	11:39 am	1.43750, 1.43750, 1.40625, 1.37500, 1.43750

(f) Instead of treating all 18 samples at once as in part (c), for each day *separately*, make retrospective R and $\bar{x}$ charts. What are your conclusions regarding process stability for each day *separately*?

(g) Find three daily estimated lip width standard deviations. How do these estimates compare to that calculated when the complete set of data is used? (See (c) above.)

(h) Would there be advantages to using subgroup standard deviations instead of subgroup ranges in parts (c) and (f) above? Explain.

35. Refer to the **Plastic Packaging** case in problem 32.

   (a) Make a control chart for the standard deviation of hole position. Is short-term variation stable?

   (b) Make a control chart for mean hole position based on subgroup standard deviations. Is process aim stable?

   (c) For each day *separately*, make charts for the standard deviation of hole position. Is short term variation stable for each day?

   (d) For each day *separately*, make charts for mean hole position (use an estimate of σ based on the subgroup standard deviations). Is process aim stable for each day?

   (e) For each of (a), (b), (c), and (d), was it helpful to use the subgroup standard deviations instead of the subgroup ranges as in problem 32? Why or why not?

36. Refer to the **Hose Skiving** case of problem 11 in the Chap. 1 exercises. The plant works two shifts/day. Five hoses were sampled every two hours from each of three production lines and skive length, y measured. Specifications for skive length are $target \pm .032$ in. The values ($x = y - target$) in the accompanying tables are in units of .001 in above target.

    (a) Explain (possibly using the notions of "rational subgrouping" and "stratification") why it would not make good sense to combine data taken at a particular time period from the three different production lines to make a single "sample." (Particularly in cases where it is only possible to select a single item from each line at a given time period, the urge to make such a "sample" is strong, and this kind of error is a common one.)

    (b) Compute the 48 sample means and ranges for the data given in Table 3.22. Then separately for lines 1, 2, and 3, make both $\bar{x}$ charts and R charts. Comment on what they indicate about the stability of the skiving process on the three lines over the two days of this study.

    (c) One *could* think about plotting all 48 sample means on a single chart, for example, plotting means from lines 1, 2, and 3 in that order at a given time period. Discuss why that is not a terribly helpful way of summarizing the data. (Will it be easier or harder to see trends for a given production line on this kind of plot compared to the separate charts of part (b)?)

37. Consider the hypothetical values in Table 3.23 on page 181 from a process where $T(t)$ is the target value, $Y(t)$ is the realized value of the characteristic of interest, $E(t) = T(t) - Y(t)$, $\Delta E(t) = E(t) - E(t-1)$, and $\Delta^2 E(t) = \Delta E(t) - \Delta E(t-1)$. A PID controller $\Delta X(t) = \kappa_1 \Delta E(t) + \kappa_2 E(t) + \kappa_3 \Delta^2 E(t)$ has been used.

    (a) What does $\Delta X(t)$ represent? Was the adjustment $\Delta X(3)$ made before or after observing $Y(3) = 0$?

    (b) Suppose the integral gain in the control algorithm is 4. What are the proportional and derivative gains?

    (c) Using your answer in (b), find the complete set of $\Delta X(t)$s.

    (d) Find the average squared error for the last 9 periods, the last 8 periods, ..., and the last 3 periods.

    (e) Make a plot of your values from (d) as follows. Label the horizontal axis with t. For $t = 4$ plot the average squared error for periods 4 through 12, for $t = 5$ plot the average squared error for periods 5 through 12, ..., and for $t = 10$ plot the average squared error for periods 10 through 12. Does the appearance of this plot give you hope that any transient or "startup" effects have been eliminated before the

TABLE 3.22. Data for problem 36

Day	Time	Line 1 Skive length	Line 2 Skive length	Line 3 Skive length
1	8:00 am	3, 2, 4, −5, 2	−17, 3, 2, 10, 4	−3, −5, 7, 10, 3
1	10:00 am	5, −4, −3, 0, −2	13, 3, −2, 12, 15	3, 5, 5, 8, 1
1	12:00 pm	−5, 5, 5, −3, 2	14, 6, 10, 5, 1	3, 6, 6, 5, 5
1	2:00 pm	−2, 5, 4, −3, 2	7, 2, 10, 16, 13	5, −2, 5, 4, 6
1	4:00 pm	−10, 2, 1, 2, 1	−15, −12, −2, −4, 0	2, 5, 4, 1, 1
1	6:00 pm	−5, −6, −3, −3, −7	−4, −6, −4, −4, 4	2, 1, 0, 1, 1
1	8:00 pm	−5, 0, −3, −3, −8	2, −5, −5, −3, −4	1, 3, 5, −6, −10
1	10:00 pm	−5, −10, 10, −9, −3	0, −1, −2, −1, 0	−7, −5, 4, 2, −9
2	8:00 am	2, 4, 1, 0, −5	15, 2, 16, 10, 14	18, 15, 5, 3, 4
2	10:00 am	−3, 3, −4, 5, 3	12, 4, −10, 10, −3	3, 2, −2, −5, 2
2	12:00 pm	−5, −7, 6, 8, −10	1, −7, 4, −5, −9	4, 2, 2, 1, 3
2	2:00 pm	3, −4, 4, 6, −3	−6, 8, −5, 18, 20	6, 5, 4, 2, 5
2	4:00 pm	−10, −7, −3, −1, −3	−2, −4, −5, −1, −3	2, 0, 1, −3, 5
2	6:00 pm	0, −1, −6, −2, 0	−2, −2, −2, −4, −2	2, −5, −7, −3, −5
2	8:00 pm	2, 4, −2, −3, 5	0, 2, −1, −1, −2	−6, −3, −10, −4, −7
2	10:00 pm	1, 0, −1, 7, −5	−1, −2, 0, −1, −1	0, −4, −7, −10, −2

last few periods and that those periods adequately represent control algorithm performance? Explain.

38. **Paper Dry Weight**. Before progressing to the collection of the data in Table 3.9, several different PID algorithms were tried. Miami University Paper Science Laboratory Research Associate Doug Hart set up the paper-making machine with 1% de-inked pulp stock to produce 20-lb bond paper. No filler was used. Then Jobe and Hart began an investigation into how to best control the dry weight variable. Twelve periods of data were obtained to benchmark the process behavior without any pump speed adjustments. (It is well known that the pump speed does affect final dry weight.) A standard setting of 4.5 (45% of maximum speed that was known to produce paper with a dry weight in the vicinity of the target of $70 \, g/m^2$) was used. Paper dry weight measurements were made at roughly 5-min intervals, and these are presented in Table 3.24 as $Y(t)$. Units are g/m^2.

   (a) Plot the measured values $Y(t)$ versus t.
   (b) Plot the errors $E(t)$ versus t.
   (c) Find the average squared error for periods 1 through 12, for periods 2 through 12,..., and for periods 10 through 12.
   (d) Make a plot of your values from (c) for $t = 1, 2, \ldots, 10$. (At time t plot the average squared error for periods t through 12.)

39. Refer to the **Paper Dry Weight** case of problem 38. Hart informed Jobe that for every 5-tick increase on the speed pump dial, paper dry weight

TABLE 3.23. Values for problem 37

Period, t	$T(t)$	$Y(t)$	$E(t)$	$\Delta E(t)$	$\Delta^2 E(t)$	$\Delta X(t)$
1	4	2	2			
2	4	1	3	1		
3	4	0	4	1	0	18
4	4	2	2	−2	−3	1
5	4	2	2	0	2	
6	4	3	1	−1	−1	
7	5	3	2	1	2	
8	5	4	1	−1	−2	
9	5	5	0	−1	0	
10	5	6	−1	−1	0	
11	5	6	−1	0	1	
12	5	6	−1	0	0	

TABLE 3.24. Values for problem 38

Time	Period, t	$T(t)$	$Y(t)$	$E(t)$
8:45	1	70	75.3	−5.3
8:50	2	70	75.8	−5.8
8:55	3	70	73.1	−3.1
9:00	4	70	72.4	−2.4
9:05	5	70	73.5	−3.5
9:10	6	70	72.8	−2.8
9:15	7	70	72.6	−2.6
9:20	8	70	71.7	−1.7
9:25	9	70	69.8	.2
9:30	10	70	66.9	3.1
9:45*	11	70	70.9	−.9
9:50	12	70	71.7	−1.7

increases about $1.5 \, \text{g}/\text{m}^2$. This means that in rough terms, to increase a dry weight by $1 \, \text{g}/\text{m}^2$, an increase of pump speed setting of about 3.33 ticks is needed.

(a) If one were to consider an "integral-only" version (a $\kappa_1 = \kappa_3 = 0$ version) of the control equation (3.52) for use with the paper-making machine, why might $\kappa_2 = 3.33$ be a natural first choice? (X is in ticks, while T and Y are in g/m^2.)

(b) The "integral-only" controller of part (a) was used for 7 time periods and paper dry weight data collected. This is summarized in Table 3.25. Fill in the $\Delta X(t)$ and $E(t)$ columns in that table for $t = 1, 2, \ldots, 8$. (The machine was running without adjustment with X set at 4.5 until 9:55. The measurements were taken far enough apart in time that the entire effect of a pump speed change ordered on the basis of data through a given period was felt at the next measuring period.)

TABLE 3.25. Values for problem 39

Time Period	t	$T(t)$	$Y(t)$	$E(t)$	$\Delta X(t)$
9:55	1	70	72.1		
10:08	2	70	70.6		
10:14	3	70	71.3		
10:25	4	70	67.1		
10:32	5	70	71.5		
10:38	6	70	70.3		
10:44	7	70	68.4		
10:50	8	70	71.7		

(c) Plot $Y(t)$ versus t.

(d) Plot $E(t)$ versus t.

(e) Find the average squared error for periods 2 through 8, for periods 3 through 8,..., for periods 6 through 8.

(f) Make a plot of your values from (e) for $t = 2, \ldots, 6$. (At time t plot the average squared error for periods t through 8.) Does the appearance of this plot give you hope that any transient or "startup" effects have been eliminated before the last few periods and that those periods adequately represent control algorithm performance?

40. Refer to the **Paper Dry Weight** case in problems 38 and 39. At 10:50 the speed pump dial was set back to 4.5 (45 %) and left there for 5 min in order to return the system to the benchmark conditions of problem 38. A new coefficient κ_2 in an integral control algorithm was adopted, and beginning at 10:55 this new adjustment algorithm was employed for 7 periods with results summarized in Table 3.26.

TABLE 3.26. Values for problem 40

Time Period	t	$T(t)$	$Y(t)$	$E(t)$	$\Delta X(t)$
10:55	1	70	72.0	-2	-3.32
11:01	2	70	71.7		
11:13	3	70	71.1		
11:19	4	70	68.8		
11:25	5	70	69.6		
11:31	6	70	71.8		
11:37	7	70	68.2		
11:43	8	70	69.7		

(a) Find the value of the new coefficient κ_2 used by Jobe and Hart. Then fill in the $E(t)$ and $\Delta X(t)$ values in the table for $t = 2, \ldots, 8$.

(b) Plot $Y(t)$ versus t.

(c) Plot $E(t)$ versus t.

(d) Find the average squared error for periods 2 through 8, for periods 3 through 8,..., and for periods 6 through 8.

(e) Make a plot of your values from (d) for $t = 2, \ldots, 6$. (At time t plot the average squared error for periods t through 8.) Does the appearance of this plot give you hope that any transient or "startup" effects have been eliminated before the last few periods and that those periods adequately represent control algorithm performance?

41. Refer to the **Paper Dry Weight** case of problems 38, 39, and 40. After making the measurement at 11:43 indicated in problem 40, the speed pump dial was again set back to 4.5 and left there for 5 min (from 11:44 to 11:49). (This was again done to in some sense return the system to the benchmark conditions.) Hart and Jobe decided to include both integral and proportional terms in a new control equation, and $\kappa_2 = 1.66$ and $\kappa_1 = .83$ were selected for use in Eq. (3.52). (The same integral control coefficient was employed, and a proportional coefficient half as large as the integral coefficient was added.) This new adjustment algorithm was used to produce the values in Table 3.27.

TABLE 3.27. Values for problem 41

Time	Period, t	$T(t)$	$Y(t)$	$E(t)$	$\Delta E(t)$	$\Delta X(t)$
11:49	1	70	70.9	$-.9$		
11:54	2	70	70.3	$-.3$	.6	0
11:59	3	70	68.8			
12:06	4	70	70.0			
12:12	5	70	69.6			
12:18	6	70	69.3			
12:24	7	70	68.4			
12:30	8	70	68.4			
12:36	9	70	69.8			

(a) Find the values of $E(t)$, $\Delta E(t)$, and $\Delta X(t)$ for periods 3 through 9.

(b) Plot $Y(t)$ versus t.

(c) Plot $E(t)$ versus t.

(d) Find the average squared error for periods 3 through 9, for periods 4 through 9,..., and for periods 7 through 9.

(e) Make a plot of your values from (d) for $t = 3, \ldots, 7$. (At time t plot the average squared error for periods t through 9.) Does the appearance of this plot give you hope that any transient or "startup" effects have been eliminated before the last few periods and that those periods adequately represent control algorithm performance?

42. Refer to the **Paper Dry Weight** case of problems 38 through 41. Yet another control algorithm was considered. κ_1 from problem 41 was halved and the coefficient κ_2 was left at 1.66. The pump speed dial was set to 4.5 at 12:37. Thereafter, the new "PI" control algorithm was used to produce the values in Table 3.28.

TABLE 3.28. Values for problem 42

Time	Period, t	$T(t)$	$Y(t)$	$E(t)$	$\Delta E(t)$	$\Delta X(t)$
12:42	1	70	66.2			
12:45	2	70	66.4			
12:51	3	70	67.2			
12:58	4	70	69.4			
1:04	5	70	69.5			
1:10	6	70	69.2			
1:16	7	70	70.1			
1:22	8	70	66.2			
1:29	9	70	71.7			

(a) Find $E(t)$ for all 9 periods and $\Delta E(t)$ and the corresponding $\Delta X(t)$ for periods 2 through 9.

(b) Plot $Y(t)$ versus t.

(c) Plot $E(t)$ versus t.

(d) Find the average squared error for periods 3 through 9, for periods 4 through 9, ..., and for periods 7 through 9.

(e) Make a plot of your values from (d) for $t = 3, \ldots, 7$. (At time t plot the average squared error for periods t through 9.) Does the appearance of this plot give you hope that any transient or "startup" effects have been eliminated before the last few periods and that those periods adequately represent control algorithm performance?

43. Refer to Example 35.

(a) Plot $Y(t)$ versus t.

(b) Plot $E(t)$ versus t.

(c) Find the average squared error for periods 4 through 11, for periods 5 through 11,..., and for periods 9 through 11.

(d) Make a plot of your values from (c) for $t = 4, \ldots, 9$. (At time t plot the average squared error for periods t through 11.) Does the appearance of this plot give you hope that any transient or "startup" effects have been eliminated before the last few periods and that those periods adequately represent control algorithm performance?

44. Refer to the **Paper Dry Weight** case and specifically the plots in problems 38d), 39f), 40e), 41e), 42e), and 43d). Which control equation seems to be best in terms of producing small average squared error?

45. Rewrite the PID control equation (3.52) so that $\Delta X(t)$ is expressed in terms of a linear combination of $E(t)$, $E(t-1)$, and $E(t-2)$, the current and two previous errors.

46. Fill levels of jelly jars are of interest. Every half hour, three jars are taken from a production line and net contents measured and recorded. The range and average of these three measurements are calculated and plotted on charts. One of these charts is intended to monitor location of the fill distribution, and the other is useful in monitoring the spread of the fill distribution.

    (a) What is the name for the chart used to monitor location of the fill-level distribution?

    (b) What is the name for the chart used to monitor spread of the fill-level distribution?

    (c) What is the name and value of the tabled constant used to make retrospective control limits for process location?

    (d) What are the names and values of the two tabled constants used to make retrospective control limits for process spread or short-term variability?

    (e) In this context, what constitutes a natural subgroup?

    (f) Give an expression for the usual estimate of process short-term variability (σ) based on an average of subgroup ranges.

47. Consider again the scenario of problem 46. Suppose that instead of ranges and averages, sample standard deviations and averages are computed and plotted:

    (a) What is the name and value of the tabled constant used to make retrospective control limits for process location?

    (b) What are the names and values of the two tabled constants used to make retrospective control limits for process spread or variability?

    (c) Give an expression for the usual estimate of process short-term variability (σ) based on an average of subgroup standard deviations.

48. Consider again the scenario of problems 46 and 47, and suppose that instead of three jars, 10 jars are sampled every half hour. Redo problems 46 and 47 with this change. For a given set of ranges or standard deviations, say which sets of retrospective control limits are wider apart with this new sample size.

49. Consider again the scenario of problems 46 and 47, and suppose that instead of plotting averages to monitor location, the decision is made to plot medians. What multiple of σ (or an estimate of this quantity) would be used to set control limits for medians around some central value in the case that $n = 3$? In the case that $n = 11$?

50. Consider drained weights of the contents of cans of Brand X green beans. Believable values for the process mean and standard deviation of these weights are 21.0 oz and 1.0 oz, respectively. Suppose that in a Brand X canning factory, 8 of these cans are sampled every hour and their net contents determined. Sample means and ranges are then computed and used to monitor stability of the filling process.

    (a) What is the name and value of the multiplier of $\sigma = 1.0$ that would be used to establish a center line for sample ranges?

    (b) What are the names and values of the multipliers of $\sigma = 1.0$ that would be used to establish upper and lower control limits for sample ranges?

    (c) What center line and control limits should be established for sample means?

51. Consider again the situation of problem 50, but suppose that instead of ranges and averages, sample standard deviations and averages are computed and plotted. Answer the questions posed in problem 50 (in the case of sample standard deviations instead of ranges).

52. Consider again the situation of problems 50 and 51, and suppose that instead of eight cans, only five cans are sampled every hour. Redo problems 50 and 51 with this change. Say which sets of control limits are wider apart with this new sample size.

53. **Electronic Card Assemblies**. In a 1995 article in *Quality Engineering*, Ermer and Hurtis discussed applications of control charting to the monitoring of soldering defects on electronic card assemblies. One assembly technology they studied was pin-in-hole (PIH) technology, which uses wave soldering to secure components to printed circuit boards after the leads of components have been inserted through holes drilled in the boards. The most common types of soldering defects encountered using this technology are "shorts" (unwanted electrical continuity between points on an assembly) and "opens" (the absence of desired electrical continuity between points on an assembly).

    Production of a particular card is done in "jobs" consisting of 24 cards. All cards from a job are tested, and a count is made of the total number defects found on the job. What type of probability model might plausibly be used to describe the number of defects found on a given job? What type of control chart might you use to monitor the production of soldering defects? Suppose that records on 132 jobs show a total of 2 defects recorded. What retrospective control limits might then be applied to the 132 different counts of defects? Does a job with any defect at all signal a lack of control?

54. **Milling Operation.** A student group studied a milling operation used in the production of a screen fixture mounting. Of primary importance was a "deviation from flatness" measurement. The units of measurement were .001 in. In the past, deviations from flatness had an average of 2.45 and a standard deviation of 1.40. What do the values of these and the fact that deviation from flatness cannot be negative suggest about the plausibility of a normal model for deviation from flatness? For what follows temporarily put aside any misgivings you might rightly have.

   (a) Set up standards given control limits for process location. (Monitoring is to be done on the basis of subgroups of size one.)

   (b) Ten consecutive mountings produced the deviation from flatness values below (units are .001 in).

   $$.5, 4.5, 2.0, 2.0, 3.0, 3.0, 2.0, 4.5, 3.0, 0.0$$

   Together with the limits in (a), use these data values to make a control chart for monitoring process aim. Has there been a process change away from the standards? Why or why not?

   (c) Find the moving ranges of adjacent observations and the mean of these 10 observations.

   (d) Make a retrospective individuals chart using the moving ranges and grand average of the 10 data values. Give the center line and control limits. What do you conclude based on this chart?

55. Refer to the **Milling Operation** case in problem 54.

   (a) For purposes of process monitoring only, let a target deviation from flatness be $\mu = 5$, and suppose the process standard deviation is $\sigma = 1.40$, as in problem 54. (In functional terms a 0 deviation from flatness is ideal.) Compute control limits for individuals based on this set of standards. Give the center line and control limits.

   (b) Plot the individuals from problem 54(b) using your new limits from (a). Does it appear that there has been a process change from standard conditions? Why or why not?

   (c) Discuss the practical meaning of the terms "stability," "shift," and "out of control" in light of parts (b) and (d) of problem 54 and part (b) above.

56. Refer to the **Lab Carbon Blank** case in problem 21 of Chap. 1. Suppose that repeated measurements of the same blank are normally distributed. For convenience, the data are repeated here.

Test number	1	2	3	4	5	6	7
Measured carbon	5.18	1.91	6.66	1.12	2.79	3.91	2.87

Test number	8	9	10	11	12	13	14
Measured carbon	4.72	3.68	3.54	2.15	2.82	4.38	1.64

(a) Find retrospective control limits and center line for the sequence of measurements. Use "3-sigma" limits.

(b) Plot the individual responses versus time and compare them to the limits found in (a). Do you detect any measurement process instability? Why or why not?

(c) Give an estimated mean measured carbon content. Give an estimated standard deviation of measured carbon content, $\overline{MR}/1.128$.

57. Refer to the **Lab Carbon Blank** case of problem 56. Suppose the nominal or "real" carbon content is 1.0.

(a) Find control limits and center line to apply to the data of problem 56. Use 3-sigma limits and $\overline{MR}$ in place of a real average range ($\overline{R}$) in the formula for retrospective limits. Make use of the nominal value 1.0 in place of $\overline{\overline{x}}$.

(b) Plot the x values from problem 56, and compare them to your limits from (a), i.e., make an individuals chart.

(c) What dilemma is revealed by your chart in (b) above and problem 56(b)? Discuss this using phrases such as "consistency of location," "shift in the mean," "off-target process," and "unstable process."

58. Refer to the **Lab Carbon Blank** case in problems 56 and 57. It is unknown whether the carbon measurements were made by the same person or by as many as 14 different people. What configuration of operators would be most effective in isolating instrument changes in the measurement of carbon content? Defend your answer in terms of the concept of "sources of variability."

59. Consider the analysis of a series of samples some time after their collection.

(a) What might be learned from an analysis based on standards given control charts?

(b) What might be learned from an analysis instead using retrospective limits on the control charts?

60. Refer to the **Paper Dry Weight** case in problem 38. Recall that the target for dry weight of 20-lb bond paper is $70\,\text{g}/\text{m}^2$. The pump speed controlling the flow of liquid pulp mixture onto the conveyor-roller mechanism was held fixed at 4.5 (45 % of maximum flow) in the production of the data in problem 38. Assume that under stable process conditions, dry weights are normally distributed. The dry weight of a sample was recorded for each of 12 consecutive samples, approximately equally spaced in time.

   (a) Find control limits to apply to the data of problem 38. Use the nominal dry weight of 70 as a target value, and employ $\overline{MR}/1.128$ as an estimate of σ. Does the process appear to be stable? Why or why not?

   (b) 100 measurements correspond to how many subgroups in the context of problem 38?

   (c) Suppose that the limits of (a) are applied to the future monitoring of individuals. About what ARL is produced if σ is as in part (a), but μ increases from its standard value by $3.5\,\text{g}/\text{m}^2$? Assume the process is stable and dry weights are normally distributed.

   (d) If completely retrospective control limits were used ($\overline{x}$ was used in place of the target value for dry weight), would your conclusion in (a) change? Why or why not?

61. **Transmission Housings.** Apple, Hammerand, Nelson, and Seow analyzed data taken from a set of 35 transmission housings. In addition to the side cover hole diameter considered in problem 4 of Section 2 of this chapter, they also examined upper borehole diameters on the transmission housings. For y the hole diameter in inches, the values below concern $x = (y - 3.5000) \times 10^4$, diameters stated in ten thousandths of an inch above 3.5000 in. Specification limits for the upper borehole diameter were $3.502 \pm .002$ in. (Below, 19 represents $y = 3.5019$, 28 represents $y = 3.5028$, etc.)

Transmission housing	1	2	3	4	5	6	7	8	9	10	11	12
Measured diameter	19	28	25	22	18	20	20	14	20	12	16	16

Transmission housing	13	14	15	16	17	18	19	20	21	22	23	24
Measured diameter	22	22	22	21	23	21	20	18	18	18	12	11

Transmission housing	25	26	27	28	29	30	31	32	33	34	35
Measured diameter	13	12	16	12	10	20	21	15	28	26	24

   (a) What is the subgroup size?

   (b) Give appropriate retrospective lower and upper control limits and center line for monitoring the hole diameters.

   (c) What does the chart using on your limits from (b) indicate about the stability of the upper borehole diameter production process? Why?

CHAPTER 4

PROCESS CHARACTERIZATION AND CAPABILITY ANALYSIS

The previous chapter dealt with tools for monitoring processes and detecting physical instabilities. The goal of using these is finding the source(s) of any process upsets and removing them, creating a process that is consistent/repeatable/predictable in its pattern of variation. When that has been accomplished, it then makes sense to summarize that pattern of variation graphically and/or in terms of numerical summary measures. These descriptions of consistent process behavior can then become the bases for engineering and business decisions about if, when, and how the process should be used.

This chapter discusses methods for characterizing the pattern of variation exhibited by a reasonably predictable system. Section 4.1 begins by presenting some more methods of statistical graphics (beyond those in Sect. 1.5) useful in process characterization. Then Sect. 4.2 discusses some "Process Capability Indices" and confidence intervals for them. Next, prediction and tolerance intervals for measurements generated by a stable process are presented in Sect. 4.3. Finally, Sect. 4.4 considers the problem of predicting the variation in output for a system simple enough to be described by an explicit equation, in terms of the variability of system inputs.

4.1 More Statistical Graphics for Process Characterization

The elementary methods of Sect. 1.5 provide a starting point for picturing the pattern of variability produced by a process. Slightly more sophisticated methods are also possible and often prove useful. Some of these are the topic of this section. After briefly reviewing the simple ideas of dot plots and stem-and-leaf diagrams, graphical tools based on the concept of distribution quantiles are discussed. The tools of quantile plots, box plots, and both empirical and theoretical Q-Q plots (probability plots) are presented.

4.1.1 Dot Plots and Stem-and-Leaf Diagrams

Two very effective tools for presenting small to moderate-sized data sets are dot plots and stem-and-leaf diagrams. A **dot plot** is made by ruling off an appropriate scale and then placing a large dot above the scale for each data point, stacking dots corresponding to points that are identical (to the indicated precision of the data). A **stem-and-leaf** diagram is made by using a vertical line (a stem) to separate the leading digits for data values from the final few (usually one or two) digits. These (sets of) final digits are stacked horizontally to form "leaves" that function like the bars of a histogram, portraying the shape of the data set. The virtue of a stem-and-leaf diagram is that it provides its picture of data set shape without loss of the exact individual data values.

Example 36 *Tongue Thickness on Machined Steel Levers. Unke, Wayland, and Weppler worked with a machine shop on the manufacture of some steel levers. The ends of these levers were machined to form tongues. Table 4.1 contains measured thicknesses for 20 tongues cut during the period the students worked on the process. The units are inches, and engineering specifications for the tongue thickness were .1775 to .1875. Figure 4.1 is a dot plot of the tongue-thickness data, and Fig. 4.2 is a corresponding stem-and-leaf diagram. On the stem-and-leaf diagram, the digits ".18X" have been placed to the left of the stem, and only the final digits recorded in Table 4.1 have been used to make the leaves. Notice also that the digits in the leaves have been sorted smallest to largest to further organize the data.*

TABLE 4.1. Measured tongue thicknesses for twenty machined steel levers (inches)

.1825	.1817	.1841	.1813	.1811
.1807	.1830	.1827	.1835	.1829
.1827	.1816	.1819	.1812	.1834
.1825	.1828	.1868	.1812	.1814

FIGURE 4.1. Dot plot of the tongue-thickness data

```
.186 | 8
.185 |
.184 | 1
.183 | 0 4 5
.182 | 5 5 7 7 8 9
.181 | 1 2 2 3 4 6 7 9
.180 | 7
```

FIGURE 4.2. Stem-and-leaf diagram of tongue thicknesses

A useful variation on the basic stem-and-leaf diagram is placing two of them back to back, using the same stem for both, with leaves for the first projecting to the left and leaves for the second projecting to the right. This is helpful when one needs to make comparisons between two different processes.

Example 37 *Heat Treating Gears.* *The article "Statistical Analysis: Mack Truck Gear Heat Treating Experiments" by P. Brezler (*Heat Treating, *November 1986) describes a company's efforts to find a way to reduce distortion in the heat treating of some gears. Two different methods were considered for loading gears into a continuous carburizing furnace, namely, laying the gears flat in stacks and hanging them from rods passing through the gear bores. Table 4.2 on page 194 contains measurements of "thrust face runout" made after heat treating 38 gears laid and 39 gears hung. Figure 4.3 shows back-to-back stem-and-leaf diagrams for the runouts. (Notice that in this display, there are two sets of stems for each leading digit, a "0-4" stem and a "5-9" stem.) The stem-and-leaf diagrams clearly show the laying method to produce smaller distortions than the hanging method.*

TABLE 4.2. Thrust face runouts for gears laid and gears hung (.0001 in)

Laid	Hung
5, 8, 8, 9, 9, 9, 9, 10, 10, 10	7, 8, 8, 10, 10, 10, 10, 11, 11, 11
11, 11, 11, 11, 11, 11, 11, 12, 12, 12	12, 13, 13, 13, 15, 17, 17, 17, 17, 18
12, 13, 13, 13, 13, 14, 14, 14, 15, 15	19, 19, 20, 21, 21, 21, 22, 22, 22, 23
15, 15, 16, 17, 17, 18, 19, 27	23, 23, 23, 24, 27, 27, 28, 31, 36

```
              Laid Runouts                      Hung Runouts

                        9 9 9 9 8 8 5 | 0 | 7 8 8
4 4 4 3 3 3 3 2 2 2 2 1 1 1 1 1 1 0 0 0 | 1 | 0 0 0 0 1 1 1 2 3 3 3
                        9 8 7 7 6 5 5 5 5 | 1 | 5 7 7 7 7 8 9 9
                                        | 2 | 0 1 1 1 2 2 2 3 3 3 3 4
                                      7 | 2 | 7 7 8
                                        | 3 | 1
                                        | 3 | 6
```

FIGURE 4.3. Back-to-back stem-and-leaf plots of thrust face runouts (10^{-4} in)

Dot plots and stem-and-leaf diagrams are primarily tools for working data analysis. Histograms are more easily understood by the uninitiated than are stem-and-leaf diagrams. As such, they are more appropriate for final reports and presentations. But because of the grouping that is usually required to make a histogram, they do not provide the kind of complete picture of a data set provided by the simple dot plots and stem-and-leaf diagrams discussed here.

4.1.2 Quantiles and Box Plots

The concept of **quantiles** is especially helpful for building descriptions of distributions. Roughly speaking, the p quantile (or $100 \times p$th percentile) of a distribution is a number such that a fraction p of the distribution lies to the left and a fraction $1 - p$ lies to the right of the number. If one scores at the .8 quantile (80th percentile) on a national college entrance exam, about 80 % of those taking the exam had lower marks, and 20 % had higher marks. Or, since 95 % of the standard normal distribution is to the left of 1.645, one can say that 1.645 is the .95 quantile of the standard normal distribution. The notation $Q(p)$ will be used in this book to stand for the p quantile of a distribution.

Notation for the p Quantile

While the idea of a distribution quantile should be more or less understandable in approximate terms, a precise definition of $Q(p)$ is needed. For example, faced with a sample consisting of the five values $4, 5, 5, 6$, and 8, exactly what should one mean by $Q(.83)$? What number should one announce as placing 83 % of the five numbers to the left and 17 % to the right? Definition 38 gives the two-part convention that will be used in this book for quantiles of data sets.

Definition 38 *For a data set consisting of n values $x_1 \leq x_2 \leq \cdots \leq x_n$ (x_i is the ith smallest data value),*

1. *for any positive integer i, if $p = (i - .5)/n$, the **p quantile** of the data set is*

$$Q\left(\frac{i-.5}{n}\right) = x_i, \quad \text{and}$$

2. *for values of p not of the form $(i - .5)/n$ for any integer i but with $.5/n < p < (n - .5)/n$, the **p quantile** of the data set is found by linear interpolation between the two values $Q((i - .5)/n)$ with corresponding $(i - .5)/n$ closest to p.*

Definition 38 defines $Q(p)$ for all p between $.5/n$ and $(n - .5)/n$. If one wants to find a particular single quantile for a data set, say $Q(p)$, one may solve the equation

$$p = \frac{i-.5}{n}$$

for i, yielding

$$i = np + .5. \tag{4.1}$$

Index i of the Ordered Data Point that is $Q(p)$

If $np + .5$ is an integer, then $Q(p)$ is simply the $(np + .5)$th smallest data point. If relationship (4.1) leads to a non-integer value for i, one interpolates between the ordered data points with indices just smaller than and just larger than $np + .5$.

It is sometimes helpful to plot $Q(p)$ as a function of p. Such a plot is called a **quantile plot** and can be made by plotting the n points $((i - .5)/n, x_i)$ and then drawing in the interpolating line segments. It gives essentially the same information about a distribution as the possibly more familiar "cumulative frequency ogive" of elementary statistics.

Example 39 *(Example 36 continued.)* Returning to the tongue-thickness example, Table 4.3 on page 196 shows the $n = 20$ ordered data values and corresponding values of $(i - .5)/n$. From the values in Table 4.3, it is clear, for example, that $Q(.425) = .1819$. But should one desire the $.83$ quantile of the tongue-thickness data set, one must interpolate appropriately between $Q(.825) = .1834$ and $Q(.875) = .1835$. Doing so,

$$\begin{aligned}
Q(.83) &= \frac{.830 - .825}{.875 - .825} Q(.875) + \left(1 - \frac{.830 - .825}{.875 - .825}\right) Q(.825) \\
&= .1(.1835) + .9(.1834), \\
&= .18341.
\end{aligned}$$

And giving a more complete summary of the entire quantile function for the tongue-thickness data set, Fig. 4.4 on page 196 is a quantile plot based on Table 4.3.

TABLE 4.3. Ordered tongue thicknesses and values of $((i - .5)/20)$

i	$p = \frac{i-.5}{20}$	$x_i = Q\left(\frac{i-.5}{20}\right)$	i	$p = \frac{i-.5}{20}$	$x_i = Q\left(\frac{i-.5}{20}\right)$
1	.025	.1807	11	.525	.1825
2	.075	.1811	12	.575	.1827
3	.125	.1812	13	.625	.1827
4	.175	.1812	14	.675	.1828
5	.225	.1813	15	.725	.1829
6	.275	.1814	16	.775	.1830
7	.325	.1816	17	.825	.1834
8	.375	.1817	18	.875	.1835
9	.425	.1819	19	.925	.1841
10	.475	.1825	20	.975	.1868

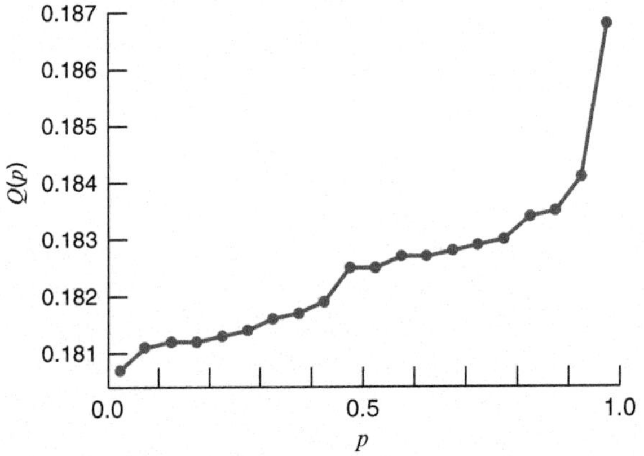

FIGURE 4.4. Quantile plot for the tongue-thickness data

Special values of p have corresponding specially named p quantiles. The .5 quantile of a distribution is the usual **median**, symbolized as $\tilde{x}$ in Sect. 3.2. $Q(.25)$ is often called the **first quartile** of a distribution and $Q(.75)$ is called the **third quartile**. And the values $Q(.1), Q(.2), \ldots, Q(.9)$ are called the **deciles** of a distribution.

Example 40 *(Examples 36 and 39 continued.) Looking carefully at Table 4.3, it is easy to see that the median of the tongue-thickness data set is the simple average of $Q(.475) = .1825$ and $Q(.525) = .1825$. That is, $\tilde{x} = Q(.5) = .1825$. Similarly, the first quartile of the data set is halfway between $Q(.225) = .1813$ and $Q(.275) = .1814$. That is, $Q(.25) = .18135$. And the third quartile of the tongue-thickness data is the mean of $Q(.725) = .1829$ and $Q(.775) = .1830$, namely, $Q(.750) = .18295$.*

Quantiles are basic building blocks for many useful descriptors of a distribution. The median is a well-known measure of location. The difference between the quartiles is a simple measure of spread called the **interquartile range.** In symbols

$$IQR = Q(.75) - Q(.25). \qquad (4.2)$$

Interquartile Range

And there are a number of helpful graphical techniques that make use of quantiles. One of these is the **box plot**.

Figure 4.5 shows a generic box plot. The box locates the middle 50 % of the distribution, with a dividing line drawn at the median. The placement of this dividing line indicates symmetry (or lack thereof) for the center part of the distribution. Lines (or "whiskers") extend out from the box to the most extreme data points that are within $1.5 IQR$ (1.5 times the box length) of the box. Any data values that fall more than $1.5 IQR$ away from the box are plotted individually and in the process identified as unusual or "outlying."

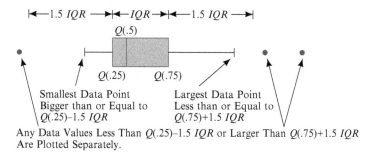

FIGURE 4.5. Generic box plot

Example 41 *(Examples 36, 39, and 40 continued.) As an illustration of the calculations necessary to implement the schematic shown in Fig. 4.5, consider the making of a box plot for the tongue thicknesses. Previous calculation has shown the median thickness to be .1825 and the quartiles of the thickness distribution to be $Q(.25) = .18135$ and $Q(.75) = .18295$. Thus, from display (4.2), the interquartile range for the thickness data set is*

$$IQR = Q(.75) - Q(.25) = .18295 - .18135 = .0016.$$

Then, since

$$Q(.25) - 1.5 IQR = .18135 - .0024 = .17895$$

and there are no data points less than .17895, the lower whisker extends to the smallest data value, namely, .1807. Further, since

$$Q(.75) + 1.5 IQR = .18295 + .0024 = .18535$$

and there is one data point larger than this sum, the value .1868 will be plotted individually, and the upper whisker will extend to the second largest data value, namely, .1841. Figure 4.6 on page 198 is a box plot for the tongue-thickness data.

It reveals some asymmetry in the central part of the data set, its relative short-tailedness to the low side, and the one very large outlying data value.

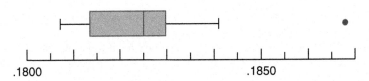

FIGURE 4.6. Box plot for the tongue-thickness data

Box plots carry a fair amount of information about distribution location, spread, and shape. They do so in a very compact way. Many of them can be placed on a single page to facilitate comparisons among a large number of distributions. The next example illustrates the comparison of three distributions using side-by-side box plots.

Example 42 *Comparing Hardness Measurement Methods. Blad, Sobatka, and Zaug did some hardness testing on a single metal specimen. They tested it on three different machines, 10 times per machine. A dial Rockwell tester, a digital Rockwell tester, and a Brinell tester were used. The Brinell hardnesses they recorded (after conversion in the case of the Rockwell readings) are given in Table 4.4.*

Figure 4.7 shows box plots for the measurements produced by the three hardness testers. It is very helpful for comparing them. It shows, among other things, the comparatively large variability and decided skewness of the Brinell machine measurements and the fact that the dial Rockwell machine seems to read consistently higher than the Digital Rockwell machine.

TABLE 4.4. Hardness values for a single specimen obtained from three different testers (Brinell hardness)

Dial Rockwell	Digital Rockwell	Brinell
536.6, 539.2, 524.4, 536.6	501.2, 522.0, 531.6, 522.0	542.6, 526.0, 520.5, 514.0
526.8, 531.6, 540.5, 534.0	519.4, 523.2, 522.0, 514.2	546.6, 512.6, 516.0, 580.4
526.8, 531.6	506.4, 518.1	600.0, 601.0

4.1.3 Q-Q Plots and Normal Probability Plots

An important application of quantiles is to the careful comparison of the shapes of two distributions through making and interpreting Q-Q **plots.** The version of the Q-Q plot that is easiest to understand is that where both distributions involved are empirical, representing data sets. The *most important* version is that where one

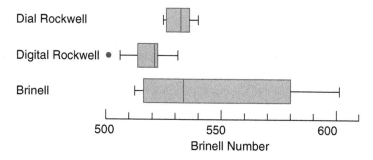

FIGURE 4.7. Box plots for hardness measurements made on three different testers

distribution is empirical and the other is a theoretical distribution. One is essentially investigating how the shape of a data set matches that of some probability distribution. This discussion will begin with the easy-to-understand (but practically less important) case of comparing shapes for two data sets and then proceed to the more important case.

Consider the two small data sets given in Table 4.5. Figure 4.8 shows dot plots for them and reveals that by most standards, they have the same shape. There are several ways one might try to quantify this fact. For one thing, the relative sizes of the gaps or differences between successive ordered data values are the same for the two data sets. That is, the gaps for the first data set are in the ratios $1:1:0:2$, and for the second data set, the ratios are $2:2:0:4$.

TABLE 4.5. Two small artificial data sets

Data set 1	Data set 2
2, 3, 4, 4, 6	5, 7, 9, 9, 13

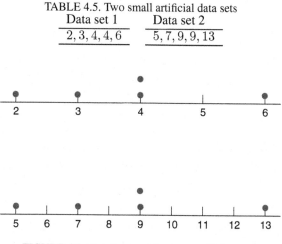

FIGURE 4.8. Dot diagrams for two small data sets

A second (and for present purposes more germane) observation is that the ordered values in the second data set are linearly related to those in the first. In fact, the values in the second data set were derived by doubling those in the first data set and adding 1. This means that the second quantile function is linearly related

to the first by
$$Q_2(p) = 2Q_1(p) + 1 .$$

Notice then that if one makes up ordered pairs of the form $(Q_1(p), Q_2(p))$ and plots them, all of the plotted points will fall on a single line. Using the values $p = .1, .3, .5, .7,$ and $.9$, one has in the present case the five ordered pairs:

$$\begin{aligned}
(Q_1(.1), Q_2(.1)) &= (2, 5) , \\
(Q_1(.3), Q_2(.3)) &= (3, 7) , \\
(Q_1(.5), Q_2(.5)) &= (4, 9) , \\
(Q_1(.7), Q_2(.7)) &= (4, 9) , \quad \text{and} \\
(Q_1(.9), Q_2(.9)) &= (6, 13) ,
\end{aligned}$$

and the scatterplot in Fig. 4.9.

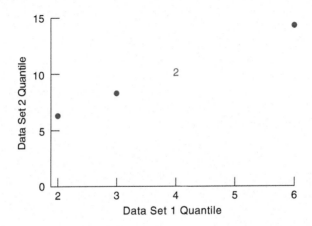

FIGURE 4.9. Q-Q plot for the data sets of Table 4.5

What is true in this highly simplified and artificial example is true in general. Equality of "shape" for two distributions is equivalent to the two corresponding quantile functions being linearly related. A way of investigating the extent to which two distributions have the same shape is to plot for suitable p, ordered pairs of the form

$$(Q_1(p), Q_2(p)) , \tag{4.3}$$

Points for a Q-Q Plot of Distributions 1 and 2

looking for linearity. Where there is perfect linearity on such a plot, equality of shape is suggested. Where there are departures from linearity, those departures can often be interpreted in terms of the relative shapes of the two distributions. Consider, for example, a modified version of the present example where the value 5 in the second data set is replaced by 1. Figure 4.10 is the Q-Q plot for this modification.

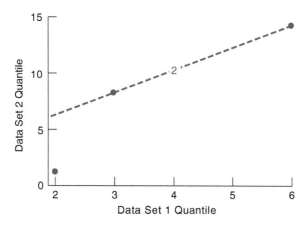

FIGURE 4.10. *Q-Q* plot after modifying one value in Table 4.5

Notice that while four of the plotted points in Fig. 4.10 fall on a single straight line, the fifth, $(2, 1)$, does not fall on that line. It is below/to the right of the line established by the rest of the points. To bring it "back on line" with the rest, it would have to be moved either up or to the left on the plot. This says that relative to the shape of data set 1, the second data set is long tailed to the low side. Equivalently, relative to the shape of data set 2, the first is short tailed to the low side.

The important version of *Q-Q* plotting where the first distribution is that of a data set and the second is a theoretical or probability distribution is usually called **probability plotting.** And the most common kind of probability plotting is **normal plotting,** where one is investigating the degree of similarity between the shape of a data set and the archetypal bell shape of the normal distribution.

The values of p typically used in making the points (4.3) for a probability plot are those corresponding exactly to the data points in hand, namely, those of the form $(i - .5)/n$ for integer i. Using such values of p, if $Q_z(p)$ is the standard normal quantile function, it follows that a normal plot can be made on regular graph paper by plotting the n points

$$\left(x_i, Q_z\left(\frac{i - .5}{n}\right)\right), \qquad (4.4)$$

Points for a Normal Plot of an x Data Set

where as in Definition 38, x_i is the ith smallest data value.

Standard normal quantiles for use in display (4.4) can, of course, be found by locating values of p in the body of a cumulative normal probability table like Table A.1 and then reading corresponding quantiles from the table's margin. Statistical packages provide "inverse cumulative probability" functions that can be used to automate this procedure. And there are approximations for $Q_z(p)$ that are quite adequate for plotting purposes. One particularly simple approximation (borrowed from *Probability and Statistics for Engineers and Scientists* by Walpole

Approximate Standard Normal p Quantile

$$Q_z(p) \approx 4.91[p^{.14} - (1-p)^{.14}],$$

which returns values that are accurate to within .01 for $.005 \leq p \leq .995$ (and to within .05 for $.001 \leq p \leq .999$).

Example 43 *(Examples 36, 39, 40, and 41 continued.)* Consider the problem of assessing how normal/bell shaped the tongue-thickness data of Table 4.1 are. A normal probability plot (theoretical Q-Q plot) can be used to address this problem. Table 4.6 shows the formation of the necessary ordered pairs, and Fig. 4.11 is the resulting normal plot. The plot might be called roughly linear, except for the point corresponding to the largest data value. In order to get that point back in line with the others, one would need to move it either to the left or up. That is, relative to the normal distribution shape, the data set is long tailed to the high side. The tongue thickness of .1868 simply does not fit into the somewhat normal-looking pattern established by the rest of the data.

TABLE 4.6. Coordinates for points of a normal plot of the tongue-thickness data

i	$x_i = Q\left(\frac{i-.5}{20}\right)$	$Q_z\left(\frac{i-.5}{20}\right)$	i	$x_i = Q\left(\frac{i-.5}{20}\right)$	$Q_z\left(\frac{i-.5}{20}\right)$
1	.1807	−1.96	11	.1825	.06
2	.1811	−1.44	12	.1827	.19
3	.1812	−1.15	13	.1827	.32
4	.1812	−.93	14	.1828	.45
5	.1813	−.76	15	.1829	.60
6	.1814	−.60	16	.1830	.76
7	.1816	−.45	17	.1834	.93
8	.1817	−.32	18	.1835	1.15
9	.1819	−.19	19	.1841	1.44
10	.1825	−.06	20	.1868	1.96

Theoretical Q-Q plotting (probability plotting) is important for several reasons. First, it helps one judge how much faith to place in calculations based on a probability distribution and suggests in what ways the calculations might tend to be wrong. For example, Fig. 4.11 suggests that if one uses a normal distribution to describe tongue thickness, the frequency of very large data values might well be underpredicted.

A second way in which probability plotting is often helpful is in providing graphical estimates of distribution parameters. For example, it turns out that if one makes a normal plot of an exactly normal distribution, the slope of the plot is the reciprocal of σ, and the horizontal intercept is μ. That suggests that for a real data set whose normal plot is fairly linear, one might infer that

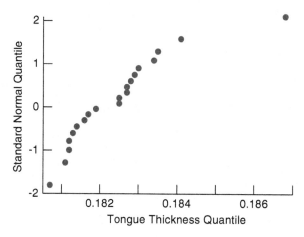

FIGURE 4.11. Normal plot of the tongue-thickness data

1. the horizontal intercept of an approximating line is a sensible estimate of the mean of the process generating the data, and

2. the reciprocal of the slope is a sensible estimate of the standard deviation of the process generating the data.

Estimates of a Mean and Standard Deviation from a Normal Plot

Example 44 *Angles of Holes Drilled by Electrical Discharge Machining (EDM). Duren, Ling, and Patterson worked on the production of some small, high-precision metal parts. Holes in these parts were being drilled using an electrical discharge machining technique. The holes were to be at an angle to one flat surface of the parts, and engineering specifications on that angle were $45° \pm 2°$. The actual angles produced were measured on 50 consecutive parts and are given in Table 4.7. (The units there are degrees and the values are in decimal form. The data were originally in degrees and minutes.) Figure 4.12 on page 204 is a normal plot of the hole angle data. Notice that the plot is fairly linear and that the horizontal intercept of an approximating line is near the sample mean $\bar{x} = 44.117$, while the slope of an approximating line is approximately the reciprocal of $s = .983$.*

TABLE 4.7. Angles of fifty holes drilled by electrical discharge machining (degrees)

46.050	45.250	45.267	44.700	44.150	44.617	43.433	44.550	44.633	45.517
44.350	43.950	43.233	45.933	43.067	42.833	43.233	45.250	42.083	44.067
43.133	44.200	43.883	44.467	44.600	43.717	44.167	45.067	44.000	42.500
45.333	43.467	43.667	44.000	44.000	45.367	44.950	45.100	43.867	43.000
42.017	44.600	43.267	44.233	45.367	44.267	43.833	42.450	44.650	42.500

The facts that (for bell-shaped data sets) normal plotting provides a simple way of approximating a standard deviation and that 6σ is often used as a measure of the intrinsic spread of measurements generated by a process together lead to

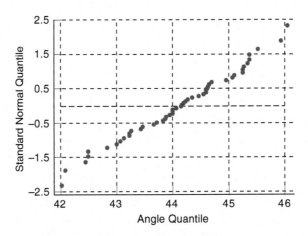

FIGURE 4.12. Normal plot of the tongue-thickness data

the common practice of basing **process capability analyses** on normal plotting. Figure 4.13 on page 206 shows a very common type of industrial form that essentially facilitates the making of a normal plot by removing the necessity of evaluating the standard normal quantiles $Q_z(p)$. (On the special vertical scale, one may simply use the plotting position p rather than $Q_z(p)$, as would be required when using regular graph paper.) After plotting a data set and drawing in an approximating straight line, 6σ can be read off the plot as the difference in horizontal coordinates for points on the line at the "$+3\sigma$" and "-3σ" vertical levels (i.e., with $p = .0013$ and $p = .9987$).

A form like the one in Fig. 4.13 encourages the *plotting* of process data (always a plus) and also allows even fairly nonquantitative people to easily estimate and develop some intuition about "the process spread." Normal plotting is certainly not the last word in process characterization, but it *is* a very important tool that can and should be used alongside some of the other (numerical) methods presented in the following sections.

Section 4.1 Exercises

1. A distributor of spices and seasonings checks moisture content of lots of pepper that it receives from a supplier. The company's specifications on moisture content are from 9 % to 12 %. The following are moisture content data for 22 incoming pepper lots (in %). The data are not listed according to production or arrival order.

    $11.5, 11.1, 11.1, 10.9, 10.6, 10.6, 10.6, 10.7, 10.7, 10.4, 10.4,$

    $10.5, 10.5, 10.3, 10.0, 10.0, 10.1, 9.7, 9.6, 9.3, 9.3, 9.3$

    (a) Make a normal probability plot for these data. Does it appear that normal distribution is an adequate model for % moisture content in lots of pepper? Why or why not?

(b) Based on the normal probability plot in (a), estimate the mean and standard deviation of % moisture content in a pepper lot.

(c) Make a dot plot for these data.

(d) Make a histogram for these data using 9.25 to 9.49, 9.5 to 9.74, 9.75 to 9.99, 10.0 to 10.24, ..., 11.5 to 11.75 as the "bins."

(e) Make a box plot for these data.

(f) Give the first quartile, median, and third quartile for these data.

(g) Find the interquartile range for this set of data.

2. Continue with the data of problem 1. It was noted that the data are not listed here according to production or arrival order.

(a) In addition to the data above, what should have been recorded to monitor (over time) the % moisture content in pepper lot production?

(b) For your conclusions and estimates in problem 1 to fairly represent the future, what must be true (over time) about the pepper lot production process?

3. A painting process was studied and determined to be stable. The variable of interest was paint thickness on a particular item at a specified location on the item. The following paint thickness data were recorded for $n = 10$ such items in some unspecified units.

$$.2, .8, 4.1, 2.4, 9.7, 9.5, 4.4, 6.2, 5.9, 1.8$$

(a) Make a normal probability plot for the data. Does it appear that a normal model for paint thickness is adequate? Why or why not?

(b) Make a box plot of the data. Do you find any outliers?

(c) Based on the normal probability plot, (graphically) estimate the 1st quartile, the 3rd quartile, and the median of paint thicknesses.

4. Below are individual values of the percentages, x, of "small" grit particles in a bulk abrasive used by a company in the production of a type of sandpaper. The company has already done extensive process monitoring and has concluded the percentage of this "small" grit particle % in lots of the material is stable. Below are the most recent data obtained from 11 samples.

$$11.0, 14.2, 13.1, 19.6, 16.7, 13.3, 14.1, 14.7, 16.5, 16.1, 13.7$$

(a) Make a normal probability plot for the data. Does it appear that a normal model for percent of small grit particles is adequate? Why or why not?

(b) Find the sample mean and sample standard deviation of x.

(c) Based on the normal probability plot (graphically), estimate the 90th percentile of percent small grit particles in a sample of this material.

206   Chapter 4. Process Characterization and Capability Analysis

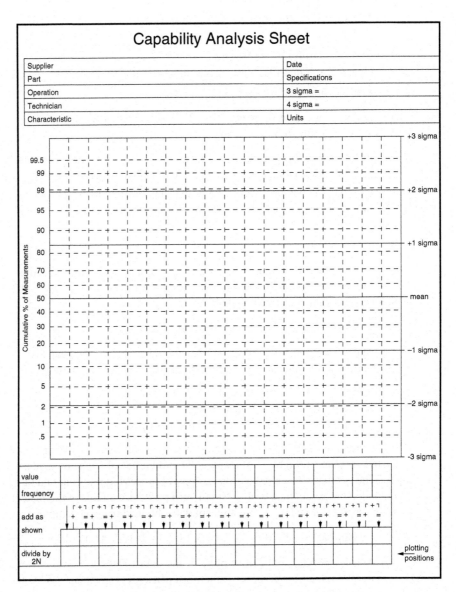

FIGURE 4.13. Capability analysis form

4.2 Process Capability Measures and Their Estimation

The methods of Sect. 4.1 can be used to picture the pattern of variation associated with a process. It is also helpful to have some numerical summary measures to quote as more or less representing/condensing the graphics. Of course, the mean and standard deviation are useful in this regard. But there are also more specialized measures that have come into common use in quality assurance circles. Some of these are the subject of this section.

This section presents some measures of process performance that are appropriate for processes that generate at least roughly normally distributed data. The "process capability" and two "capability ratios," C_p and C_{pk}, are discussed, and confidence intervals for them are presented. But before introducing these, it is important to offer a disclaimer: unless a normal distribution makes sense as a description of process output, these measures are of dubious relevance. Further, the confidence interval methods presented here for estimating them are completely unreliable unless a normal model is appropriate. So the normal plotting idea presented in the last section is a very important prerequisite for these methods.

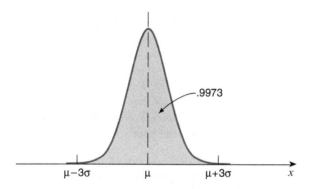

FIGURE 4.14. Normal distribution

The large majority of a normal distribution is located within three standard deviations of its mean. Figure 4.14 illustrates this elementary point. In light of this fact, it makes some sense to say that 6σ is a measure of process spread

Process Capability

and to call 6σ the **process capability** for a stable process generating normally distributed measurements.

The fact that there are elementary statistical methods for estimating the standard deviation of a normal distribution implies that it is easy to give confidence limits for the process capability. That is, if one has in hand a sample of n observations with corresponding sample standard deviation s, then confidence limits for 6σ are simply

Confidence Limits for 6σ

$$6s\sqrt{\frac{n-1}{\chi^2_{\text{upper}}}} \quad \text{and/or} \quad 6s\sqrt{\frac{n-1}{\chi^2_{\text{lower}}}}, \tag{4.5}$$

where χ^2_{upper} and χ^2_{lower} are upper and lower percentage points of the χ^2 distribution with $n-1$ degrees of freedom. If the first limit in display (4.5) is used alone as a lower confidence bound for 6σ, the associated confidence level is the probability that a χ^2_{n-1} random variable takes a value less than χ^2_{upper}. If the second limit in display (4.5) is used alone as an upper confidence bound for 6σ, the associated confidence is the probability that a χ^2_{n-1} random variable exceeds χ^2_{lower}. If both limits in display (4.5) are used to make a two-sided interval for the process capability, the associated confidence level is the probability that a χ^2_{n-1} random variable takes a value between χ^2_{lower} and χ^2_{upper}.

Example 45 *Process Capability for the Angles of EDM-Drilled Holes (Example 44 revisited). Figure 4.12 shows the angle data of Table 4.7 to be reasonably described by a normal distribution. As such, it makes sense to consider estimating the process capability for the angles at which holes are drilled. Recall that the students' data had $n = 50$ and $s = .983$. From the approximation provided with Table A.3 or a statistical package, the .05 and .95 quantiles of the χ^2 distribution for $\nu = n - 1 = 49$ degrees of freedom are, respectively, 33.93 and 66.34. Thus, from display (4.5), the interval with end points*

$$6(.983)\sqrt{\frac{50-1}{66.34}} \quad \text{and} \quad 6(.983)\sqrt{\frac{50-1}{33.93}},$$

that is,

$$5.07° \quad \text{and} \quad 7.09°$$

is a 90% confidence interval for the process capability. (One is "90% sure" that the process spread is at least 5.07° and no more than 7.09°.)

Where there are both an upper specification U and a lower specification L for measurements generated by a process, it is common to compare process variability to the spread in those specifications. One way of doing this is through **process capability ratios**. And a popular process capability ratio is

Chapter 4. Process Characterization and Capability Analysis

Capability Ratio C_p

$$C_p = \frac{U - L}{6\sigma}. \quad (4.6)$$

When this measure is 1, the process output will fit more or less exactly inside specifications *provided the process mean is exactly on target at* $(U + L)/2$. When C_p is larger than 1, there is some "breathing room" in the sense that a process would not need to be perfectly aimed in order to produce essentially all measurements inside specifications. On the other hand, where C_p is less than 1, no matter how well a process producing normally distributed observations is aimed, a significant fraction of the output will fall outside specifications.

The very simple form of Eq. (4.6) makes it clear that once one knows how to estimate 6σ, one may simply divide the known difference in specifications by confidence limits for 6σ in order to derive confidence limits for C_p. That is, lower and upper confidence limits for C_p are, respectively,

Confidence Limits for C_p

$$\frac{(U-L)}{6s}\sqrt{\frac{\chi^2_{\text{lower}}}{n-1}} \quad \text{and/or} \quad \frac{(U-L)}{6s}\sqrt{\frac{\chi^2_{\text{upper}}}{n-1}}, \quad (4.7)$$

where again χ^2_{upper} and χ^2_{lower} are upper and lower percentage points of the χ^2 distribution with $n - 1$ degrees of freedom. If the first limit in display (4.7) is used alone as a lower confidence bound for C_p, the associated confidence level is the probability that a χ^2_{n-1} random variable exceeds χ^2_{lower}. If the second limit in display (4.7) is used alone as an upper confidence bound for C_p, the associated confidence is the probability that a χ^2_{n-1} random variable is less than χ^2_{upper}. If both limits in display (4.7) are used to make a two-sided interval for C_p, the associated confidence level is the probability that a χ^2_{n-1} random variable takes a value between χ^2_{lower} and χ^2_{upper}.

Example 46 *(Examples 44 and 45 continued.)* Recall from Example 44 in Sect. 4.1 that the engineering specifications on the angles for the EDM-drilled holes were $45° \pm 2°$. That means that for this situation, $U - L = 4°$. So, since from before one is 90 % confident that 6σ is between $5.07°$ and $7.09°$, one can be 90 % confident that C_p is between

$$\frac{4}{7.09} \quad \text{and} \quad \frac{4}{5.07},$$

that is, between

$$.56 \quad \text{and} \quad .79.$$

Of course, this same result could have been obtained beginning directly with expressions (4.7) rather than starting from the limits (4.5) for 6σ.

C_p is more a measure of process *potential* than it is a measure of current performance. Since the process aim is not considered in the computation of C_p, it is possible for a misaimed process with very small intrinsic variation to have a huge value of C_p and yet currently be turning out essentially no product in specifications. C_p measures only "what could be" were the process perfectly aimed. This is not necessarily an undesirable feature of C_p, but it is one that users need to understand.

Another process capability index that does take account of the process mean (and is more a measure of current process performance than of potential performance) is one commonly known as C_{pk}. This measure can be described in words as "the number of 3σs that the process mean is to the good side of the closest specification." For example, if $U - L$ is 10σ and μ is 4σ below the upper specification, then C_{pk} is $4\sigma/3\sigma = 1.33$. On the other hand, if $U - L$ is 10σ and μ is 4σ above the upper specification, then C_{pk} is -1.33.

In symbols,

Capability Index C_{pk}

$$C_{pk} = \min\left\{\frac{U-\mu}{3\sigma}, \frac{\mu-L}{3\sigma}\right\} = \frac{U-L-2\left|\mu-\frac{U+L}{2}\right|}{6\sigma}. \qquad (4.8)$$

This quantity will be positive as long as μ is between L and U. It will be large if μ is between L and U (and close to neither L nor U), and $U - L$ is large compared to σ. C_{pk} is never larger than C_p, i.e.,

$$C_{pk} \leq C_p,$$

and the two are equal only when μ is exactly at the mid-specification $(L+U)/2$.

Making a confidence interval for C_{pk} is more difficult than making one for C_p. The best currently available method is only appropriate for large samples and provides a real confidence level that only approximates the nominal one. The method is based on the natural single-number estimate of C_{pk},

Estimate of C_{pk}

$$\widehat{C_{pk}} = \min\left\{\frac{U-\bar{x}}{3s}, \frac{\bar{x}-L}{3s}\right\} = \frac{U-L-2\left|\bar{x}-\frac{U+L}{2}\right|}{6s}. \qquad (4.9)$$

Then confidence limits for C_{pk} are

Confidence Limits for C_{pk}

$$\widehat{C_{pk}} \pm z\sqrt{\frac{1}{9n} + \frac{\widehat{C_{pk}}^2}{2n-2}}. \qquad (4.10)$$

If z is the p quantile of the standard normal distribution ($z = Q_z(p)$), a single one of the two limits in display (4.10) is an approximately $p \times 100\,\%$ (lower or upper) confidence bound for C_{pk}. If both limits are used to make a two-sided interval, then the approximate confidence level is $(2p-1) \times 100\,\%$.

Example 47 *(Examples 44, 45, and 46 continued.)* The 50 EDM hole angles have sample mean $\bar{x} = 44.117$. Then from relationship (4.9),

$$\widehat{C_{pk}} = \min\left\{\frac{47 - 44.117}{3(.983)}, \frac{44.117 - 43}{3(.983)}\right\} = \min\{.98, .38\} = .38.$$

So, for example, since the .95 quantile of the standard normal distribution is 1.645, an approximate 95% lower confidence bound for C_{pk} is from expression (4.10)

$$.38 - 1.645\sqrt{\frac{1}{9(50)} + \frac{(.38)^2}{2(50) - 2}} = .28.$$

One can be in some sense approximately 95% sure that C_{pk} for the angles in the EDM drilling process is at least .28.

Overreliance upon process capability measures can be justly criticized. Critics have correctly noted that

1. 6σ, C_p, and C_{pk} have unclear relevance when a process distribution is not normal,

2. "one-number summaries" like those discussed here can leave much unsaid about what a process is doing or even the shape of a distribution of measurements it is generating, and

3. really going to work tuning a process, monitoring for and removing upsets, and determining what it is really "capable" of doing involve much more than the simple estimation of 6σ or one of the measures C_p or C_{pk}.

In addition to these objections, the capability ratios C_p and C_{pk} depend upon specifications that are sometimes subject to unannounced change (even arbitrary change). This makes it difficult to know from one reporting period to the next what has happened to process variability if estimates of C_p or C_{pk} are all that are provided. It thus seems that for purposes of comparisons across time, if any of the measures of this section are to be used, the simple process capability 6σ is most attractive.

Despite these issues, the measures of this section are very popular. Provided one understands their limitations and simply views them as one of the many tools for summarizing process behavior, they have their place. But the wise engineer will not assume that computing and reporting one of these figures is in any way the last word in assessing process performance.

Section 4.2 Exercises

1. Consider again problem 1 of Sect. 4.1 and the data given there. Assume pepper lot production has a stable lot percent moisture content, and percent

moisture content is approximately normally distributed. Acceptable lot percent moisture content has been specified as between 9 % and 12 %.

(a) Give limits that you are 95 % sure will bracket 6σ where σ is the lot-to-lot standard deviation of percent moisture content.

(b) Does your interval in (a) suggest 6σ is less than $U - L$? How does your answer relate to the potential quality of the pepper lot production process (with respect to percent moisture content)?

(c) Give a single-number estimate of C_{pk} for percent measured moisture content in the lots of pepper. Interpret your estimate.

(d) Give a single-number estimate of C_p for percent measured moisture content in the lots of pepper. Interpret your estimate.

(e) Give 95 % confidence limits for a capability index that measures *(not potential but)* current pepper lot production process performance.

(f) Give 95 % confidence limits for a capability index that reflects what the pepper lot process performance could be if the percent moisture content per lot could be centered. Is the process centered? Why or why not?

2. Often, capability indices are used as "quick" reflections of process quality and as measures of quality improvement (comparing indices over time). This can sometimes be misleading.

(a) Consider the capability index C_{pk}. Assuming a process has the same μ and σ, how might one "artificially" increase C_{pk} even though the process has not improved at all? Consider the capability ratio C_p. Again, assuming a process has the same μ and σ, how might one "artificially" increase C_p even though the process potential has not improved at all?

(b) Answer (a) with respect to *estimated* C_{pk} and C_p when $\bar{\bar{x}}$ and $\hat{\sigma}$ do not change.

3. Lengths of steel shafts are measured by a laser gauge that produces a coded voltage proportional to shaft length (above some reference length). In the units produced by the gauge, shafts of a particular type have length specifications 2300 ± 20. Below are measured lengths of $n = 10$ such shafts. Assume the shafts came from a physically stable process.

$$2298, 2301, 2298, 2289, 2291, 2290, 2287, 2280, 2289, 2290$$

(a) Using a normal distribution model for shaft length, give 90 % confidence limits for a process capability index that measures process *potential*.

(b) Using a normal distribution model for shaft length, give 90 % confidence limits for a process capability index that measures current process performance.

(c) Based on your answers to (a) and (b) what do you conclude about the current and potential process performance?

4. The ratio of an upper confidence limit for a process capability 6σ to a lower confidence limit for the quantity can be taken as a measure of how much one can learn about it based on a given size sample. For a sample of size $n = 30$ and 95 % confidence, to within what factor (multiple) does this ratio indicate 6σ can be estimated?

5. Consider again problem 3 of Sect. 4.1 and the data given there. A painting process was investigated and determined to be stable with an approximately normal distribution. The variable of interest was paint thickness. Acceptable paint thicknesses are between 4 and 6 units.

   (a) Using the sample standard deviation from the data, give a 90 % confidence interval for 6σ, where σ is the standard deviation of paint thickness from all items (at the specified location on the items).

   (b) Using the standard deviation and sample average from the data, would the current painting process be considered "acceptable" or of high quality? Why?

6. Consider again problem 4 of Sect. 4.1 and the data given there. Recall that the percentages (x) of "small" grit particles in bulk abrasive used by a company in the production of a type of sandpaper are of interest. The company has already done extensive process monitoring and has concluded the percentage of this "small" grit particle is stable. Specifications are from 13.0 % to 16.6 %.

   (a) Using the sample mean and standard deviation, estimate the potential process capability.

   (b) Using the sample mean and standard deviation, estimate the current process performance as expressed in a capability index.

   (c) Give two-sided 95 % confidence intervals for the indices in (a) and (b).

   (d) Briefly comment on the current process performance and potential performance.

4.3 Prediction and Tolerance Intervals

The methods of the previous section and those of elementary statistics can be used to characterize the pattern of variation produced by a stable process through the estimation of process *parameters*. Another approach is to provide intervals likely to contain either the next measurement from the process or a large portion of

214   Chapter 4. Process Characterization and Capability Analysis

all additional values it might generate. Such intervals are the subject of this section. The so-called prediction interval and tolerance interval methods for normal processes are considered first. Then some simple "nonparametric" methods are considered. These can be used to produce prediction and tolerance intervals for any stable process, regardless of whether or not the process distribution is normal.

4.3.1 Intervals for a Normal Process

The usual confidence interval methods presented in elementary statistics courses concern the estimation of process parameters like the mean μ and standard deviation σ. The estimation methods for 6σ, C_p, and C_{pk} presented in Sect. 4.2 essentially concern the empirical approximation of interesting functions of these process parameters. A completely different approach to process characterization is to use data, not to approximate process parameters or functions of them but rather to provide intervals in some sense representing where *additional observations* are likely to fall. Two different formulations of this approach are the making of prediction intervals and the making of tolerance intervals.

A **prediction interval** based on a sample of size n from a stable process is an interval thought likely to contain a single additional observation drawn from the process. Suppose one makes a 90 % prediction interval. The associated confidence guarantee is that if the whole business of "selecting a sample of n, making the corresponding interval, observing an additional value and checking to see if the additional value is in the interval" is repeated many times, about 90 % of the repetitions will be successful.

Where it is sensible to assume that one is sampling from a normal process, a very simple formula can be given for prediction limits. That is, prediction limits for a single additional observation from a normal distribution are

Normal
Distribution
Prediction
Limits

$$\bar{x} - ts\sqrt{1 + \frac{1}{n}} \quad \text{and/or} \quad \bar{x} + ts\sqrt{1 + \frac{1}{n}}, \qquad (4.11)$$

where t is a quantile of the t distribution with $\nu = n - 1$ associated degrees of freedom. If t is the p quantile and the first limit in display (4.11) is used alone as a lower prediction bound, the associated confidence level is $p \times 100\,\%$. Similarly, if t is the p quantile and the second limit in display (4.11) is used alone as an upper prediction bound, the associated confidence level is $p \times 100\,\%$. And if t is the p quantile and both limits in display (4.11) are used as the end points of a two-sided prediction interval, the associated prediction confidence level is $(2p - 1) \times 100\,\%$.

Example 48 Predicting the Angle of an Additional EDM-Drilled Hole (Examples 44, 45, 46, and 47 revisited). *The normal plot of Fig. 4.12 shows the angle data of Table 4.7 to be reasonably normal looking. Capability figures for the EDM drilling process were estimated in Examples 45, 46, and 47. Here consider the prediction of a single additional angle measurement. Either from the t distribution quantiles given in Table A.2 or from a statistical package, the $p = .95$ quantile* *of the t distribution with $\nu = 50 - 1 = 49$ degrees of freedom is 1.6766. Then recalling that for the angle data $\bar{x} = 44.117$ and $s = .983$, formulas (4.11) show that a 90% two-sided prediction interval for a single additional angle has end points*

$$44.117 - 1.6766(.983)\sqrt{1 + \frac{1}{50}} \quad \text{and} \quad 44.117 + 1.6766(.983)\sqrt{1 + \frac{1}{50}},$$

that is,

$$42.45° \quad \text{and} \quad 45.78°.$$

One can be in some sense 90% sure that a single additional angle generated by the stable EDM drilling process will be between 42.45° and 45.78°.

A second formulation of the problem of producing an interval locating where additional observations from a process are likely to fall is that of making a **tolerance interval** for some (typically large) fraction of the process distribution. A tolerance interval for a fraction p of a distribution is a data-based interval thought likely to bracket at least that much of the distribution. Suppose one uses a formula intended to produce 95% tolerance intervals for a fraction $p = .90$ of a distribution. Then the associated confidence guarantee is that if the whole business of "selecting a sample of n, computing the associated interval and checking to see what fraction of the distribution is in fact bracketed by the interval" is repeated many times, about 95% of the intervals will bracket at least 90% of the distribution.

Where it is sensible to assume one is sampling from a normal process distribution, very simple formulas can be given for tolerance limits. That is, using constants τ_1 or τ_2 given in Tables A.6, a one-sided tolerance limit for a fraction p of an entire normal process distribution is

$$\boxed{\bar{x} - \tau_1 s \quad \text{or} \quad \bar{x} + \tau_1 s,} \tag{4.12}$$

One-sided Normal Distribution Tolerance Limits

while a two-sided tolerance interval for a fraction p of such a distribution can be made using end points:

$$\boxed{\bar{x} - \tau_2 s \quad \text{and} \quad \bar{x} + \tau_2 s.} \tag{4.13}$$

Two-sided Normal Distribution Tolerance Limits

The constant τ_1 or τ_2 may be chosen (from Table A.6.2 or A.6.1, respectively) to provide a 95% or a 99% confidence level.

Example 49 *(Example 48 continued.) Rather than as before predicting a single additional EDM-drilled angle, consider the problem of announcing an interval likely to contain 95% of the angle distribution. In fact, for purposes of illustration, consider the making of both a 99% one-sided lower tolerance bound for 95%*

of all additional angles and a 99 % two-sided tolerance interval for 95 % of all additional angles.

Beginning with the one-sided problem, the first formula in display (4.12) and the $n = 50$, $p = .95$, and 99 % confidence level entry of Table A.6.2 produce the lower tolerance bound

$$44.117 - 2.269(.983) = 41.89°.$$

One can be "99 % sure" that at least 95 % of all angles are $41.89°$ or larger.

In a similar fashion, using the formulas (4.13) and the $n = 50$, $p = .95$, and 99 % confidence level entry of Table A.6.1, one has the end points

$$44.117 - 2.580(.983) = 41.58° \quad \text{and} \quad 44.117 + 2.580(.983) = 46.65°.$$

That is, one can be in some sense 99 % sure that at least 95 % of all angles are in the interval $(41.58, 46.65)$.

It is instructive to compare the second interval in Example 48 to the 95 % prediction interval obtained earlier, namely, $(42.45, 45.78)$. The tolerance interval is clearly larger than the prediction interval. This is typical of what happens using common (large) confidence levels for tolerance intervals. A tolerance interval is simply designed to do a more ambitious task than a corresponding prediction interval. That is, a prediction interval aims to locate a single additional measurement while a tolerance interval intends to locate most of *all* additional observations. It is therefore not surprising that the tolerance interval would need to be larger.

4.3.2 Intervals Based on Maximum and/or Minimum Sample Values

The prediction and tolerance intervals prescribed by displays (4.11), (4.12), and (4.13) are very definitely normal distribution intervals. If a normal distribution is not a good description of the stable process data-generating behavior of a system under consideration, the confidence guarantees associated with these formulas are null and void. If measurements x are not normal, on occasion it is possible to find a "transformation" $g(\cdot)$ such that transformed measurements $g(x)$ are normal. When this can be done, one can then simply find prediction or tolerance intervals for $g(x)$ and then "untransform" the end points of such intervals (using the inverse function $g^{-1}(\cdot)$) to provide prediction or tolerance intervals for raw values x. This approach to making intervals really amounts to finding a convenient scale upon which to express the variable of interest when the original one turns out to be inconvenient.

A second approach to making prediction and tolerance intervals when a process distribution does not seem to be normal is to use limits that carry the same confidence level guarantee for *any* (continuous) stable process distribution. Such limits can be based on minimum and/or maximum values in a sample. Because their applicability is not limited to the normal "parametric family" of distributions, these limits are sometimes called **nonparametric limits**.

If one has a sample of n measurement from a stable process, the most obvious of all statistical intervals based on those measurements are

$$(\min x_i, \infty), \qquad (4.14)$$

$$(-\infty, \max x_i), \qquad (4.15)$$

and

$$(\min x_i, \max x_i). \qquad (4.16)$$

One-sided Prediction or Tolerance Interval (Lower Bound)

One-sided Prediction or Tolerance Interval (Upper Bound)

Two-sided Prediction or Tolerance Interval

It turns out that any of these intervals can be used as either a prediction interval for a single additional observation from a process or as a tolerance interval for a fraction p of the process distribution.

Where either of the one-sided intervals (4.14) or (4.15) is used as a prediction interval, the associated prediction confidence level is

$$\frac{n}{n+1}. \qquad (4.17)$$

Prediction Confidence of Intervals (4.15) and (4.14)

Where the two-sided interval (4.16) is used, the associated prediction confidence level is

$$\frac{n-1}{n+1}. \qquad (4.18)$$

Prediction Confidence of Interval (4.16)

Where either of the one-sided intervals (4.14) or (4.15) is used as a tolerance interval for a fraction p of the output from a stable process, the associated confidence level is

$$1 - p^n. \qquad (4.19)$$

Confidence Level of Tolerance Intervals (4.14) and (4.15)

And where the two-sided interval (4.16) is used as a tolerance interval, the associated confidence level is

$$1 - p^n - n(1-p)p^{n-1}. \qquad (4.20)$$

Confidence Level of Tolerance Interval (4.16)

Example 50 *Prediction and Tolerance Intervals for Tongue Thicknesses of Machined Levers (Example 36 revisited). The normal plot in Fig. 4.11 shows the tongue-thickness data of Table 4.1 to be long tailed to the high side and clearly not adequately described as approximately normal. As such, the normal distribution formulas (4.11) through (4.13) are not appropriate for making prediction or* tolerance intervals for tongue thicknesses. But if one assumes that the machining process represented by those data is stable, the methods represented in formulas (4.14) through (4.20) can be used.

Consider first the problem of announcing a two-sided prediction interval for a single additional tongue thickness. Reviewing the data of Table 4.1, it is easy to see that for the sample of size $n = 20$ represented there,

$$\min x_i = .1807 \quad \text{and} \quad \max x_i = .1868.$$

So, in view of displays (4.16) and (4.18), the interval with end points

$$.1807 \text{ in} \quad \text{and} \quad .1868 \text{ in}$$

can be used as a prediction interval for a single additional tongue thickness with associated prediction confidence

$$\frac{20 - 1}{20 + 1} = .905 = 90.5\,\%.$$

One can in some sense be 90 % sure that an additional tongue thickness generated by this machining process would be between .1807 in and .1868 in.

As a second way of expressing what the data say about other tongue thicknesses, consider the making of a two-sided tolerance interval for 90 % of all tongue thicknesses. The method represented in displays (4.16) and (4.20) implies that the interval with end points

$$.1807 \text{ in} \quad \text{and} \quad .1868 \text{ in}$$

has associated confidence level

$$1 - (.9)^{20} - 20(1 - .9)(.9)^{19} = .608 = 60.8\,\%.$$

One can be only about 61 % sure that 90 % of all tongue thicknesses generated by the machining processes are between .1807 in and .1868 in.

The prediction and confidence interval methods presented here are a very small fraction of those available. In particular, there are methods specifically crafted for other families of process distributions besides the normal family. The reader is referred to the book *Statistical Intervals: A Guide for Practitioners* by Hahn and Meeker for a more comprehensive treatment of the many available methods, should the ones presented in this section not prove adequate for his or her purposes.

Section 4.3 Exercises

1. Consider again problems 1 from Sects. 4.1 and 4.2 and the data given in Sect. 4.1. Pepper lot production has a stable lot percent moisture content

that is approximately normally distributed. Acceptable lot percent moisture content is between 9 % and 12 %.

  (a) Give limits that you are "95 % sure" will bracket the measured moisture content from the next sampled lot.

  (b) Among the $n = 22$ lots tested, the smallest and largest measured moisture contents are, respectively, 9.3 % and 11.5 %. How "sure" can one be that 95 % of all additional lots will have measured contents between these two values?

  (c) Give limits that you are "95 % sure" will bracket the percent moisture content in 95 % of all pepper lots.

2. Consider again problem 3 of Sect. 4.2 and the data given there. Lengths of steel shafts are measured by a laser gauge that produces a coded voltage proportional to shaft length (above some reference length). In the units produced by the gauge, shafts of a particular type have length specifications 2300 ± 20. Assume the $n = 10$ shafts measured to produce the data in Sect. 4.2 came from a physically stable process.

  (a) How confident are you that an 11th shaft produced by this process would measure at least 2280? (Give a numerical confidence level without relying on a normal distribution assumption.)

  (b) Assuming shaft lengths are normally distributed, give two-sided limits that you are 95 % sure contain at least 99 % of all measured shaft lengths.

  (c) Suppose (that shaft lengths are normally distributed and) you purchase one additional shaft. Construct an interval that will include the length of the one you buy with 95 % confidence.

3. Consider again problem 3 of Sect. 4.1 and problem 5 of Sect. 4.2 and the data given in Sect. 4.1. Recall that a painting process was investigated and determined to be stable and producing an approximately normal distribution of thicknesses.

  (a) Using the data in Sect. 4.1, find limits that you are "95 % sure" contain 99 % of thickness values.

  (b) Suppose your normal probability analysis detected a strong departure from a normal distribution. If you use .2 and 9.7 as limits for an additional measured thickness, what level of confidence would you have?

  (c) Suppose your normal probability analysis had detected a strong departure from a normal distribution. If you use .2 and 9.7 as limits for 90 % of all paint thicknesses, what level of confidence would you have?

4. Consider again problem 4 of Sect. 4.1 and problem 6 of Sect. 4.2 and the data given in Sect. 4.1. Recall that percentages (x) of "small" grit particles in bulk abrasive lots used by a company in the production of a certain type of sandpaper are of interest and that the company has already done extensive process monitoring and has concluded that these are stable with an approximately normal distribution. Specifications are from 13.0 % to 16.6 %.

   (a) Find limits that you are "99 % sure" will bracket the lot percent of "small" grit particles from the next sample.
   (b) Find limits that you are "99 % sure" will bracket the lot percent "small" grit particles from 90 % of all samples.
   (c) Give the respective names for the intervals constructed in (a) and (b).
   (d) Find a limit L such that you are "95 % sure" at least 95 % of all samples will have at least $L\%$ "small" grit particles.

4.4 Probabilistic Tolerancing and Propagation of Error

The methods of the previous three sections have had to do with characterizing the pattern of variation associated with a stable process on the basis of a sample from that process. There are occasions where one needs to predict the pattern of variation associated with a stable system *before* such data are available. (This is quite often the case in engineering *design* contexts, where one must choose between a number of different possible designs for a process or product without having many systems of each type available for testing.)

Where the product or process of interest can be described in terms of a relatively simple equation involving the properties of some components or system inputs, and information is available on variabilities of the components or inputs, it is often possible to do the necessary prediction. This section presents methods for accomplishing this task. There is first a brief discussion of the use of simulations. Then a simple result from probability theory concerning the behavior of linear combinations of random variables is applied to the problem. And finally, a very useful method of approximation is provided for situations where the exact probability result cannot be invoked.

In abstract terms, the problem addressed in this section can be phrased as follows. Given k random system inputs $X, Y, \ldots, Z$, an output of interest U, and the form of a function g giving the exact value of the output in terms of the inputs,

$$U = g(X, Y, \ldots, Z), \tag{4.21}$$

how does one infer properties of the random variable U from properties of $X, Y, \ldots, Z$? For particular joint distributions for the inputs and fairly simple functions g, the methods of multivariate calculus can sometimes be invoked to find formulas for the distribution of U. But problems that yield easily to such an approach are rare, and much more widely applicable methods are needed for engineering practice.

One quite general tool for this problem is that of **probabilistic simulations**. These are easily accomplished using widely available statistical software. What one does is to use (pseudo-)random number generators to produce many (say n) realizations of the vector $(X, Y, \ldots, Z)$. Upon plugging each of these realizations into the function g, one obtains realizations of the random variable U. Properties of the distribution of U can then be inferred in approximate fashion from the empirical distribution of these realizations. This whole program is especially easy to carry out when it is sensible to model the inputs $X, Y, \ldots, Z$ as independent. Then the realizations of the inputs can be generated separately from the k marginal distributions. (The generation of realizations of *dependent* variables $X, Y, \ldots, Z$ is possible but beyond the scope of this discussion.)

Example 51 *Approximating the Distribution of the Resistance of an Assembly of Three Resistors. The "laws" of physics often provide predictions of the behavior of simple physical systems. Consider the schematic of the assembly of three resistors given in Fig. 4.15 on page 222. Elementary laws of physics lead to the prediction that if $R_1, R_2,$ and R_3 are the resistances of the three resistors in the schematic, then the resistance of the assembly will be*

$$R = R_1 + \frac{R_2 R_3}{R_2 + R_3}.$$

Suppose that one is contemplating the use of such assemblies in a mass-produced product. And suppose further that resistors can be purchased so that R_1 has mean 100 Ω and standard deviation 2 Ω and that both R_2 and R_3 have mean 200 Ω and standard deviation 4 Ω. What can one then predict about the variability of the assembly resistance?

Table 4.8 on page 222 holds some simple simulation code and output for the open-source R *statistical package, based on an assumption that the three resistances are independent and normally distributed. Ten thousand simulated values of R have been created. Figure 4.16 on page 223 shows the histogram that results.*

It is evident from Table 4.8 that assembly resistances are predicted to average on the order of 200.0 Ω *and to have a standard deviation on the order of* 2.5 Ω. *And Fig. 4.16 indicates that the distribution of assembly resistances is roughly bell shaped. Of course, rerunning the simulation would produce slightly different results. But for purposes of obtaining a quick, rough-and-ready picture of predicted assembly variability, this technique is very convenient and powerful.*

For the very simplest of functions g relating system inputs to the output U, it is possible to provide expressions for the mean and variance of U in terms of the means and variances of the inputs. Consider the case where g is linear in the

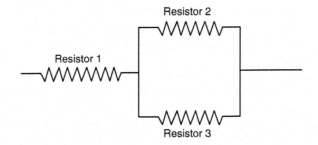

FIGURE 4.15. An assembly of 3 resistors

TABLE 4.8. Code and output for the resistance simulation
```
> R1<-c(rnorm(10000,mean=100,sd=2))
> R2<-c(rnorm(10000,mean=200,sd=4))
> R3<-c(rnorm(10000,mean=200,sd=4))
> R<-R1+(R2*R3/(R2+R3))
> summary(R)
 Min. 1st Qu. Median Mean 3rd Qu. Max.
 191.3 198.3 200.0 200.0 201.6 209.7
> mean(R)
[1] 199.9601
> sd(R)
[1] 2.466333
> hist(R)
```

inputs, that is, for constants $a_0, a_1, \ldots, a_k$, suppose that

$$U = a_0 + a_1 X + a_2 Y + \cdots + a_k Z. \tag{4.22}$$

Then if the variables $X, Y, \ldots, Z$ are independent with respective means $\mu_X, \mu_Y, \ldots, \mu_Z$ and variances $\sigma_X^2, \sigma_Y^2, \ldots, \sigma_Z^2$, U has mean

Mean of a Linear Function of k Random Variables

$$\mu_U = a_0 + a_1 \mu_X + a_2 \mu_Y + \cdots + a_k \mu_Z \tag{4.23}$$

and variance

Variance of a Linear Function of k Independent Random Variables

$$\sigma_U^2 = a_1^2 \sigma_X^2 + a_2^2 \sigma_Y^2 + \cdots + a_k^2 \sigma_Z^2. \tag{4.24}$$

These facts represented in displays (4.22) through (4.24) are not directly applicable to problems like that in Example 51, since there the assembly resistance is not linear in either R_2 or R_3. But they *are* directly relevant to many engineering problems involving geometrical dimensions. That is, often geometrical variables of interest on discrete parts or assemblies of such parts are sums and differences of more fundamental variables. Clearances between shafts and ring bearings are differences between inside diameters of the bearings and the shaft

Chapter 4. Process Characterization and Capability Analysis 223

Histogram of R

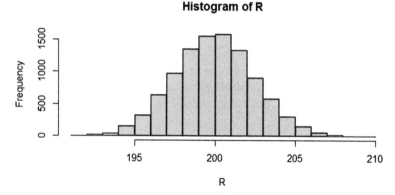

FIGURE 4.16. Histogram for 10,000 simulated assembly resistances, R

diameters. Thicknesses of five-ply sheets of plywood are sums of thicknesses of the individual layers. Tongue thicknesses of steel levers machined on both sides are the original steel bar thicknesses minus the depths of cut on both sides of the bar and so on.

Example 52 *Choosing a Box Size in a Packaging Problem. Miles, Baumhover, and Miller worked with a company that was having a packaging problem. The company bought cardboard boxes nominally 9.5 in in length intended to hold 4 units of a product they produced, stacked side by side in the boxes. They were finding that many boxes were unable to accommodate the full 4 units, and a sensible figure was needed for a new target dimension on the boxes.*

The students measured the thicknesses of 25 units of product. They found that these had $\bar{x} = 2.577$ in and $s = .061$ in. They also measured several of the nominally 9.5-in boxes and found their (inside) lengths to have mean $\bar{x} = 9.556$ in and $s = .053$ in. Consider applying the results (4.23) and (4.24) to the problem of finding a workable new target dimension for the inside length of boxes ordered by this company.

Let X_1, X_2, X_3, and X_4 be the thicknesses of 4 units to be placed in a box and Y be the inside length of the box. Then the clearance or "head space" in the box is

$$U = Y - X_1 - X_2 - X_3 - X_4.$$

This simple relationship is illustrated in Fig. 4.17 on page 224.

Based on the students' measurements, a plausible model for the variables Y, X_1, X_2, X_3, X_4 is one of the independence where Y has mean to be chosen and standard deviation .053 and each of the X variables has mean 2.577 and standard deviation .061. Then, since U is of form (4.22), display (4.24) implies that U has

$$\sigma_U^2 = 1^2 \sigma_Y^2 + (-1)^2 \sigma_{X_1}^2 + (-1)^2 \sigma_{X_2}^2 + (-1)^2 \sigma_{X_3}^2 + (-1)^2 \sigma_{X_4}^2,$$

that is,

$$\sigma_U^2 = (.053)^2 + 4(.061)^2 = .0177,$$

so that

$$\sigma_U = .133 \text{ in}.$$

One might then hope for at least approximate normality of U and reason that if the mean of U were set at $3\sigma_U$, essentially all of the boxes would be able to hold the required 4 units of product (few values of U would be negative). Again remembering that U is of form (4.22), display (4.23) implies that

$$\mu_U = \mu_Y - \mu_{X_1} - \mu_{X_2} - \mu_{X_3} - \mu_{X_4} = \mu_Y - 4(2.577).$$

So setting $\mu_U = 3\sigma_U = 3(.133)$, one has

$$3(.133) = \mu_Y - 4(2.577),$$

that is, one wants

$$\mu_Y = 10.707 \text{ in}.$$

(Then given that nominally 9.5-in boxes were running with mean inside lengths of 9.556 in, it might be possible to order boxes with nominal lengths .056 in below the value for μ_Y found above without creating packing problems.)

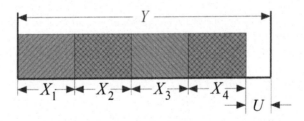

FIGURE 4.17. Schematic for a packaging problem

Example 52 is a very nice example of a real "probabilistic tolerancing" problem. Such problems can be effectively attacked using the relationships (4.23) and (4.24). It is evident from the example that if the standard deviation of the head space (namely, $\sigma_U = .133$ in) is unacceptably large, then either the uniformity of the thicknesses of the units of product or the uniformity of the inside lengths of the boxes will need substantial improvement. (In fact, setting $\sigma_Y = 0$ and recalculating σ_U will show the reader that the potential for variance reduction associated with the box length is small. It is product uniformity that will require attention.)

The simple exact probability result represented in displays (4.22) through (4.24) is not of direct help where U is nonlinear in one or more of $X, Y, \ldots, Z$. But it suggests how one might proceed to develop approximate formulas for the mean and variance of U for even nonlinear g. That is, provided g is smooth near the point $(\mu_X, \mu_Y, \ldots, \mu_Z)$ in k-dimensional space, if the point $(x, y, \ldots, z)$ is not too far from $(\mu_X, \mu_Y, \ldots, \mu_Z)$, a first-order multivariate Taylor expansion of g implies that

$$\begin{aligned}g(x,y,\ldots,z) &\approx g(\mu_X,\mu_Y,\ldots,\mu_Z) + g'_x(\mu_X,\mu_Y,\ldots,\mu_Z)(x-\mu_X) \\ &+ g'_y(\mu_X,\mu_Y,\ldots,\mu_Z)(y-\mu_Y) + \cdots + g'_z(\mu_X,\mu_Y,\ldots,\mu_Z)(z-\mu_Z)\end{aligned} \quad (4.25)$$

where the subscripted g' functions are the partial derivatives of g. Now the function on the right of approximation (4.25) is linear in the variables $x, y, \ldots, z$. So plugging the random variables $X, Y, \ldots, Z$ into approximation (4.25) and then applying the probability result indicated in displays (4.22) through (4.24), one can arrive at approximations for μ_U and σ_U^2. These turn out to be

$$\mu_U \approx g(\mu_X, \mu_Y, \ldots, \mu_Z) \quad (4.26)$$

Approximate Mean of a Function of k Random Variables

and

$$\sigma_U^2 \approx \left(\frac{\partial g}{\partial x}\right)^2 \sigma_X^2 + \left(\frac{\partial g}{\partial y}\right)^2 \sigma_Y^2 + \cdots + \left(\frac{\partial g}{\partial z}\right)^2 \sigma_Z^2, \quad (4.27)$$

Approximate Variance of a Function of k Independent Random Variables

where the partial derivatives indicated in display (4.27) are evaluated at the point $(\mu_X, \mu_Y, \ldots, \mu_Z)$. (The notation for the partial derivatives used in display (4.27) is more compact but less complete than the g' notation used on the right of approximation (4.25). The same partials are involved.) The formulas (4.26) and (4.27) are often called **the propagation of error formulas** in that they provide a simple approximate view of how "error" or variation "propagates" through a function g. And it is also worth noting that the exact result (4.24) for linear g is (upon realizing that $a_1, a_2, \ldots, a_k$ are the partial derivatives of a linear g) essentially a special case of relationship (4.27).

Example 53 *Uncertainty in the Measurement of Viscosity of S.A.E. no. 10 Oil. One technique for measuring the viscosity of a liquid is to place it in a cylindrical container and determine the force needed to turn a cylindrical rotor of nearly the same diameter as the container at a given velocity. If F is the force, D_1 is the diameter of the rotor, L is the length of the rotor, D_2 is the inside diameter of the container, and ν is the velocity at which the rotor surface moves, then the implied viscosity is*

$$\eta = \frac{F(D_2 - D_1)}{\pi \nu D_1 L} = \frac{F}{\pi \nu L}\left(\frac{D_2}{D_1} - 1\right).$$

Suppose that one wishes to measure the viscosity of S.A.E. no. 10 oil and the basic measurement equipment available has precision adequate to provide standard deviations for the variables $F, D_1, D_2, L,$ and ν,

$$\sigma_F = .05\text{N}, \sigma_{D_1} = \sigma_{D_2} = \sigma_L = .05\text{ cm} \quad \text{and} \quad \sigma_\nu = 1\text{ cm/sec}.$$

Further, suppose that approximate values for the quantities F, D_1, D_2, L, and ν are

$$F \approx 151\,\text{N}, D_1 \approx 20.00\,\text{cm}, D_2 \approx 20.50\,\text{cm}, L \approx 20.00\,\text{cm}, \text{ and } \nu \approx 30\,\text{cm/sec}.$$

These approximate values will be used as means for the variables and formula (4.27) employed to find an approximate standard deviation to use in describing the precision with which the viscosity can be determined.

To begin with, the partial derivatives of η with respect to the various measured quantities are

$$\frac{\partial \eta}{\partial F} = \frac{(D_2 - D_1)}{\pi \nu D_1 L}, \quad \frac{\partial \eta}{\partial D_1} = \frac{F}{\pi \nu L}\left(-\frac{D_2}{D_1^2}\right), \quad \frac{\partial \eta}{\partial D_2} = \frac{F}{\pi \nu D_1 L},$$

$$\frac{\partial \eta}{\partial L} = -\frac{F(D_2 - D_1)}{\pi \nu D_1 L^2}, \quad \text{and} \quad \frac{\partial \eta}{\partial \nu} = -\frac{F(D_2 - D_1)}{\pi \nu^2 D_1 L}.$$

And it is straightforward to check that if the approximate values of F, D_1, D_2, L, and ν are plugged into these formulas, then (in the appropriate units)

$$\frac{\partial \eta}{\partial F} = 1.326 \times 10^{-5}, \quad \frac{\partial \eta}{\partial D_1} = -4.106 \times 10^{-3}, \quad \frac{\partial \eta}{\partial D_2} = 4.005 \times 10^{-3},$$

$$\frac{\partial \eta}{\partial L} = -1.001 \times 10^{-4}, \quad \text{and} \quad \frac{\partial \eta}{\partial \nu} = -6.676 \times 10^{-5}.$$

Then from expression (4.27), it is apparent that an approximate variance for η is

$$\sigma_\eta^2 \approx (1.326 \times 10^{-5})^2(.05)^2 + (-4.106 \times 10^{-3})^2(.05)^2 + (4.005 \times 10^{-3})^2(.05)^2 \\ + (-1.001 \times 10^{-4})^2(.05)^2 + (-6.676 \times 10^{-5})^2(1)^2,$$

and doing the arithmetic and taking the square root, one finds

$$\sigma_\eta \approx 2.9 \times 10^{-4}\,\text{N sec/cm}^2.$$

Making use of relationship (4.26), this standard deviation of measurement accompanies a "true" or mean measured viscosity of about

$$\eta = \frac{151(20.50 - 20.00)}{\pi 30(20)(20)} = 20.0 \times 10^{-4}\,\text{N sec/cm}^2.$$

There are several practical points that need to be made about the methods presented here before closing this section. First, remember that formulas (4.27) and (4.27) are only approximations (based on the linearization of g at the point $(\mu_X, \mu_Y, \ldots, \mu_Z)$). It is a good idea to cross-check results one gets using these propagation of error formulas with results of simulations (a different kind of approximation). For example, Table 4.9 holds R code and output for a simulation of 10,000 viscosities and shows substantial agreement with the calculations in Example 53.

A reasonable question is "If one is going to do simulations anyway, why bother to do the hard work to use the propagation of error formulas?" One answer lies in important extra insight into the issue of variance transmission provided by formula (4.27). Formula (4.27) can be thought of as providing a partition of the variance of the output U into separate parts attributable to the various inputs individually. That is, each of the terms on the right side of formula (4.27) is related to a single one of the inputs $X, Y, \ldots, Z$ and can be thought of as that variable's impact on the variation in U. Comparing these can lead to the identification of the biggest source(s) of (unwanted) variability in a system and allow consideration of where engineering resources might best be invested in order to try and reduce the size of σ_U. To get these kinds of insights from simulations would require the comparison of many different simulations using various hypothetical values of the standard deviations of the inputs.

Propagation of Error and Variance Partitioning

Example 54 *Partitioning Variance in Measured Viscosity (Example 53 continued). Returning to the approximation for σ_η^2 and displaying some of the intermediate arithmetic, one has*

$$\left(\frac{\partial \eta}{\partial F}\right)^2 \sigma_F^2 = (1.326 \times 10^{-5})^2 (.05)^2 = 4.4 \times 10^{-13},$$

$$\left(\frac{\partial \eta}{\partial D_1}\right)^2 \sigma_{D_1}^2 = (-4.106 \times 10^{-3})^2 (.05)^2 = 4.21 \times 10^{-8},$$

$$\left(\frac{\partial \eta}{\partial D_2}\right)^2 \sigma_{D_2}^2 = (4.005 \times 10^{-3})^2 (.05)^2 = 4.01 \times 10^{-8},$$

$$\left(\frac{\partial \eta}{\partial L}\right)^2 \sigma_L^2 = (-1.001 \times 10^{-4})^2 (.05)^2 = 2.51 \times 10^{-11},$$

TABLE 4.9. R Code and output for the viscosity simulation

```
> F<-c(rnorm(10000,mean=151,sd=.05))
> D1<-c(rnorm(10000,mean=20,sd=.05))
> D2<-c(rnorm(10000,mean=20.5,sd=.05))
> L<-c(rnorm(10000,mean=20,sd=.05))
> v<-c(rnorm(10000,mean=30,sd=1))
>
> Eta<-F*((D2/D1)-1)/(pi*v*L)
> summary(Eta)
   Min.    1st Qu.    Median     Mean    3rd Qu.     Max.
0.0008432 0.0018080 0.0020040 0.0020060 0.0022020 0.0030430
> mean(Eta)
[1] 0.002006421
> sd(Eta)
[1] 0.000291604
```

and
$$\left(\frac{\partial \eta}{\partial \nu}\right)^2 \sigma_\nu^2 = (-6.676 \times 10^{-5})^2 (1)^2 = 4.46 \times 10^{-9}.$$

It is then evident from these values that the biggest contributors to the variance of η are the two diameter measurements. (Next in order of importance is the velocity measurement, whose contribution is an order of magnitude smaller.) The single most effective method of improving the precision of the viscosity measurement would be to find a more precise method of measuring the diameters of the rotor and cylinder.

It is instructive to note that although the standard deviation of the length measurement is exactly the same as those of the two diameter measurements (namely, 05 cm), the length measurement contribution to σ_η^2 is much smaller than those of the two diameter measurements. This is because the partial derivative of η with respect to L is much smaller than those with respect to D_1 and D_2. That is, the contributions to the overall variance involve not only the variances of the inputs but the "gains" or rates of change of the output with respect to the inputs. This is only sensible. After all, if g is constant with respect to an input variable, whether or not it varies should have no impact on the output variation. On the other hand, if g changes rapidly with respect to an input, any variation in that variable will produce substantial output variation.

A final caution regarding the methods of this section is that one should not expect the impossible from them. They are tools for predicting the pattern of variation *within a particular model*. But even the best of equations we use to describe physical phenomena are only approximations to reality. They typically ignore variables whose effects on a response of interest are "small." They are often good only over limited ranges of the inputs (and of other variables that do not even appear in the equations) and so on. So the kinds of predictions that have been illustrated in this section should be thought of as typically producing *underpredictions* of the variability that will be seen, should one observe a number of realizations of the output variable over a period of time.

Section 4.4 Exercises

1. The force opposing the motion of a block of weight W moving across a flat horizontal surface is $F = kW$ where k is the coefficient of kinetic friction specific to the block and the surface. Suppose that in an application of this simple physical relationship, W has a mean $\mu_W = 10$ lb and standard deviation $\sigma_W = .2$ lb, and k varies with a mean of $\mu_k = .3$ and standard deviation $\sigma_k = .01$.

    (a) Find the expected force opposing the motion of a block.
    (b) Find the standard deviation of force necessary to accelerate the block.

(c) Find "2-sigma" limits on either side of the expected force necessary to accelerate the block.

2. Sheets of book paper have mean thickness of .01 in and standard deviation .0001 in. A particular book will have 200 sheets of this paper in it. (Ignore the cover for this problem.) Let x_i correspond to thickness of the ith page for any one copy of the book.

   (a) Which function correctly models thickness of a book copy: $200x$ or $x_1 + x_2 + \cdots + x_{200}$? Why?
   (b) Find the average book thickness.
   (c) Find the standard deviation of book thickness.
   (d) Within how many inches from the average book thickness should one expect about 95 % of the book copies to be?

3. Consider two holes (A and B) on a mass-produced part. On a coordinate system with origin at the ideal position of hole A, ideal A is at $(0,0)$, and ideal B is at $(5,0)$. Suppose holes A and B are not exactly positioned because of imprecision in the manufacturing process. Hence, actually A is located at (x_1, y_1) and hole B is actually at $(5 + x_2, y_2)$. The variables x_1 and x_2 correspond to horizontal positioning errors, and the variables y_1 and y_2 are vertical errors. Defining $u = x_2 - x_1$ and $v = y_2 - y_1$, the actual distance between the positions of holes A and B is $D = \sqrt{(5+u)^2 + v^2}$.

   (a) A design engineer believes the holes can be drilled independently where x_1 and x_2 have standard deviations $\sigma_{x_1} = \sigma_{x_2} = .01$. What does this person then expect will be the standard deviation of $u = x_2 - x_1$?
   (b) Suppose that u and v can be described as independent variables with 0 means and standard deviations $\sigma_u = \sigma_v = .02$. Find an approximate standard deviation for D.
   (c) $n = 20$ parts were sampled and D measured. For these measurements $\overline{D} = 5.0017$, and sample standard deviation of D was .0437. A normal probability plot suggested a normal model for D, and the hole drilling process was determined to be stable. Give end points of an interval that you are "90 % sure" will contain the distance between hole positions for a single additional part.

4.5 Chapter Summary

When a process is behaving consistently, it makes sense to try to characterize or describe its behavior in quantitative terms. This chapter has discussed methods for this enterprise. Graphical methods including the important tool of normal plotting were discussed first. Then measures of process capability and their estimation were considered in the second section. Prediction and tolerance intervals were presented next, as means of projecting outputs of a stable process based on a sample. Finally, propagation of error and simulations were introduced as tools for engineering design that can sometimes be used to predict variation in system performance from variation in component characteristics.

4.6 Chapter 4 Exercises

1. **Weld Pulls.** The following scenario and data are used by GE aircraft engines as part of a data analysis demonstration. A shop uses spot welding to join two pieces of an assembly. Welds have been failing in the field, and some have been observed to pop apart in a 280° F electrostatic paint oven. Process-monitoring efforts are going to be applied to these welds, and as a part of a preliminary "snapshot" of current process performance, 25 weld strengths are measured for each of two machines. (Specifications on weld strength are that an individual weld button should hold up under a 1100 psi pull without tearing. No upper specification limit is used, but manufacturing personnel believe that "blue welds" with strengths larger than 1800 psi are brittle and difficult to finish because of excessive dimpling.) The strength data are below in psi.

Machine 1					Machine 2				
1368	1129	1020	1157	1531	1187	1862	1821	1713	1887
1022	1195	1288	1220	1792	1110	1376	1871	1315	1498
1313	1764	989	1666	1643	1206	1736	1904	1873	1208
1703	1764	1952	1706	2004	1696	1307	1965	1305	1744
1135	1946	1105	1502	1629	1358	1215	1551	1369	1375

   (a) For the machine 1 weld strengths, make a dot diagram, a stem-and-leaf plot, and a frequency table, where the first category begins with 900 psi and the last category ends with 2100 psi. (Use six categories of equal length.) Make the relative frequency histogram corresponding to your frequency table.

   (b) Redo (a) for machine 2.

   (c) Make back-to-back stem-and-leaf plots for the two machines.

   (d) Make back-to-back relative frequency histograms for the two machines.

   (e) Suppose engineering management decides that since current production goals do not require full use of both welders, all welding will be done with the better of the two machines. (The other machine will not be used.) Which machine should be used? Why?

2. Refer to the **Weld Pull** case in problem 1.

    (a) Find the 25 values of $(i - .5)/25$ for $i = 1, 2, \ldots, 25$ and (by ordering the observations from smallest to largest) find the corresponding quantiles of the machine 1 data.

    (b) Find the .10, .25, .50, and .90 quantiles of the machine 1 data.

    (c) Do you expect your answers in part (b) to be exactly the corresponding quantiles for *all* welds made by machine 1? Why or why not?

    (d) Find the 25 standard normal quantiles $Q_z(\frac{i-.5}{25})$ for $i = 1, 2, \ldots, 25$.

    (e) Use your answers to (a) and (d) to make a normal plot for the machine 1 pull strengths. Does it appear the pull strengths are coming from a normal distribution? Why or why not?

    (f) Apply the natural logarithm (ln) transformation to each of the raw strength quantiles from (a). (The results are quantiles of the distribution of log strengths.)

    (g) Plot standard normal quantiles in (d) versus the log strength quantiles from (f). Does it appear that the log strength distribution is normal? Why or why not?

3. Refer to the **Weld Pull** case of problems 1 and 2.

    (a) Redo problem 2 for machine 2.

    (b) Make side-by-side box plots for the original data from problem 1. (Make one for data from machine 1 and one for data from machine 2.)

    (c) Making use of the facts of the case given in problem 1 and your graph from (b), compare machine 1 and machine 2 weld qualities.

4. **Oil Field Production.** Geologists and engineers from a large oil company considered drilling new wells in a field where 64 wells had previously been drilled. The oil production figures for each of the 64 wells were available for analysis and are given in Table 4.10 (units of the data are 1000 barrels).

TABLE 4.10. Data for problem 4

217.1	43.4	79.5	82.2	56.4	36.6	12.0	12.1*
53.2	69.5	26.9	35.1	49.4	64.9	28.3	20.1*
46.4	156.5	13.2	47.6	44.9	14.8	104.9	30.5*
42.7	34.6	14.7	54.2	34.6	17.6	44.5	7.1*
50.4	37.9	32.9	63.1	92.2	29.1	10.3*	10.1*
97.7	12.9	196.0	69.8	37.0	61.4	37.7*	18.0*
103.1	2.5	24.9	57.4	58.8	38.6	33.7*	3.0*
51.9	31.4	118.2	65.6	21.3	32.5	81.1*	2.0*

Twelve of the wells were "completed" using a different technique than used for the other 52 wells. (The process of completing a well involves stimulation of the rock formation to draw the last "hard to get" oil.) The data values with an "*" correspond to the 12 wells that were completed using the second or alternative method.

Knowledge of the .10, .50, and .90 quantiles of a field's production distribution is very useful to geologists and engineers as they decide whether it is economically feasible to drill again.

(a) Make a stem-and-leaf plot of oil production for the 64 wells. Then make back-to-back stem-and-leaf plots for the two groups of wells with different completion methods. Does it appear there is a difference in the distributions of total production for the two groups? Explain.

(b) Make side-by-side box plots for the two groups of wells with different completion methods. Compare the two distributions based on these plots.

(c) Find and plot the 52 points that make up a normal probability plot for the oil production of the wells completed by the first method. Does it appear that there is any serious departure from the normal distribution shape in these data? Explain.

(d) Find and graph the 12 points that make up a normal probability plot for the oil production of the wells completed by the alternative method. Does it appear there is any serious departure from the normal distribution shape in these data? Explain.

(e) Find and graph the 64 points that make up a normal probability plot for the oil production of the whole set of wells. Does it appear there is any serious departure from the normal distribution shape in these data? Explain.

5. Refer to the **Oil Field Production** case in problem 4.

(a) Find the .10, .50, and .90 quantiles of the standard normal distribution, $Q_z(.1)$, $Q_z(.5)$, and $Q_z(.9)$.

Note that the p quantile of a normal distribution with mean μ and standard deviation σ is $\mu + \sigma Q_z(p)$.

(b) Find the sample mean and sample standard deviation of production for the 52 wells completed by the first method. Use these, a normal distribution assumption, the formula above, and your answer to (a), to estimate the .1, .5, and .9 quantiles of the production distribution for this field (under the first completion method).

(c) Find directly the .10, .50, and .90 quantiles of the production data for the first well completion method (represented by the 52 wells).

(d) Find the sample mean and sample standard deviation of production for the 12 wells completed by the alternative method. Use these, a normal distribution assumption, the formula above, and your answer to (a) to estimate the .1, .5, and .9 quantiles of the production distribution for this field (under the alternative completion method).

(e) Find directly the .10, .50, and .90 quantiles of the production data for the alternative well completion method (represented by the 12 wells).

(f) For the first completion method, which set of estimates do you recommend for the .10, .50, and .90 quantiles, the set from (b) or the one from (c)? Why?

(g) For the alternative completion method, which set of estimates do you recommend for the .10, .50, and .90 quantiles, the set from (d) or the one from (e)? Why?

6. Refer to the **Oil Field Production** case in problems 4 and 5. Apply the natural log (ln) transformation to every data value in problem 4. Redo (b)–(g) in problem 5 for the transformed data (the log productions).

7. Consider the following small ordered data set for the variable x.

$$1, 4, 5, 10, 11$$

(a) Apply the natural logarithm (ln) transformation to each of the data values. Does the order of the values change as the xs are transformed to $\ln(x)$s?

(b) Find the .30, .50, and .70 quantiles of both the x distribution and the $\ln(x)$ distribution.

(c) Let y be the .50 quantile of the $\ln(x)$ distribution. Find $\exp(y)$.

(d) Let w be the .30 quantile of the $\ln(x)$ distribution. Find $\exp(w)$.

(e) What relationship do you see between quantiles of the x distribution and quantiles of the $\ln(x)$ distribution?

8. **Part Hardness.** Measured hardness values for eight heat-treated steel parts produced on a single day were (in units of mm)

$$3.175,\ 3.200,\ 3.100,\ 3.200, 3.150,\ 3.100,\ 3.100,\ \text{and}\ 3.175.$$

(a) What must have been true about the heat-treating process on the day of data collection before an estimate of C_{pk} derived from these data could possibly have any practical relevance or use?

(b) What other quantities (besides raw data values like those above) are needed in order to estimate C_{pk}? Where should these quantities come from?

The analysts that obtained the data above had previously done a gauge R&R study on the hardness measurement process. Their conclusion was that for this method of hardness measurement, the repeatability standard deviation is $\sigma \approx .044$ mm.

(c) Suppose that hardness specifications were $\pm.150$ mm around an ideal value. Is the gauging method used here adequate to check conformance to such specifications? Explain. (Even if there were no reproducibility variance, would it be adequate?)

(d) Compare the sample standard deviation of the eight measurements above to the estimate of σ developed in the gauge R&R study. Based on this comparison, do you see clear evidence of any real hardness differences in the eight parts?

(e) Formula (4.8) involves μ, σ, U, and L. If these parameters refer to "true" (as opposed to values measured with error) hardnesses for parts, how optimistic are you about getting a sensible estimate of C_{pk} using the data in this problem? Explain.

(f) If one uses the hardness values from this problem to estimate "σ" and set up a monitoring scheme for part hardness (say a chart for individuals), will your scheme allow for any "natural manufacturing variability" as part of all-OK conditions? Explain in light of your answer to (d). Do you see this as necessarily good or bad if the eight parts used in this study fairly represent an entire day's production of such parts?

9. Refer to the **Oil Field Production** case in problems 4, 5, and 6.

(a) In problems 6b) and 6c), you were asked to find estimates of population .10, .50, and .90 quantiles of the log production distribution (for the first completion method) using two different methods. Exponentiate those estimates (i.e., plug them into the $\exp(\cdot)$ function). These exponentiated values can be treated as estimates of quantiles of the raw production distribution. Compare them to the values obtained in problems 5b) and 5c).

(b) In problems 6d) and 6e), you were asked to find estimates of population .10, .50, and .90 quantiles of the log production distribution (for the alternative completion method) using two different methods. Exponentiate those estimates (i.e., plug them into the $\exp(\cdot)$ function). These exponentiated values can be treated as estimates of quantiles of the raw production distribution. Compare them to the values obtained in problems 5d) and 5e).

(c) In all, you should have three different sets of estimates in part (a). Which set do you find to be most credible? Why?

(d) In all, you should have three different sets of estimates in part (b). Which set do you find to be most credible? Why?

10. Refer to the **Oil Field Production** case in problems 4 and 6. Consider the 52 wells completed using the first method, and assume that oil production from wells like those on the field in question is approximately normally distributed.

   (a) Find a two-sided interval that has a 95 % chance of containing oil production values from 95 % of all wells like the 52.

   (b) Find a two-sided interval that has a 95 % chance of containing the total oil production of the next well drilled in the same manner as the 52.

   (c) Find a lower bound, B, such that one can be 99 % sure that 90 % of all wells drilled in the same manner as the 52 will produce B or more barrels of oil.

11. Refer to the **Oil Field Production** case in problems 4, 6, and 10. Redo (a)–(c) in problem 10 assuming that the logarithm of total oil production for a well like the 52 is normally distributed. (Do the computations called for in (a) through (c) on the log production values, and then exponentiate your results to get limits for raw production values.)

12. Refer to problems 10 and 11. Which of the two sets of intervals produced in problems 10 and 11 do you find to be more credible? Why?

13. Refer to the **Oil Field Production** case in problems 10 and 11. Redo (a)–(c) in problem 10 and then problem 11 for wells like the 12 completed using the alternative method. Which set of intervals (those derived from the original data or from the transformed data) do you find to be more credible? Why?

14. Refer to the **Oil Field Production** case in problems 10, 11, and 13. One important feature of the real problem has been ignored in making projections of the sort in problems 10, 11, and 13. What is this? (Hint: Is the "independent random draws from a fixed population" model necessarily sensible here? When using a straw to drink liquid from a glass, how do the first few "draws" compare to the last few in terms of the amount of liquid obtained per draw?)

15. **Sheet Metal Saddle Diameters.** Eversden, Krouse, and Compton investigated the fabrication of some sheet metal saddles. These are rectangular pieces of sheet metal that have been rolled into a "half tube" for placement under an insulated pipe to support the pipe without crushing the insulation. Three saddle sizes were investigated. Nominal diameters were 3.0 in, 4.0 in, and 6.5 in. Twenty saddles of each size were sampled from production. Measured diameters (in inches) are in Table 4.11.

   (a) Make side-by-side box plots for the data from the three saddle sizes.

   (b) Make a frequency table for each of the three saddle sizes. (Use categories of equal length. For the 3-in saddle data, let 2.95 in be the lower limit of the first category and 3.20 in be the upper limit of the

TABLE 4.11. Data for problem 15

3- in saddles				4- in saddles			
3.000	3.031	2.969	3.063	4.625	4.250	4.250	4.313
3.125	3.000	3.000	3.125	4.313	4.313	4.313	4.250
3.000	3.063	3.000	3.063	4.313	4.313	4.125	4.063
3.031	2.969	2.969	3.094	4.094	4.125	4.094	4.125
3.063	3.188	3.031	2.969	4.156	4.156	4.156	4.125

6.5- in saddles			
6.625	6.688	6.406	6.438
6.469	6.469	6.438	6.375
6.500	6.469	6.375	6.375
6.469	6.406	6.375	6.469
6.469	6.438	6.500	6.563

last category, and employ five categories. For the 4- in saddle data, let 4.05 in be the lower limit of the first category and 4.65 in be the upper limit of the last category, and use five categories. For the 6.5- in saddle data, let 6.35 in be the lower limit of the first category and 6.70 in be the upper limit of the last category and use five categories.)

(c) Make the relative frequency histograms corresponding to the three frequency tables requested in (b).

(d) Supposing specification limits are $\pm.20$ in around the respective nominal diameters, draw in the corresponding specifications on the three relative frequency histograms made in (c). What do you conclude about the saddle-making process with respect to meeting these specifications?

(e) Make a quantile plot for each saddle size.

(f) Find the .25, .50, and .75 quantiles for data from each saddle size. Give the corresponding IQR values.

16. Refer to the **Saddle diameters** case in problem 15.

    (a) Find the 20 values $(i - .5)/20$ for $i = 1, 2, \ldots, 20$, and (by ordering values in each of the data sets from smallest to largest) find the corresponding quantiles of the three data sets in problem 15.

    (b) Find the standard normal quantiles $Q_z(\frac{i-.5}{20})$ for $i = 1, 2, \ldots, 20$. Then for each of the three data sets from problem 15, plot the standard normal quantiles versus corresponding diameter quantiles.

    (c) Draw in straight lines summarizing your graphs from (b). Note that $Q_z(.25) \approx -.67$, $Q_z(.50) = 0$, and $Q_z(.75) \approx .67$, and from your lines drawn on the plots, read off diameters corresponding to these standard normal quantiles. These values can function as additional estimates of quantiles of the diameter populations. Give the corresponding estimated IQRs.

(d) Suppose the object is to make inferences about quantiles of all saddles of a given size. For each saddle size, which set of estimates do you find to be more credible, those from problem 15f) or those from (c) above? Why?

17. Refer to the **Saddle diameters** case in problem 15. Assume diameters for each saddle size are approximately normally distributed.

(a) For each saddle size, find a two-sided interval that you are 95 % sure will include 90 % of all saddle diameters.

(b) For each saddle size, find a two-sided interval that you are 95 % sure will contain the diameter of the next saddle of that type fabricated.

(c) For each saddle size, find a numerical value so that you are 99 % sure that 90 % of all saddles of that type have diameters at least as big as your value.

(d) For each saddle size, find a numerical value so that you are 95 % sure the next saddle of that type will have a diameter at least as big as your value.

18. Refer to the **Saddle diameters** case in problems 15 and 17. Consider first (for the nominally 3- in saddles) the use of the two-sided statistical interval

$$(\min x_i, \max x_i).$$

(a) Thought of as an interval hopefully containing 90 % of all diameters (of nominally 3- in saddles), what confidence should be attached to this interval?

(b) Thought of as an interval hopefully containing the next diameter (of a nominally 3- in saddle), what confidence should be attached to this interval?

Consider next (for the nominally 3- in saddles) the use of the one-sided statistical interval:

$$(\min x_i, \infty).$$

(c) How sure are you that 90 % of all diameters (of nominally 3- in saddles) lie in this interval?

(d) How sure are you that the next diameter (of a nominally 3- in saddle) will lie in this interval?

19. Refer to problems 17 and 18.

(a) The intervals requested in 17 and 18 are based on a mathematical model of "independent random draws from a fixed universe." Such a model makes sense if the saddle-forming process is physically stable. Suppose the data in problem 17 are in fact listed in the order of fabrication (read left to right and then top to bottom for each size) of the saddles. Investigate (using a retrospective Shewhart X chart with σ estimated by $\overline{MR}/1.128$ as in Sect. 3.2.3) the stability of the fabrication process (one saddle size at a time). Does your analysis indicate any problems with the relevance of the basic "stable process" model assumptions? Explain.

(b) The intervals requested in problem 17 are based on a normal distribution model, while the ones in problem 18 are not. Which set of intervals seems most appropriate for the nominally 3- in saddles? Why?

20. **Case hardening in Red Oak.** Kongable, McCubbin, and Ray worked with a sawmill on evaluating and improving the quality of kiln-dried wood for the use in furniture, cabinetry, flooring, and trim. Case hardening (one of the problems that arises in drying wood) was the focus of their work. Free water evaporates first during the wood-drying process until about 30 % (by weight) of the wood is water. Bound water then begins to leave cells, and the wood begins to shrink as it dries. Stresses produced by shrinkage result in case hardening. When case-hardened wood is cut, some stresses are eliminated, and the wood becomes distorted. To test for case hardening of a board, prongs are cut in the board, and a comparison of the distances between the prongs before and after cutting reveals the degree of case hardening. A decrease indicates case hardening, no change is ideal, and an increase indicates reverse casehardening. The engineers sampled 15 dried 1- in red oak boards, and cut out prongs from each board. Distances between the prongs were measured and are recorded in Table 4.12. (Units are inches.)

TABLE 4.12. Data for problem 20

Board	Before	After	Board	Before	After
1	.80	.58	9	.80	.64
2	.79	.29	10	.79	.27
3	.77	.50	11	.79	.79
4	.79	.77	12	.80	.80
5	.90	.55	13	.79	.41
6	.77	.36	14	.80	.72
7	.90	.57	15	.79	.37
8	.80	.55			

(a) The data here are most appropriately thought of as which of the following: (i) two samples of $n = 15$ univariate data points or (ii) a single sample of $n = 15$ bivariate data points? Explain.

(b) One might wish to look at either the "before" or "after" width distribution by itself. Make dot plots useful for doing this.

(c) As an alternative to the dot plots of part (b), make two box plots.

(d) Make normal probability plots for assessing how bell shaped the "before" and "after" distributions appear to be. Comment on the appearance of these plots.

A natural means of reducing the before-and-after measurements to a single univariate data set is to take differences (say *after* − *before*).

(e) Suppose that the intention was to mark all sets of prongs to be cut at a distance of .80 in apart. Argue that despite this good intention, the *after* − *before* differences are probably a more effective measure of case hardening than the "after" measurements alone (or the "after" measurements minus .80). Compute these 15 values.

(f) Make a dot plot and a box plot for the differences. Mark on these the ideal difference. Do these plots indicate that case hardening has taken place?

(g) Make a normal probability plot for the differences. Does this measure of case hardening appear to follow a normal distribution for this batch of red oak boards? Explain.

(h) If there were no case hardening, what horizontal intercept would you expect to see for a line summarizing your plot in (g)? Explain.

(i) When one treats the answers in parts (a) through (h) above as characterizations of "case hardening," one is really making an implicit assumption that there are no unrecognized assignable/nonrandom causes at work. (For example, one is tacitly assuming that the boards represented in the data came from a single lot that was processed in a consistent fashion, etc.) In particular, there is an implicit assumption that there were no important time-order-related effects. Investigate the reasonableness of this assumption (using a retrospective Shewhart X chart with σ estimated by $\overline{MR}/1.128$ as in Sect. 3.2.3) supposing that the board numbers given in the table indicate order of sawing and other processing. Comment on the practical implications of your analysis.

21. Refer to the **Case hardening** case in problem 20. Assume both the "after" values, and the *after* − *before* differences are approximately normal.

    (a) Find a two-sided interval that you are 99 % sure will contain the next "after" measurement for such a board.

    (b) Find a two-sided interval that you are 95 % sure contains 95 % of all "after" measurements.

    (c) Find a two-sided interval that you are 95 % sure will contain the next difference for such a board.

(d) Find a two-sided interval that you are 99 % sure contains 99 % of all differences for such boards.

22. Refer to the **Case hardening** case in problem 20 and in particular to the $after - before$ differences, x. Consider the use of the two-sided statistical interval
$$(\min x_i, \max x_i).$$

(a) Thought of as an interval hopefully containing the next difference, what confidence should be attached to this interval?

(b) Thought of as an interval hopefully containing 95 % of all differences, what confidence should be attached to this interval?

(c) Thought of as an interval hopefully containing 99 % of all differences, what confidence should be attached to this interval?

23. Refer to the **Case hardening** case in problems 20, 21, and 22 and in particular to the $after-before$ differences. The prediction interval in problem 21c) relies on a normal distribution assumption for the differences, while the corresponding interval in problem 22a) does not. Similarly, the tolerance intervals in problems 21b) and 21d) rely on a normal distribution assumption for the differences, while the corresponding intervals in problems 22b) and 22c) do not.

(a) Which set of intervals (the one in problem 21 or the one in problem 22) is most appropriate? Why?

(b) Compare the intervals (based on the differences) found in problem 21 with those from problem 22 in terms of lengths and associated confidence levels. When a normal distribution model is appropriate, what does its use provide in terms of the practical effectiveness of statistical intervals?

(c) If one were to conclude that differences cannot be modeled as normal and were to find the results of problem 22 to be of little practical use, what other possibility remains open for finding prediction and tolerance intervals based on the differences?

24. **Bridgeport Numerically Controlled Milling Machine.** Field, Lorei, Micklavzina, and Stewart studied the performance of a Bridgeport numerically controlled milling machine. Positioning accuracy was of special concern. Published specifications for both x and y components of positioning accuracy were "$\pm.001$ in" One of the main problems affecting positioning accuracy is "backlash." (Backlash is the inherent play that a machine has when it stops movement or reverses direction of travel in a given plane.) The group conducted an experiment aimed at studying the effects of backlash.

A series of holes was reamed, moving the machine's head in the x direction only. Then a series of holes was reamed, moving the machine's head in the y direction only. (The material used was a 1/4-in-thick acrylic plate. It was first spot faced with a 3/8-in drill bit to start and position the hole. The hole was then drilled with a regular 15/64-in twist drill. A 1/4-in reamer was finally used to improve hole size and surface finish. This machining was all done at 1500 rpm and a feed rate of 4 in/min.) The target x distance between successive holes in the first set was 1.25 in, while the target y distance between successive holes in the second set was .75 in Table 4.13 contains the measured distances between holes (in inches).

It is not completely obvious how to interpret positioning accuracy specifications like the "*nominal* $\pm$ *.001 in*" ones referred to above. But it is perhaps sensible to apply them to distances between successive holes like the ones made by the students.

TABLE 4.13. Data for problem 24

x Movement					y Movement				
1.2495	1.2485	1.2505	1.2495	1.2505	.7485	.7490	.7505	.7500	.7480
1.2490	1.2495	1.2505	1.2500	1.2505	.7515	.7495	.7500	.7490	.7505
1.2495	1.2525	1.2500	1.2520	1.2505	.7490	.7525	.7500	.7490	.7490
1.2485	1.2510	1.2480	1.2480	1.2495	.7510	.7485	.7490	.7485	.7495
1.2505	1.2490	1.2505	1.2500	1.2505	.7480	.7500	.7500	.7485	.7490
1.2500	1.2515	1.2500	1.2495	1.2495	.7500	.7490	.7500	.7510	.7505

(a) Make a box plot for the x distances. Indicate on the plot the ideal value of 1.250 in and the specifications $L = 1.249$ in and $U = 1.251$ in.

(b) Make a box plot for the y distances. Indicate on the plot the ideal value of .750 in and the specifications $L = .749$ in and $U = .751$ in.

(c) Is there any clear indication in your box plots from (a) and (b) that x positioning is better than y positioning (or vice versa)? Explain.

(d) Make a normal plot for the x distances. Is it sensible to treat x distances as normally distributed? Why?

(e) From the plot in (d), estimate the mean and the standard deviation of x distances. Explain how you got your answer.

(f) Make a normal plot for the y distances. Is it sensible to treat y distances as normally distributed? Why?

(g) From the plot in (f), estimate the mean and the standard deviation of y distances. Explain how you got your answer.

25. Refer to the **Bridgeport Numerically Controlled Milling Machine** case in problem 24. Assume both the x and y distances are approximately normally distributed.

(a) Make a two-sided interval that you are 95 % sure contains 99 % of x distances between such reamed holes (nominally spaced 1.250 in apart on a horizontal line).

(b) Does your interval in (a) "sit within" the specification limits for such x distances? Is this circumstance appealing? Why or why not?

(c) Make a two-sided interval that you are 99 % sure contains 95 % of y distances between such reamed holes (nominally spaced .750 in apart on a vertical line).

(d) Does your interval in (c) "sit within" the specification limits for such y distances? Is this circumstance appealing? Why or why not?

(e) In light of the results of problem 24d) and problem 24f), are the intervals made in (a) and (c) above credible? Explain.

26. **Hose Cleaning.** Delucca, Rahmani, Swanson, and Weiskircher studied a process used to ensure that some industrial hoses are free of debris. Specifications were that the inside surfaces of these were to carry no more than 44 mg of contaminant per square meter of hose surface. (The hoses are cleaned by blowing air through them at high pressure.)

Periodically, five of these hoses are tested by rinsing them with trichloroethylene, filtering the liquid, and recovering solids washed out in the cleaning fluid. The data in Table 4.14 are milligrams of solids per m^2 of inside hose surface from 13 such tests.

This scenario and data set have a number of interesting features. For one, specifications here are inherently one-sided, so that two-sided capability measures like C_p and C_{pk} do not make much sense in this context. For another, it is obvious from a plot of the observed contamination levels against sample number, or from a plot of sample standard deviations against sample means, that the variability in measured contamination level increases with mean contamination level.

Consider first the matter of clear dependence of standard deviation on mean. In a circumstance like this (and particularly where data range over several orders of magnitude), it is often helpful to conduct an analysis not on the raw data scale but on a logarithmic scale instead. Notice that on a logarithmic scale, the upper specification for solids washed out in a test is $\ln(44) = 3.78$ ($\ln \mathrm{mg/m^2}$).

(a) Replace the raw data by their natural logarithms. Then compute 13 subgroup means and standard deviations.

(b) Make retrospective $\bar{x}$ and s charts based on the transformed data. Working on the log scale, is there evidence of process instability in either of these charts? Would an analysis on the original scale of measurement look any more favorable (in terms of process stability)?

TABLE 4.14. Data for problem 26

Sample	Contamination levels
1	45.00, 47.77, 145.43, 31.84, 45.01
2	24.00, 33.37, 22.87, 27.89, 21.46
3	13.50, 17.75, 13.34, 9.87, 15.42
4	3.0, 5.21, 1.82, 6.97, 11.13
5	8.62, 4.91, 18.42, 6.16, 6.58
6	231.02, 440.32, 136.24, 379.77, 171.78
7	257.00, 207.18, 240.09, 213.93, 389.62
8	107.57, 101.40, 133.49, 141.50, 92.56
9	51.00, 47.72, 59.45, 53.75, 46.51
10	85.00, 58.40, 52.30, 60.50, 46.84
11	44.00, 45.66, 83.30, 47.31, 66.13
12	88.37, 44.35, 35.65, 146.78, 37.50
13	59.30, 55.67, 62.52, 33.66, 34.96

(c) In light of your answer to (b), explain why it does not make sense to try to state process capabilities based on the data presented here.

Suppose that after some process analysis and improvements in the way hoses are cleaned, control charts for logarithms of measured contaminations show no signs of process instability. Suppose further, that then combining ten samples of five measured log contamination rates to produce a single sample of size $n = 50$, one finds $\bar{x} = 3.05$ and $s = 2.10$ for (logged) contamination rates and a normal model to be a good description of the data.

(d) Find an interval that (assuming continued process stability) you are 95 % sure will include log contamination levels for 95 % of all future tests of this type. Transform this interval into one for contamination levels measured on the original scale.

(e) Find an interval that (assuming continued process stability) you are 99 % sure will contain the next measured log contamination level. Transform this interval into one for the next contamination level measured on the original scale.

(f) Give a 95 % two-sided confidence interval for the process capability, 6σ, measured on the log scale.

Although C_p and C_{pk} are not relevant in problems involving one-sided specifications, there are related capability indices that can be applied. In cases where there is only an upper engineering specification U, the measure

$$CPU = \frac{U - \mu}{3\sigma}$$

can be used, and in cases where there is only a lower engineering specification L, there is the corresponding measure

$$CPL = \frac{\mu - L}{3\sigma}.$$

As it turns out, the formula (4.10) used to make lower confidence bounds for C_{pk} can be applied to the estimation of CPU or CPL (after replacing $\widehat{C_{pk}}$ by $\widehat{CPU} = \dfrac{U - \bar{x}}{3s}$ or by $\widehat{CPL} = \dfrac{\bar{x} - L}{3s}$) as well.

(g) Find and interpret the estimate of CPU corresponding to the description above of log contamination levels obtained after process improvement.

(h) Give a 95 % lower confidence bound for CPU.

(i) Suppose management raises the upper specification limit to 54 mg/m^2. How does this change your answers to (g) and (h)?

(j) When comparing estimated capability ratios from different time periods, what does part (i) show must be true in order to allow a sensible comparison?

27. **Drilling Depths.** Deford, Downey, Hahn, and Larsen measured drill depths in a particular type of pump housing. Specifications on the depth for each hole measured were $1.29 \pm .01$ in. Depth measurements for 24 holes were taken and recorded in the order of drilling in Table 4.15.

(a) Find the average moving range for these data (as in Sect. 3.2.3) and estimate σ by $\overline{MR}/1.128$. Use this estimate of the process short-term variability to make a retrospective "3-sigma" Shewhart X chart for these data. Is there evidence of drilling process instability on this chart?

TABLE 4.15. Data for problem 27

Hole	Measured depth	Hole	Measured depth
1	1.292	13	1.292
2	1.291	14	1.290
3	1.291	15	1.292
4	1.291	16	1.291
5	1.291	17	1.290
6	1.291	18	1.291
7	1.290	19	1.290
8	1.291	20	1.291
9	1.292	21	1.290
10	1.291	22	1.290
11	1.291	23	1.291
12	1.290	24	1.291

(b) In light of part (a), it perhaps makes sense to treat the measured depths as a single sample of size $n = 24$ from a stable process. Do so and give 90 % lower confidence bounds for both C_p and C_{pk}.

(c) Give a 95 % two-sided confidence interval for the process capability 6σ.

(d) Find an interval that you are 95 % sure contains 90 % of all depth measurements for holes drilled like these.

(e) The validity of all the estimates in this problem depends upon the appropriateness of a normal distribution model for measured hole depth. The fact that the gauging seems to have been fairly crude relative to the amount of variation seen in the hole depths (only three different depths were ever observed) makes a completely satisfactory investigation of the reasonableness of this assumption impossible. However, do the best you can in assessing whether normality seems plausible. Make a normal probability plot for the data and discuss how linear the plot looks (making allowance for the fact that the gauging is relatively crude).

28. Refer to the **Journal Diameters** case of problem 16 in the Chap. 3 exercises. Suppose that after some attention to control charts and process behavior, engineers are able to bring the grinding process to physical stability. Further, suppose that a sample of $n = 30$ diameters then has sample mean $\bar{x} = 44.97938$, sample standard deviation $s = .00240$, and a fairly linear normal probability plot. In the following, use the mid-specification of 44.9825 as a target diameter.

(a) Give 95 % lower confidence bounds for both C_p and C_{pk}.

(b) Which estimated index from (a) reflects potential process performance? Which one summarizes current performance? Explain.

(c) If (say because of heavy external pressure to show "quality improvement," real or illusory) the lower specification for these diameters was arbitrarily lowered and the upper specification was arbitrarily raised, what would happen to C_p and C_{pk}?

(d) In light of your answer to part (c), what must be clarified when one is presenting (or interpreting someone else's presentation of) a series of estimated C_p's or C_{pk}'s?

(e) Find a two-sided interval that you are 99 % sure will include the next measured journal diameter from the process described in problem 16 of Chap. 3. What mathematical assumptions support the making of this interval?

(f) Find a two-sided interval that you are 95 % sure contains 99 % of all measured journal diameters.

29. Refer to the **U-bolt Threads** case of problem 20 in the Chap. 3 exercises.

    (a) Make retrospective $\bar{x}$ and R control charts for the thread-length data. Is there evidence of process instability in these means and ranges?

    (b) In light of part (a), it makes sense to treat the $5 \times 18 = 90$ data points in problem 20 of Chap. 3 as a single sample from a stable process. You may check that the grand sample mean for these data is $\bar{x} = 11.13$ and the grand sample standard deviation is $s = 1.97$. (The units are .001 in above nominal length.) Compare this latter figure to $\bar{R}/d_2$, and note that as expected there is substantial agreement between these figures.

    (c) Specifications for thread lengths are given in problem 20h) of Chap. 3. Use these and $\bar{x}$ and s from part (b) and give estimates of C_p and C_{pk}.

    (d) Find a 95 % lower confidence bound for C_{pk}.

    (e) Find a 90 % two-sided confidence interval for C_p. What mathematical model assumptions support the making of this interval?

    (f) Which of the two quantities, C_p or C_{pk}, is a better indicator of the actual performance of the U-bolt thread-making process? Why?

    (g) Find a 99 % two-sided confidence interval for the process short-term standard deviation, σ.

    (h) Find a two-sided interval that you are 99 % sure contains 95 % of all thread lengths for bolts from this process.

    (i) What mathematical model assumptions support the making of the inferences in parts (d), (e), (g), and (h)? How would you check to see if those are sensible in the present situation?

    (j) Use a statistical package to help you make the check you suggested in part (i), and comment on your result.

30. An engineer plans to perform two different tests on a disk drive. Let X be the time needed to perform test 1 and Y be the time required to perform test 2. Further, let μ_X and μ_Y be the respective means of these random variables and σ_X and σ_Y be the standard deviations. Of special interest is the total time required to make both tests, $X + Y$.

    (a) What is the mean of $X + Y$, $\mu_{X+Y} = E(X + Y)$ (in terms of μ_X and μ_Y)?

    (b) What is required in terms of model assumptions in order to go from σ_X and σ_Y to a standard deviation for $X + Y$, $\sigma_{X+Y} = \sqrt{\text{Var}(X + Y)}$?

    (c) Make the assumption alluded to in (b), and express σ_{X+Y} in terms of σ_X and σ_Y.

Suppose that in order to set work standards for testing disk drives of a certain model, the engineer runs both test 1 and test 2 on n such drives. Test 1 time requirements have sample mean $\bar{x}$ and sample standard deviation s_x, while test 2 time requirements have sample mean $\bar{y}$ and sample standard deviation s_y.

(d) How would you use the information from the n tests to estimate μ_{X+Y}? (Give a formula.)

(e) How would you use the information from the n tests to estimate the function of σ_X and σ_Y that you gave as an answer in part (c)? (Give a formula.)

(f) Suppose that in the engineer's data, there is a fairly large (positive) sample correlation between the observed values of X and Y. (For example, drives requiring an unusually long time to run test 1 also tended to require an unusually long time to run test 2.) Explain why you should then not expect your answer to (e) to serve as a sensible estimate of σ_{X+Y}. In this circumstance, if one has access to the raw data, what is a safer way to estimate σ_{X+Y}? (Hint: Suppose you can reconstruct the n sums $x + y$.)

(g) Would the possibility alluded to in part (f) invalidate your answer to part (d)? Explain.

31. Let X be the inside length, Y be the inside width, and Z be the inside height of a box. Suppose $\mu_X = 20, \mu_Y = 15$, and $\mu_Z = 12$, while $\sigma_X = .5, \sigma_Y = .25$, and $\sigma_Z = .3$. (All units are inches.) Assume inside height, inside width, and inside length are "unrelated" to one another.

(a) Find the mean area of the inside bottom of the box.

(b) Approximate the standard deviation of the area of the inside bottom of the box.

(c) Find the mean inside volume of the box.

(d) Approximate the standard deviation of the inside volume of the box.

32. **Impedance in a Micro-Circuit.** In their article "Robust Design Through Optimization Techniques," which appeared in *Quality Engineering* in 1994, Lawson and Madrigal modeled impedance (Z) in a thin film redistribution layer as

$$Z = \left(\frac{87.0}{\sqrt{\epsilon + 1.41}} \right) \ln \left(\frac{5.98A}{.80B + C} \right)$$

where A is an insulator thickness, B is a line width, C is a line height, and ϵ is the insulator dielectric constant. ϵ is taken to be known as 3.10, while A, B, and C are treated as random variables (as they would be in the manufacture of such circuits) with means $\mu_A = 25, \mu_B = 15$, and $\mu_C = 5$ and standard deviations $\sigma_A = .3333, \sigma_B = .2222$, and $\sigma_C = .1111$. (Units for A, B, and C are 10^{-6} in. The article does not give the units of Z.)

(a) Find an approximate mean and standard deviation of impedance for this type of device. Assume A, B, and C are independent.

(b) Variation in which of the variables is likely to be the largest contributor to variation in impedance of manufactured devices of this type? Explain.

33. **TV Electron Guns.** In their article, "Design Evaluation for Reduction in Performance Variation of TV Electron Guns," which appeared in *Quality Engineering* in 1992, Ranganathan, Chowdhury, and Seksaria reported on their efforts to improve consistency of cutoff voltages for TV electron guns. Cutoff voltage (Y) can be modeled in terms of several geometric properties of an electron gun as

$$Y = \frac{KA^3}{(B+D)(C-D)}$$

where K is a known constant of proportionality, A is the diameter of an aperture in the gun's first grid, B is the distance between that grid and the cathode, C is the distance between that grid and a second grid, and D is a measure of lack of flatness of the first grid. The quality team wished to learn how variation in A, B, C, and D was ultimately reflected in variation in Y. (In particular, it was of interest to know the relative importance of the lack of flatness variable. At the beginning of the study, an expensive selective assembly process was being used to try to compensate for problems with consistency in this variable.)

(a) Find an expression for an approximate standard deviation of cutoff voltage assuming A, B, C, and D are independent.

(b) In a manufacturing context such as the present one, which of the terms in an expression like you produced for part (a) represent target values set in product design, and which represent item-to-item manufacturing variation? How might realistic values for these latter terms be obtained?

The following means and standard deviations are consistent with the description of the case given in the article. (Presumably for reasons of corporate security, complete details are not given in the paper, but these values are consistent with everything that is said there. In particular, no units for the dimensions A through D are given.)

Dimension	Mean	Standard deviation
A	.70871	.00443
B	.26056	.00832
C	.20845	.02826
D	0	.00865

(c) Using the means and standard deviations given above and your answer to (a), approximate the fractions of variance in cutoff voltage attributable to variability in the individual dimensions A through D.

(d) In light of your answer to (c), does it seem that the company's expensive countermeasures to deal with variation in lack of flatness in the first grid are justified? Where should their variance-reduction efforts be focused next? Explain.

(e) What advantages does the method of this problem have over first measuring cutoff voltage and then dissecting a number of complete electron guns, in an effort to see which dimensions most seriously affect the voltage?

34. The heat conductivity of a circular cylindrical bar of diameter D and length L, connected between two constant temperature devices of temperatures T_1 and T_2, that conducts C calories in τ seconds is

$$\lambda = \frac{4CL}{\pi(T_1 - T_2)\tau D^2}.$$

In a particular laboratory determination of λ for brass, the quantities C, L, T_1, T_2, τ, and D can be measured with means and standard deviations approximately as follows.

Variable	Mean	Standard Deviation	$\frac{\partial \lambda}{\partial \text{variable}}$
C	240 cal	10 cal	.000825
L	100 cm	.1 cm	.199
T_1	100° C	1° C	−.00199
T_2	0° C	1° C	.00199
τ	600 sec	1 sec	.000332
D	1.6 cm	.1 cm	−.249

(The units of the partial derivatives are the units of λ, namely, cal/(cm)(sec)(° C) divided by the units of the variable in question.)

(a) Find an approximate standard deviation for a realized heat conductivity value from a single determination.

(b) In this experimental setup, which of the variables do you expect to contribute most to the variation in experimentally determined heat conductivity values? Explain.

35. An output voltage V_{out} in an audio circuit is a function of an input voltage V_{in} and gains N and K supplied, respectively, by a transformer and an amplifier via the relationship

$$V_{\text{out}} = V_{\text{in}} N K.$$

Many such circuits are to be built and used in circumstances where V_{in} may be somewhat variable.

(a) Considering manufacturing variability in N and K and variation in V_{out} attributable to environmental factors, how would you predict the mean of V_{out} without testing any of these complete circuits? How would you predict the amount of variability to be seen in V_{out} without testing any of these complete circuits?

(b) What assumption might you add to your answers to (a) in order to predict the fraction of output voltages from these circuits that will fall in the range 99 V to 100 V?

(c) Why might it be "not so smart" to rely entirely on calculations like those alluded to in (a) and (b) to the complete exclusion of product testing?

CHAPTER 5

EXPERIMENTAL DESIGN AND ANALYSIS FOR PROCESS IMPROVEMENT PART 1: BASICS

The first four chapters of this book provide tools for bringing a process to physical stability and then characterizing its behavior. The question of what to do if the resulting picture of the process is not to one's liking remains. This chapter and the next present tools for addressing this issue. That is, Chaps. 5 and 6 concern statistical methods that support intelligent process experimentation and can provide guidance in improvement efforts.

Section 5.1 presents "one-way" methods for the analysis of experimental data. These treat r samples in ways that do not refer to any special structure in the process conditions leading to the observations. They are thus among the most widely applicable of statistical methods. Then Sect. 5.2 considers the analysis of two-way complete factorial data sets and introduces some basic concepts needed in the study of systems where several different factors potentially impact a response variable. Finally, Sect. 5.3 is a long one discussing p-way factorial analysis, with particular emphasis on those complete factorial studies where each factor appears at two levels (known as the 2^p *factorial studies*).

5.1 One-Way Methods

Figure 5.1 is useful for helping one think about experimental design and data analysis. The proverbial "black box" represents a process to be studied. Into that process go many noisy/variable inputs, and out of the process comes at least one noisy output of interest, y. At the time of experimentation, there are "knobs" on the process that the experimenter can manipulate, variables $x_1, x_2, x_3, \ldots$ whose settings or "levels" are under the investigator's control. The engineer can choose values for the variables and observe one or more values of y, choose another set of values for the variables and observe one or more additional values of y, and so on. The object of such data collection and subsequent data analysis is to figure out how the black box/process responds to changes in the knob settings (i.e., how the process output depends upon the variables $x_1, x_2, x_3, \ldots$). Armed with such knowledge, one can then try to

1. optimize the choice of values for the variables $x_1, x_2, x_3, \ldots$ in terms of maximizing (or minimizing) y or,

2. identify those variables that have the largest effects on y, with the goal of prescribing very careful future supervision of those, in order to reduce variation in y.

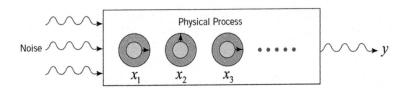

FIGURE 5.1. Black box with noisy inputs, "knobs" $x_1, x_2, x_3, \ldots$, and noisy output y

This section presents methods of data analysis that do not depend upon any specifics of the nature of the variables $x_1, x_2, x_3, \ldots$ nor on the pattern of changes one makes in them when collecting data. (In particular, there is no assumption that the variables $x_1, x_2, x_3, \ldots$ are quantitative. Nor does one need to suppose that any specific number of the variables are changed in any specific way in order to use the basic methods of this section.) All that one assumes is that there are r different sets of experimental conditions under consideration.

The presentation begins with a discussion of a "one-way" normal model for data from r different sets of conditions and an estimator of variance in that model. Then a method of making confidence intervals for linear combinations of the r means involved is presented.

5.1.1 The One-Way Normal Model and a Pooled Estimator of Variance

Example 55 is typical of many engineering experiments. Fairly small samples are taken to represent r different sets of process conditions, and one must use the limited information provided by the r samples to make comparisons.

Example 55 *Strengths of Solder Joints.* The paper "Fracture Mechanism of Brass/Sn-Pb-Sb Solder Joints and the Effect of Production Variables on the Joint Strength," by Tomlinson and Cooper, appeared in Journal of Materials Science *in 1986 and contains data on shear strengths of solder joints (units are megapascals). Table 5.1 gives part of the Tomlinson/Cooper data for six sets of process conditions (defined in terms of the cooling method employed and the amount of antimony in the solder).*

The data in Table 5.1 comprise $r = 6$ samples of common size $m = 3$, representing six different ways of making solder joints. In terms of the conceptualization of Fig. 5.1, there are $p = 2$ process "knobs" that have been turned in the collection of these data, the cooling method knob and the % antimony knob. Section 5.2 introduces methods of analysis aimed at detailing separate effects of the two factors. But to begin, we simply think of the data (and summary statistics) in Table 5.1 as generated by six different unstructured sets of process conditions.

TABLE 5.1. Shear strengths and some summary statistics for $r = 6$ different soldering methods (MPa)

Method (i)	Cooling	Sb (% weight)	Strength, y	$\bar{y}_i$	s_i
1	H$_2$O quench	3	18.6, 19.5, 19.0	19.033	.451
2	H$_2$O quench	5	22.3, 19.5, 20.5	20.767	1.419
3	H$_2$O quench	10	15.2, 17.1, 16.6	16.300	.985
4	Oil quench	3	20.0, 20.9, 20.4	20.433	.451
5	Oil quench	5	20.9, 22.9, 20.6	21.467	1.250
6	Oil quench	10	16.4, 19.0, 18.1	17.833	1.320

In order to make statistical inferences from r small samples like those represented in Table 5.1, it is necessary to adopt some model for the data generation process. By far the most widely used, most convenient, and most easily understood such mathematical description is the **one-way normal** (or Gaussian) model. In words, this model says that the r samples in hand come from r normal distributions *with possibly different means* $\mu_1, \mu_2, \ldots, \mu_r$, *but a common standard deviation* σ. Figure 5.2 on page 254 is a graphical representation of these assumptions.

One-Way Normal Model

It is helpful to also have a statement of the basic one-way Gaussian model assumptions in terms of symbols. The one that will be used in this text is that with y_{ij} the jth observation in sample i (from the ith set of process conditions),

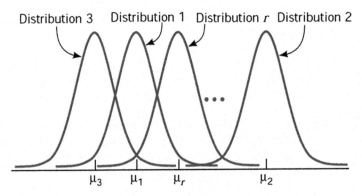

FIGURE 5.2. r normal distributions with a common standard deviation

One-Way Model Equation

$$y_{ij} = \mu_i + \epsilon_{ij}, \tag{5.1}$$

for $\mu_1, \mu_2, \ldots, \mu_r$ (unknown) means and $\epsilon_{11}, \ldots, \epsilon_{1n_1}, \epsilon_{21}, \ldots, \epsilon_{2n_2}, \ldots, \epsilon_{r1}, \ldots, \epsilon_{rn_r}$ independent normal random variables with mean 0 and (unknown) standard deviation σ. $n_1, n_2, \ldots, n_r$ are the sample sizes, and the means $\mu_1, \mu_2, \ldots, \mu_r$ and the standard deviation σ are the parameters of the model that must be estimated from data.

It is always wise to examine the plausibility of a mathematical model before using it to guide engineering decisions. So how might one investigate the appropriateness of the "constant variance, normal distributions" model represented by Eq. (5.1)? Where the sample sizes $n_1, n_2, \ldots, n_r$ are moderate to large (say on the order of at least 6 or 7), one option is to make r normal plots (possibly on the same set of axes) for the different samples. One uses the normal plotting method of Sect. 4.1 and hopes to see reasonably linear plots with roughly the same slope—linearity indicating normality and equal slopes indicating constant variance.

While making r different normal plots is a viable option where sample sizes are big enough, it is not really very helpful in contexts like Example 55 where sample sizes are extremely small. About all that can be done for very small sample sizes is to examine "residuals" in much the same way one studies residuals in regression analysis. That is, for the one-way model (5.1), the ith sample mean $\overline{y}_i$ is a kind of "predicted" or "fitted" response for the ith set of process conditions. One might then (for those samples with $n_i > 1$) compute and examine **residuals**

Residual for Data Point y_{ij}

$$e_{ij} = y_{ij} - \overline{y}_i. \tag{5.2}$$

The motivation for doing so is that in light of the model (5.1), e_{ij} is an approximation for ϵ_{ij}, and the e_{ij} might thus be expected to look approximately like Gaussian random variation. (Actually, when sample sizes vary, a slightly more sophisticated analysis would consider **standardized residuals,** which in this context are essentially the e_{ij} divided by $\sqrt{(n_i - 1)/n_i}$. But for the sake of brevity, this refinement will not be pursued here.) So, for example, if model (5.1) is to be considered completely appropriate, a normal plot of the e_{ij} ought to look reasonably linear.

Example 56 *(Example 55 continued.)* *Considering again the soldering study, Table 5.2 contains $18 = 6 \times 3$ residuals computed according to formula (5.2). Figure 5.3 is a normal plot of the 18 residuals of Table 5.2. It is quite linear and raises no great concerns about the appropriateness of an analysis of the soldering data based on the one-way normal model.*

TABLE 5.2. Residuals for the soldering data (MPa)

Method (i)	Residuals $e_{ij} = y_{ij} - \overline{y}_i$
1	$18.6 - 19.033 = -.433$, $19.5 - 19.033 = .467$, $19.0 - 19.033 = -.033$
2	$22.3 - 20.767 = 1.533$, $19.5 - 20.767 = -1.267$, $20.5 - 20.767 = -.267$
3	$15.2 - 16.300 = -1.100$, $17.1 - 16.300 = .800$, $16.6 - 16.3 = .300$
4	$20.0 - 20.433 = -.433$, $20.9 - 20.433 = .467$, $20.4 - 20.433 = -.033$
5	$20.9 - 21.467 = -.567$, $22.9 - 21.467 = 1.433$, $20.6 - 21.467 = -.867$
6	$16.4 - 17.833 = -1.433$, $19.0 - 17.833 = 1.167$, $18.1 - 17.833 = .267$

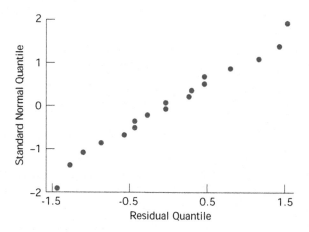

FIGURE 5.3. Normal plot of residuals in the soldering study

After considering the appropriateness of the basic model (5.1), an estimate of the parameter σ should be found. This is the (supposedly constant) standard deviation of observations y from any fixed set of process conditions. Any one of the sample standard deviations $s_1, s_2, \ldots, s_r$ could serve as an estimator of σ. But a better option is to somehow combine all r of these into a single "pooled" estimator. Usually, one computes a kind of weighted average of the r sample variances and then takes a square root in order to get back to the original units of observation. That is, a pooled estimator of σ^2 is

Pooled Estimator of σ^2

$$s^2_{\text{pooled}} = \frac{(n_1 - 1)s_1^2 + (n_2 - 1)s_2^2 + \cdots + (n_r - 1)s_r^2}{(n_1 - 1) + (n_2 - 1) + \cdots + (n_r - 1)}, \qquad (5.3)$$

or setting $n = n_1 + n_2 + \cdots + n_r$ and abbreviating the word "pooled" to the letter "P," one can rewrite display (5.3) as

Pooled Estimator of σ^2

$$s^2_{\text{P}} = \frac{(n_1 - 1)s_1^2 + (n_2 - 1)s_2^2 + \cdots + (n_r - 1)s_r^2}{n - r}. \qquad (5.4)$$

A corresponding pooled estimator of σ is thus

Pooled Estimator of σ

$$s_{\text{P}} = \sqrt{s^2_{\text{P}}}. \qquad (5.5)$$

Example 57 *(Examples 55 and 56 continued.)* *Table 5.1 gives the $r = 6$ sample variances for the soldering methods. These may be combined according to formula (5.4) to produce the pooled sample variance:*

$$\begin{aligned} s^2_{\text{P}} &= \frac{(3-1)(.451)^2 + (3-1)(1.419)^2 + \cdots + (3-1)(1.320)^2}{18 - 6}, \\ &= 1.116, \end{aligned}$$

so that

$$s_{\text{P}} = \sqrt{1.116} = 1.056 \text{ MPa}.$$

This pooled estimate of σ is meant to represent the amount of variability seen in shear strengths for any one of the six conditions included in the Tomlinson and Cooper study.

The pooled estimator (5.3) is sometimes called the "error mean square" (MSE) where people emphasize ANOVA in the analysis of experimental data. This text does not emphasize ANOVA or the significance tests that it facilitates. Rather, it concentrates on confidence intervals and graphical displays in the analysis of experimental data. And from this point of view, the present language is probably more natural.

5.1.2 Confidence Intervals for Linear Combinations of Means

The estimator s_P is a measure of background noise or observed experimental variability against which differences in the sample means $\bar{y}_1, \bar{y}_2, \ldots, \bar{y}_r$ must be judged. The particular form of the estimator specified in displays (5.3) and (5.4) allows the development of a number of simple formulas of statistical inference. For one thing, it supports the making of simple confidence intervals for linear combinations of the means $\mu_1, \mu_2, \ldots, \mu_r$.

That is, for constants $c_1, c_2, \ldots, c_r$, consider the problem of estimating the **linear combination of means**

$$L = c_1\mu_1 + c_2\mu_2 + \cdots + c_r\mu_r, \qquad (5.6)$$

based on samples from process conditions $1, 2, \ldots, r$. A number of important quantities are of this form. For example, where all c_i's are 0 except one which is 1, L is just a particular mean of interest. And where all c_i's are 0 except one that is 1 and one that is -1, L is a difference in two particular means of interest. And as this chapter proceeds, it will become evident that there are other more complicated Ls that are useful in the analysis of data from process-improvement experiments.

A natural estimator for L specified in display (5.6) is

Estimator of L

$$\hat{L} = c_1\bar{y}_1 + c_2\bar{y}_2 + \cdots + c_r\bar{y}_r, \qquad (5.7)$$

obtained by replacing each population mean by its corresponding sample version. And as it turns out, one can use $\hat{L}$ as the basis of a confidence interval for L. That is, confidence limits for L are

Confidence Limits for L

$$\hat{L} \pm ts_P\sqrt{\frac{c_1^2}{n_1} + \frac{c_2^2}{n_2} + \cdots + \frac{c_r^2}{n_r}}, \qquad (5.8)$$

where t is a quantile of the t distribution with $n-r$ associated degrees of freedom. (n continues to stand for the total of the r sample sizes.) If t is the p quantile and only one of the limits indicated in display (5.8) is used, the associated confidence level is $p \times 100\,\%$. If both limits in formula (5.8) are employed to make a two-sided confidence interval, the associated confidence is $(2p-1) \times 100\,\%$.

Two important special cases of formula (5.8) are those where L is a single mean and where L is a difference of means. That is, confidence limits for the ith mean response μ_i are

Confidence Limits for a Single Mean

$$\bar{y}_i \pm ts_P\sqrt{\frac{1}{n_i}}. \qquad (5.9)$$

Confidence Limits for a Difference in Means

And confidence limits for the difference in the ith and i'th means, $\mu_i - \mu_{i'}$, are

$$\bar{y}_i - \bar{y}_{i'} \pm t s_\text{P} \sqrt{\frac{1}{n_i} + \frac{1}{n_{i'}}}. \tag{5.10}$$

Example 58 *(Examples 55, 56, and 57 continued.)* Return again to the soldering problem, and first consider an individual method mean strength, μ_i. Associated with the pooled standard deviation s_P are $n - r = 18 - 6 = 12$ degrees of freedom. So, for example, finding that the .975 quantile of the t_{12} distribution is 2.179, two-sided 95 % confidence limits for any single mean strength in the soldering study are from display (5.9):

$$\bar{y}_i \pm 2.179(1.056)\sqrt{\frac{1}{3}}, \quad \text{that is,} \quad \bar{y}_i \pm 1.328 \text{ MPa}.$$

In some sense, the $\bar{y}_i$ values in Table 5.1 are each good to within 1.328 MPa as representing their respective long-run mean solder joint strengths.

Next, consider the comparison of any two mean strengths. A way to make such a comparison is through a confidence interval for their difference, as indicated in display (5.10). Again using the fact that the .975 quantile of the t_{12} distribution is 2.179, two-sided 95 % confidence limits for any particular difference in mean strengths are

$$\bar{y}_i - \bar{y}_{i'} \pm 2.179(1.056)\sqrt{\frac{1}{3} + \frac{1}{3}}, \quad \text{that is,} \quad \bar{y}_i - \bar{y}_{i'} \pm 1.879 \text{ MPa}.$$

For example, $\mu_1 - \mu_4$ (which represents the difference in mean strengths for water-quenched and oil-quenched joints when 3 % antimony is used in the solder) can be estimated with 95 % confidence as

$$(19.033 - 20.433) \pm 1.879, \quad \text{that is,} \quad -1.400 \text{ MPa} \pm 1.879 \text{ MPa}.$$

The fact that the uncertainty reflected by the ± 1.879 MPa figure is larger in magnitude than the difference between $\bar{y}_1$ and $\bar{y}_4$ says that evidence in the data that methods 1 and 4 produce different joint strengths is not overwhelming. (The confidence interval includes 0.)

Finally, as an example of using the general form of the confidence interval formula given in display (5.8), consider estimation of

$$L = \frac{1}{3}(\mu_4 + \mu_5 + \mu_6) - \frac{1}{3}(\mu_1 + \mu_2 + \mu_3)$$

$$= -\frac{1}{3}\mu_1 - \frac{1}{3}\mu_2 - \frac{1}{3}\mu_3 + \frac{1}{3}\mu_4 + \frac{1}{3}\mu_5 + \frac{1}{3}\mu_6,$$

the difference between the average of the oil-quench mean strengths and the average of the water-quench mean strengths. First, from formula (5.7)

$$\hat{L} = -\frac{1}{3}(19.033 + 20.767 + 16.300) + \frac{1}{3}(20.433 + 21.467 + 17.833) = 1.211.$$

Then, based on the fact that each c_i is $\pm\frac{1}{3}$ and thus has square $\frac{1}{9}$, formula (5.8) gives 95% two-sided confidence limits for L of the form

$$1.211 \pm 2.179(1.056)\sqrt{6\left(\frac{1}{9}\right)\left(\frac{1}{3}\right)}, \quad \text{that is,} \quad 1.211 \text{ MPa} \pm 1.085 \text{ MPa}.$$

This shows that (at least on average across different amounts of antimony) the oil-quenched joints are detectably stronger than water-quenched joints. (The interval includes only positive values.)

It is important to realize that the confidence level associated with an interval with end points (5.8) (or one of the specializations (5.9) and (5.10)) is an *individual* confidence level, pertaining to a single interval at a time. For example, if one uses formula (5.9) r times, to estimate each of $\mu_1, \mu_2, \ldots, \mu_r$ with 90% confidence, one is 90% confident of the first interval, *separately* 90% confident of the second interval, *separately* 90% confident of the third interval, and so on. One is *not* 90% confident that all r of the intervals are correct. But there are times when it is desirable to be able to say that one is simultaneously "90% sure" of a whole collection of inferences.

If one wishes to make several confidence intervals and announce an overall or simultaneous confidence level, one simple approach is to use the **Bonferroni Inequality**. This inequality says that if l intervals have associated individual confidence levels $\gamma_1, \gamma_2, \ldots, \gamma_l$, then the confidence that should be associated with them simultaneously or as a group, say γ, satisfies

$$\gamma \geq 1 - ((1 - \gamma_1) + (1 - \gamma_2) + \cdots + (1 - \gamma_l)). \quad (5.11)$$

Bonferroni's Lower Bound on Overall Confidence

(This says that the "unconfidence" associated with a group of inferences is no worse than the sum of the individual "unconfidences," $1 - \gamma \leq (1 - \gamma_1) + (1 - \gamma_2) + \cdots + (1 - \gamma_l)$.)

Example 59 *(Examples 55 through 58 continued.)* Consider a simple use of the Bonferroni inequality (5.11) in the solder joint strength study. The $r = 6$ individual 95% confidence intervals for the means μ_i of the form $\bar{y}_i \pm 1.328$ taken as a group have simultaneous confidence level at least 70%, since

$$1 - 6(1 - .95) = .7.$$

If one wanted to be more sure of intervals for the $r = 6$ means, one could instead make (much wider) 99% individual intervals and be at least 94% confident that all six intervals cover their respective means.

The Bonferroni idea is a simple all-purpose tool that covers many different situations. It is particularly useful when there are a relatively few quantities to be estimated and individual confidence levels are large (so that the lower bound for the joint or simultaneous confidence is not so small as to be practically useless). But it is also somewhat crude, and a number of more specialized methods have been crafted to produce exact simultaneous confidence levels for estimating particular sets of Ls. When these are applicable, they will produce narrower intervals (or equivalently, higher stated simultaneous confidence levels) than those provided by the Bonferroni method. (The reader is referred to Sect. 7.3 of Vardeman and Jobe's *Basic Engineering Data Collection and Analysis* for discussion of methods of making simultaneous intervals for a set of means or for all differences in a set of means.)

Section 5.1 Exercises

1. For the one-way normal model of this section, describe the distributional aspects (means, variances, and distribution) of observations y_{ij}.

2. s_P is the pooled sample standard deviation for an experiment for comparing r different experimental conditions. What does this statistic estimate in the context of the one-way normal model?

3. Suppose $r = 3$ conditions in an experiment have sample sizes and produce sample standard deviations. What is the corresponding value of s_P?

Condition 1	Condition 2	Condition 3
$n_1 = 4$	$n_2 = 3$	$n_3 = 5$
$s_1 = 2$	$s_2 = 6$	$s_3 = 4$

4. If approximately 95 % confidence limits for μ_1 are $\bar{y}_1 \pm 3$, while 95 % confidence limits for μ_2 in the same study are $\bar{y}_2 \pm 4$, find 95 % confidence limits for the difference $\mu_1 - \mu_2$. (All limits are based on s_P. Assume the appropriate t multiplier is about 2.0.)

5. Consider an experiment with $r = 4$ conditions in which samples of size $m = 4$ are obtained under each condition. Further, suppose that $s_P = 3$. 95 % confidence limits for each μ_i are of the form $\bar{y}_i \pm \Delta$.

   (a) What is the numerical value of Δ?

   (b) Will you be 95% confident that all 4 intervals include the respective μ_i's? Why or why not? If not, your confidence in all intervals *simultaneously* is at least how much?

6. **Air Filters.** Tests of effectiveness of $r = 5$ different operator cab air filtration systems (meant to protect agricultural workers operating large mobile pesticide spraying machines) produced ratios y of outside-cab-particle-counts to inside-cab-particle-counts with summary statistics below. (Large ratios y are good.) Suppose that the variation in the ratios y can be

Chapter 5. Experimental Design and Analysis for Process Improvement   261

described by a single standard deviation appropriate for any fixed design and that the y values for each design are approximately normally distributed.

Design 1	Design 2	Design 3	Design 4	Design 5
$n_1 = 3$	$n_2 = 2$	$n_3 = 1$	$n_4 = 3$	$n_5 = 4$
$\bar{y}_1 = 50$	$\bar{y}_2 = 40$	$\bar{y}_3 = 65$	$\bar{y}_4 = 35$	$\bar{y}_5 = 55$
$s_1 = 7.2$	$s_2 = 6.5$	$s_3 = 0$	$s_4 = 5.0$	$s_5 = 7.5$

(a) Find s_P.

(b) Find 95 % confidence limits for the average ratio for design 4.

(c) Find 95 % confidence limits for the difference between the average ratio for design 4 and the average ratio for design 1 (design 4 minus design 1).

(d) Let $\mu_1, \mu_2, \mu_3, \mu_4,$ and μ_5 be the long run average y for the five designs. Suppose designs 2 and 4 use filters supplied by Company X and designs 3 and 5 use filters supplied by Company Y. Give an L (linear combination of design means) that might be used to compare the performance of the two filter companies.

(e) The numbers in the table above for a given design might summarize multiple measurements on a single prototype or summarize individual measurements on several prototypes. Which of these two possibilities is better in practical terms? Why?

7. Continue with the **Air Filter** problem.

   (a) Give a 90 % confidence interval for the L from problem 6(d).

   (b) How many possible differences $\mu_i - \mu_j$ (for $i \neq j$) are there for the five designs? (Do not count $\mu_1 - \mu_2$ and $\mu_2 - \mu_1$ as different comparisons of two means.)

   (c) An analyst wanted to make a set of intervals for all the differences $\mu_i - \mu_j$ (for $i \neq j$) and be 90 % confident that all intervals simultaneously cover their respective differences $\mu_i - \mu_j$. What confidence level for each individual interval would guarantee the 90 % figure overall?

5.2 Two-Way Factorials

It is not yet apparent how the analysis tools of the previous section address the facts that most often process-improvement experiments involve several factors and that separating their influences is a primary issue. This section begins to show how these vital matters can be handled, taking the case of $p = 2$ factors as a starting point.

The section begins with a discussion of two-way factorial data structures and graphical and qualitative analyses of these. Then the concepts of main effects and interactions are defined, and the estimation of these quantities is considered. Finally, there is a brief discussion of fitting simplified or reduced models to balanced two-way factorial data.

5.2.1 Graphical and Qualitative Analysis for Complete Two-Way Factorial Data

This section concerns situations where on the black box of Fig. 5.1 there are two "knobs" under the control of an experimenter. That is, two-factor experimentation is treated. So for ease of communication, let Factor A and Factor B be two generic names for factors that potentially impact some process output of interest, y. In cases where levels of Factors A and B are defined in terms of values of quantitative process variables x_1 and x_2, respectively, the tool of multiple regression analysis provides a powerful method of data analysis. That tool is applied to multifactor process-improvement problems in Sect. 6.2 of this book. But here, methods of analysis that can be used even when one or both of Factors A and B is qualitative will first be considered.

The most straightforward analyses of two-factor studies are possible in cases where Factor A has levels $1, 2, \ldots, I$, Factor B has levels $1, 2, \ldots, J$, and every possible combination of a level of A and a level of B is represented in the data set. Such a data set will be called an $I \times J$ **complete factorial** data set. Figure 5.4 illustrates the fact that the $I \times J$ different sets of process conditions in a complete factorial study can be laid out in a rectangular two-way table. Rows correspond to levels of Factor A, columns correspond to levels of Factor B, and the factorial is "complete" in the sense that there are data available corresponding to all $I \times J$ "cells" in the table.

Corresponding to Fig. 5.4, let

y_{ijk} = the kth observation from level i of Factor A and level j of Factor B,

$$\bar{y}_{ij} = \frac{1}{n_{ij}} \sum_k y_{ijk}$$

= the sample mean from level i of Factor A and level j of Factor B,

and

$$s_{ij}^2 = \frac{1}{n_{ij} - 1} \sum_k (y_{ijk} - \bar{y}_{ij})^2$$

= the sample variance from level i of Factor A and level j of Factor B,

Chapter 5. Experimental Design and Analysis for Process Improvement   263

	Factor B			
	Level 1	Level 2		Level J
Level 1	cell 1,1	cell 1,2	• • •	cell 1,J
Level 2	cell 2,1	cell 2,2	• • •	cell 2,J
	• • •	• • •		• • •
Level I	cell I,1	cell I,2	• • •	cell I,J

(Factor A on vertical axis)

FIGURE 5.4. $r = I \times J$ combinations in a two-way factorial study

where n_{ij} is the sample size corresponding to level i of A and level j of B. Except for the introduction of (i, j) double subscripting to recognize the two-factor structure, there is nothing new in these formulas. One is simply naming the various observations from a data structure like that indicated in Fig. 5.4 and the corresponding $r = I \times J$ sample means and variances.

Example 60 *Two-Way Analysis of Solder Joint Strengths (Examples 55 through 59 revisited).* *Table 5.3 on page 264 is essentially a repeat of Table 5.1, giving the joint strength data of Tomlinson and Cooper and corresponding sample means and standard deviations. What is new here is only that in place of naming $r = 6$ sets of process conditions with indices $i = 1, 2, 3, 4, 5,$ and 6, double subscripts (i, j) corresponding to the 2×3 different combinations of $I = 2$ different cooling methods and $J = 3$ different amounts of antimony are used.*

Figure 5.5 on Page 264 shows the six sample means and standard deviations of Table 5.3 laid out in a 2×3 table, with rows corresponding to levels of Factor A (cooling method) and columns corresponding to levels of Factor B (% Sb).

Most basically, data from a complete two-way factorial study are simply observations from $r = I \times J$ different sets of process conditions, and all of the material from the previous section can be brought to bear (as it was in Example 55) on their analysis. But in order to explicitly acknowledge the two-way structure, it is common to not only double subscript the samples (for level of A and level of B) but to also double subscript the theoretical mean responses as well, writing μ_{ij} instead of simply the μ_i used in Sect. 5.1. And so, using the obvious subscript no-

TABLE 5.3. Shear strengths and summary statistics for $I \times J = 2 \times 3$ combinations of cooling method and amount of antimony (MPa)

Factor A Cooling	i	Factor B Sb (% weight)	j	Strength, y	$\bar{y}_{ij}$	s_{ij}
H$_2$O quench	1	3	1	18.6, 19.5, 19.0	19.033	.451
H$_2$O quench	1	5	2	22.3, 19.5, 20.5	20.767	1.419
H$_2$O quench	1	10	3	15.2, 17.1, 16.6	16.300	.985
Oil quench	2	3	1	20.0, 20.9, 20.4	20.433	.451
Oil quench	2	5	2	20.9, 22.9, 20.6	21.467	1.250
Oil quench	2	10	3	16.4, 19.0, 18.1	17.833	1.320

		Factor B Sb % Weight		
		3	5	10
Factor A Cooling Method	H$_2$O	19.033 .451	20.767 1.419	16.300 .985
	Oil	20.433 .451	21.467 1.250	17.833 1.320

FIGURE 5.5. Sample means and standard deviations from Table 5.3 (MPa)

tation for ϵs the one-way model assumptions (5.1) are rewritten for the two-way factorial context as

Two-Way Model Equation

$$y_{ijk} = \mu_{ij} + \epsilon_{ijk},$$

for $\mu_{11}, \mu_{12}, \ldots, \mu_{1J}, \mu_{21}, \ldots, \mu_{2J}, \ldots, \mu_{I1}, \ldots, \mu_{IJ}$ (unknown) means and $\epsilon_{111}, \ldots, \epsilon_{11n_{11}}, \epsilon_{121}, \ldots, \epsilon_{12n_{12}}, \ldots, \epsilon_{IJ1}, \ldots, \epsilon_{IJn_{IJ}}$ independent normal random variables with mean 0 and (unknown) standard deviation σ.

Finding interpretable patterns in how the means μ_{ij} change with i and j (with level of Factor A and level of Factor B) is a primary goal in a two-way factorial analysis. A very effective first step in achieving that goal is to make a plot of the $I \times J$ sample means $\bar{y}_{ij}$ versus (say) level of Factor B, connecting points having a common level of (say) Factor A with line segments. Such a plot is usually called an **interaction plot** (although the terminology is not terribly descriptive or helpful). It is useful to indicate on such a plot the precision with which the mean responses are known. This can be done by using confidence limits for the μ_{ij} to make **error bars** around the sample means. The formulas (5.9) can be used for this purpose as long as one bears in mind that the confidence level associated with them is an individual one, not one that applies to the entire figure simultaneously.

Interaction Plot

Example 61 (*Example 60 continued.*) *Figure 5.6 is an interaction plot for the solder joint strength data. In Example 58, 95 % confidence limits for the six method means μ_{ij} were found to be of the form $\bar{y}_{ij} \pm 1.328$ MPa, and this 1.328 MPa figure has been used to make the error bars in the figure.*

Figure 5.6 gives a helpful summary of what the data say about how cooling method and amount of antimony impact solder joint strength. That is, there are strong hints that (1) a large amount of antimony in the solder is not good (in terms of producing large joint strength), (2) oil-quenched joints are stronger than water-quenched joints, and (3) patterns of response to changes in a given factor (A or B) are consistent across levels of the other factor (B or A). But these conclusions are somewhat clouded by the relatively large error bars on the plot (indicating uncertainty in knowledge about long-run mean joint strengths). The indicated uncertainty does not seem large enough to really draw into question the importance of amount of antimony in determining joint strength. But exactly how the two different cooling methods compare is perhaps somewhat murkier. And the extent to which a change in antimony levels possibly produces different changes in mean joint strength for the two different cooling methods is nearly completely clouded by the "experimental noise level" pictured by the error bars.

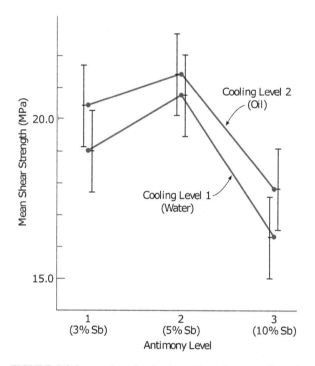

FIGURE 5.6. Interaction plot for the solder joint strength study

In two-way factorial experiments where it turns out that one has good precision (small error bars) for estimating all $I \times J$ means, an interaction plot in the style of Fig. 5.6 can be all that is really required to describe how the two factors impact y. But where one has learned less about the individual means, finer/more quantitative analyses are helpful. And the next subsection discusses such analyses, both because they can be important in two-factor studies and because they set the pattern for what is done in the analysis of p-way factorial data for $p > 2$.

5.2.2 Defining and Estimating Effects

Figure 5.6 for the solder joint strength data hints at "an effect" of cooling method, Factor A, on y. If one can make a sensible quantitative definition of exactly what the effect of a level of Factor A might mean, the possibility exists of doing inference (giving a confidence interval) for it. To those ends, notations for row and column averages of both μ_{ij}'s and corresponding $\bar{y}_{ij}$'s are needed. These are indicated in Fig. 5.7 for the specific case of the solder study and detailed in general in formulas (5.12) through (5.17).

		Factor B Sb % Weight			
		3	5	10	
Factor A Cooling Method	H₂O	$\bar{y}_{11}$	$\bar{y}_{12}$	$\bar{y}_{13}$	$\bar{y}_{1.}$
		μ_{11}	μ_{12}	μ_{13}	$\mu_{1.}$
	Oil	$\bar{y}_{21}$	$\bar{y}_{22}$	$\bar{y}_{23}$	$\bar{y}_{2.}$
		μ_{21}	μ_{22}	μ_{23}	$\mu_{2.}$
		$\bar{y}_{.1}$	$\bar{y}_{.2}$	$\bar{y}_{.3}$	$\bar{y}_{..}$
		$\mu_{.1}$	$\mu_{.2}$	$\mu_{.3}$	$\mu_{..}$

FIGURE 5.7. Two-way layout of $\bar{y}_{ij}, \bar{y}_{i.}, \bar{y}_{.j}, \bar{y}_{..}, \mu_{ij}, \mu_{i.}, \mu_{.j}, \mu_{..}$ for the 2×3 solder joint study

Let

Average of Row i Sample Means

$$\bar{y}_{i.} = \frac{1}{J} \sum_j \bar{y}_{ij} = \text{the simple average of the row } i \text{ sample means} \quad (5.12)$$

and correspondingly

$$\mu_{i.} = \frac{1}{J} \sum_j \mu_{ij} = \text{the simple average of the row } i \text{ theoretical means.}$$

$$(5.13)$$

Similarly, take

Average of Column j Sample Means

$$\bar{y}_{.j} = \frac{1}{I} \sum_i \bar{y}_{ij} = \text{the simple average of the column } j \text{ sample means}$$

$$(5.14)$$

Chapter 5. Experimental Design and Analysis for Process Improvement

and correspondingly

$$\mu_{\cdot j} = \frac{1}{I} \sum_i \mu_{ij} = \text{the simple average of the column } j \text{ theoretical means.}$$

(5.15)

Finally, use the notation

$$\overline{y}_{\cdot \cdot} = \frac{1}{IJ} \sum_{i,j} \overline{y}_{ij} = \text{the average of all } I \times J \text{ cell sample means} \quad (5.16)$$

Average of All $I \times J$ Sample Means

and correspondingly

$$\mu_{\cdot \cdot} = \frac{1}{IJ} \sum_{i,j} \mu_{ij} = \text{the average of all } I \times J \text{ theoretical means.} \quad (5.17)$$

Example 62 *(Example 60 continued.)* *Averaging $\overline{y}_{ij}$ values across rows, down columns, and over the whole table summarized in Fig. 5.5 produces the $\overline{y}_{i\cdot}, \overline{y}_{\cdot j}$ values and the value of $\overline{y}_{\cdot \cdot}$ displayed in Fig. 5.8 along with the six cell means. (Since the long-run mean strengths μ_{ij} are not known, it is not possible to present an analog of Fig. 5.8 giving numerical values for the $\mu_{i\cdot}, \mu_{\cdot j},$ and $\mu_{\cdot \cdot}$.)*

		Factor B Sb % Weight			
		3	5	10	
Factor A	H₂O	$\overline{y}_{11} = 19.033$	$\overline{y}_{12} = 20.767$	$\overline{y}_{13} = 16.300$	$\overline{y}_{1\cdot} = 18.700$
Cooling Method	Oil	$\overline{y}_{21} = 20.433$	$\overline{y}_{22} = 21.467$	$\overline{y}_{23} = 17.833$	$\overline{y}_{2\cdot} = 19.911$
		$\overline{y}_{\cdot 1} = 19.733$	$\overline{y}_{\cdot 2} = 21.117$	$\overline{y}_{\cdot 3} = 17.067$	$\overline{y}_{\cdot \cdot} = 19.306$

FIGURE 5.8. Cell, marginal, and overall means from the data in Table 5.3 (MPa)

The row and column means in Fig. 5.8 suggest a way of measuring the direct or main effects of Factors A and B on y. One might base comparisons of levels of Factor A on row averages of mean responses and comparisons of levels of Factor B on column averages of mean responses. This thinking leads to definitions of main effects and their estimated or fitted counterparts.

Definition 63 *The (theoretical)* **main effect of Factor A at level** i *in a complete $I \times J$ two-way factorial is*

$$\alpha_i = \mu_{i\cdot} - \mu_{\cdot \cdot}.$$

Definition 64 *The (estimated or)* **fitted main effect of Factor A at level** i *in a complete $I \times J$ two-way factorial is*

$$a_i = \overline{y}_{i\cdot} - \overline{y}_{\cdot \cdot}.$$

Definition 65 *The (theoretical)* **main effect of Factor B at level j** *in a complete $I \times J$ two-way factorial is*

$$\beta_j = \mu_{\cdot j} - \mu_{\cdot\cdot}.$$

Definition 66 *The (estimated or)* **fitted main effect of Factor B at level j** *in a complete $I \times J$ two-way factorial is*

$$b_j = \bar{y}_{\cdot j} - \bar{y}_{\cdot\cdot}.$$

A main effects are row averages of cell means minus a grand average, while B main effects are column averages of cell means minus a grand average. And a very small amount of algebra makes it obvious that differences in main effects of a factor are corresponding differences in row or column averages. That is, from the definitions

$$\alpha_i - \alpha_{i'} = \mu_{i\cdot} - \mu_{i'\cdot} \quad \text{and} \quad a_i - a_{i'} = \bar{y}_{i\cdot} - \bar{y}_{i'\cdot}, \tag{5.18}$$

while

$$\beta_j - \beta_{j'} = \mu_{\cdot j} - \mu_{\cdot j'} \quad \text{and} \quad b_j - b_{j'} = \bar{y}_{\cdot j} - \bar{y}_{\cdot j'}. \tag{5.19}$$

Example 67 *(Example 60 continued.)* Some arithmetic applied to the row and column average means in Fig. 5.8 shows that the fitted main effects of cooling method and antimony content for the data of Table 5.3 are

$$a_1 = 18.700 - 19.30\bar{5} = -.60\bar{5} \quad \text{and} \quad a_2 = 19.91\bar{1} - 19.30\bar{5} = .60\bar{5}$$

and

$$b_1 = 19.73\bar{3} - 19.30\bar{5} = .42\bar{7}, \; b_2 = 21.11\bar{6} - 19.30\bar{5} = 1.81\bar{1}, \quad \text{and}$$
$$b_3 = 17.06\bar{6} - 19.30\bar{5} = -2.23\bar{8}.$$

More decimals have been displayed in Example 67 than are really justified on the basis of the precision of the original data. This has been done for the purpose of pointing out (without clouding the issue with roundoff error) that $a_1 + a_2 = 0$ and $b_1 + b_2 + b_3 = 0$. These relationships are no accident. It is an algebraic consequence of the form of Definitions 63 through 66 that

$$\sum_i a_i = 0 \text{ and } \sum_j b_j = 0, \text{ and similarly } \sum_i \alpha_i = 0 \text{ and } \sum_j \beta_j = 0.$$
(5.20)

Both fitted and theoretical main effects of any factor sum to 0 over all possible levels of that factor. Notice that, in particular, relationships (5.20) imply that when a factor has only two levels, the two (theoretical or fitted) main effects must have the same magnitude but opposite signs.

The main effects of Factors A and B do not in general "tell the whole story" about how means μ_{ij} depend upon i and j. Figure 5.9 specifies two hypothetical

Chapter 5. Experimental Design and Analysis for Process Improvement   269

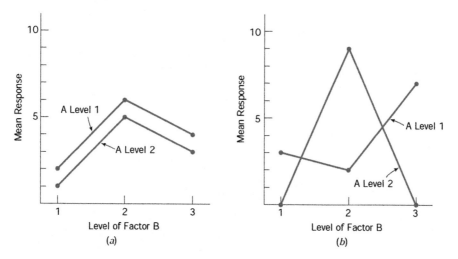

FIGURE 5.9. Two hypothetical sets of means μ_{ij} for 2×3 factorials that share the same row and column averages

FIGURE 5.10. Interaction plots for the two sets of means in Fig. 5.9

sets of means μ_{ij} for 2×3 factorials that share the same row and column averages and therefore the same main effects. Interaction plots for the two sets of means are given in Fig. 5.10.

The qualitative characters of the two plots in Fig. 5.10 are substantially different. The first graph represents a situation that is fundamentally simpler than the second. On the first graph, as one changes level of Factor B, the same change in mean response is produced regardless of whether one is looking at the first level of Factor A or at the second level. In some sense, it is legitimate to think about how Factor B impacts the mean response as independent of the level of A under discussion. The same is not true for the second plot. On the second graph, what happens to mean response when one changes level of B depends strongly on what level of A is being considered. That implies, for example, that if one were interested in maximizing mean response, the preferred level of B would depend on the level of A being used. No simple blanket recommendation like "level 2 of Factor B is best" can be made in situations like that pictured on the second plot.

A way of describing the feature of the first plot in Fig. 5.10 that makes it simple is to say that the plot exhibits **parallelism** between the profiles (across levels of B) of mean response for various levels of A. It is thus important to be able to measure the degree to which a set of means departs from the kind of simple parallelism seen on the first plot.

It turns out that parallelism on an interaction plot of μ_{ij}s is equivalent to the possibility that for all (i, j), combinations

$$\mu_{ij} = \mu_{..} + \alpha_i + \beta_j.$$

This is exactly the eventuality that the main effects (and the grand mean) completely summarize the μ_{ij}. Departures from this simple state of affairs can then be measured by taking the difference between the left- and right-hand sides of this simple ideal relationship. That is, one is led to two more definitions.

Definition 68 *The (theoretical)* ***interaction of Factor A at level*** i ***and Factor B at level*** j *in a complete* $I \times J$ *two-way factorial is*

$$\alpha\beta_{ij} = \mu_{ij} - (\mu_{..} + \alpha_i + \beta_j).$$

Definition 69 *The (estimated or)* ***fitted interaction of Factor A at level*** i ***and Factor B at level*** j *in a complete* $I \times J$ *two-way factorial is*

$$ab_{ij} = \overline{y}_{ij} - (\overline{y}_{..} + a_i + b_j).$$

To the extent that parallelism or lack thereof can be clearly identified on an interaction plot like Fig. 5.6, data-based examination of the possibility of important AB interactions in a two-factor study can proceed graphically. But a more quantitative look at the issue must begin with computation of the fitted interactions defined in Definition 69. In this regard, it is worth noting that there is an alternative "computational" formula for fitted interactions that is sometimes useful. That is,

Computational Formula for Fitted Interactions

$$ab_{ij} = \overline{y}_{ij} - \overline{y}_{i.} - \overline{y}_{.j} + \overline{y}_{..}.$$

Example 70 *(Example 60 continued.)* Returning again to the solder joint strength example, Table 5.4 organizes calculations of the six fitted interactions derived from the data of Table 5.3.

It is worth noting that the largest of the fitted interactions in Table 5.4 is smaller than even the smallest fitted main effect calculated earlier. That is numerical evidence that the lack of parallelism in Fig. 5.6 is in some sense smaller than the gap between the oil- and water-quench profiles and the differences between the observed strengths for the different amounts of antimony.

Once again, more decimal places are displayed in Table 5.4 than are really justified on the basis of the precision of the original data. This has been done so that it can be clearly seen that $ab_{11} + ab_{21} = 0, ab_{12} + ab_{22} = 0$, $ab_{13} + ab_{23} = 0, ab_{11} + ab_{12} + ab_{13} = 0$, and $ab_{21} + ab_{22} + ab_{23} = 0$. These relationships are not special to the particular data set. Fitted interactions sum to

TABLE 5.4. Calculation of the fitted interactions for the solder joint strength study

i	j	$\bar{y}_{ij}$	$\bar{y}_{..}+a_i+b_j$	$ab_{ij}=\bar{y}_{ij}-(\bar{y}_{..}+a_i+b_j)$
1	1	$19.03\bar{3}$	$19.30\bar{5}+(-.60\bar{5})+.42\bar{7}=19.12\bar{7}$	$-.09\bar{4}$
1	2	$20.76\bar{6}$	$19.30\bar{5}+(-.60\bar{5})+1.81\bar{1}=20.51\bar{1}$	$.25\bar{5}$
1	3	16.300	$19.30\bar{5}+(-.60\bar{5})+(-2.23\bar{8})=16.46\bar{1}$	$-.16\bar{1}$
2	1	$20.43\bar{3}$	$19.30\bar{5}+.60\bar{5}+.42\bar{7}=20.33\bar{8}$	$.09\bar{4}$
2	2	$21.46\bar{6}$	$19.30\bar{5}+.60\bar{5}+1.81\bar{1}=21.72\bar{2}$	$-.25\bar{5}$
2	3	$17.83\bar{3}$	$19.30\bar{5}+.60\bar{5}+(-2.23\bar{8})=17.67\bar{2}$	$.16\bar{1}$

0 down any column or across any row in any two-way factorial. It is an algebraic consequence of Definitions 68 and 69 that

$$\sum_i ab_{ij}=0 \quad \text{and} \quad \sum_j ab_{ij}=0, \quad \text{and similarly that}$$

$$\sum_i \alpha\beta_{ij}=0 \quad \text{and} \quad \sum_j \alpha\beta_{ij}=0. \tag{5.21}$$

To go beyond simply computing single-number estimates of (main and interaction) effects, to making confidence intervals for these, it is only necessary to realize that effects are linear combinations of the means μ_{ij} (Ls). And not surprisingly, the fitted effects are the corresponding linear combinations of the sample means $\bar{y}_{ij}$ (the corresponding $\hat{L}$s). That implies (repeating formula (5.8) in two-way factorial/double subscript notation) that

$$\hat{L} \pm t s_P \sqrt{\frac{c_{11}^2}{n_{11}}+\cdots+\frac{c_{1J}^2}{n_{1J}}+\frac{c_{21}^2}{n_{21}}+\cdots+\frac{c_{IJ}^2}{n_{IJ}}}, \tag{5.22}$$

Confidence Limits for L in Two-Way Notation

can be used to make confidence intervals for the main effects and interactions. The only real question in applying formula (5.22) to the estimation of an effect is what form the sum of squared coefficients over sample sizes takes. That is, one needs to know how to compute the sum:

$$\sum_{i,j} \frac{c_{ij}^2}{n_{ij}} \tag{5.23}$$

that goes under the root in formula (5.22). It is possible to derive formulas for these sums where the Ls involved are two-way factorial effects or differences in such effects. Table 5.5 on page 272 collects the very simple formulas that are appropriate when data are **balanced** (all n_{ij} are equal to some number m). Table 5.6 on page 272 gives the more complicated formulas for the quantities (5.23) needed when the n_{ij} vary.

Example 71 *(Example 60 continued.)* *Once again consider the solder joint strength example and now the problem of making confidence intervals for the various factorial effects, beginning with the interactions $\alpha\beta_{ij}$. The fitted interactions ab_{ij} are collected in Table 5.4. Use of the formula (5.22) allows one to*

TABLE 5.5. Balanced data formulas for the quantities $\sum_{i,j} \frac{c_{ij}^2}{n_{ij}}$ needed to make confidence intervals for effects in two-way factorials (All $n_{ij} = m$)

L	$\hat{L}$	$\sum_{i,j} \frac{c_{ij}^2}{n_{ij}}$
$\alpha\beta_{ij}$	ab_{ij}	$\frac{(I-1)(J-1)}{mIJ}$
α_i	a_i	$\frac{I-1}{mIJ}$
$\alpha_i - \alpha_{i'}$	$a_i - a_{i'}$	$\frac{2}{mJ}$
β_j	b_j	$\frac{J-1}{mIJ}$
$\beta_j - \beta_{j'}$	$b_j - b_{j'}$	$\frac{2}{mI}$

TABLE 5.6. General formulas for the quantities $\sum_{i,j} \frac{c_{ij}^2}{n_{ij}}$ needed to make confidence intervals for effects in two-way factorials

L	$\hat{L}$	$\sum_{i,j} \frac{c_{ij}^2}{n_{ij}}$
$\alpha\beta_{ij}$	ab_{ij}	$\left(\frac{1}{IJ}\right)^2 \left(\frac{(I-1)^2(J-1)^2}{n_{ij}} + (I-1)^2 \sum_{j' \neq j} \frac{1}{n_{ij'}} + (J-1)^2 \sum_{i' \neq i} \frac{1}{n_{i'j}} + \sum_{i' \neq i, j' \neq j} \frac{1}{n_{i'j'}} \right)$
α_i	a_i	$\left(\frac{1}{IJ}\right)^2 \left((I-1)^2 \sum_j \frac{1}{n_{ij}} + \sum_{i' \neq i, j} \frac{1}{n_{i'j}} \right)$
$\alpha_i - \alpha_{i'}$	$a_i - a_{i'}$	$\frac{1}{J^2} \left(\sum_j \frac{1}{n_{ij}} + \sum_j \frac{1}{n_{i'j}} \right)$
β_j	b_j	$\left(\frac{1}{IJ}\right)^2 \left((J-1)^2 \sum_i \frac{1}{n_{ij}} + \sum_{i, j' \neq j} \frac{1}{n_{ij'}} \right)$
$\beta_j - \beta_{j'}$	$b_j - b_{j'}$	$\frac{1}{I^2} \left(\sum_i \frac{1}{n_{ij}} + \sum_i \frac{1}{n_{ij'}} \right)$

associate "plus or minus values" with these estimates. In Example 57, the pooled estimate of σ from the joint strength data was found to be $s_P = 1.056$ MPa with $\nu = 12$ associated degrees of freedom. Since the data in Table 5.3 are balanced factorial data with $I = 2$, $J = 3$, and $m = 3$, using the first line of Table 5.5, it follows that 95 % two-sided confidence limits for the interaction $\alpha\beta_{ij}$ are

$$ab_{ij} \pm 2.179(1.056)\sqrt{\frac{(2-1)(3-1)}{3(2)(3)}}, \quad \text{that is} \quad ab_{ij} \pm .767 \text{ MPa}.$$

Notice then that all six of the intervals for interactions (centered at the values in Table 5.4) contain positive numbers, negative numbers, and 0. It is possible that the $\alpha\beta_{ij}$ are all essentially 0, and correspondingly the lack of parallelism on Fig. 5.6 is no more than a manifestation of experimental error. By the standard of these 95% individual confidence limits, the magnitude of the uncertainty associated with any fitted interaction exceeds that of the interaction itself and the apparent lack of parallelism is "in the noise range."

Next, using the second row of Table 5.5, it follows that 95% two-sided confidence limits for the cooling method main effects are

$$a_i \pm 2.179(1.056)\sqrt{\frac{2-1}{3(2)(3)}}, \quad \text{that is} \quad a_i \pm .542 \text{ MPa}.$$

This is in accord with the earlier more qualitative analysis of the joint strength data made on the basis of Fig. 5.6 alone. The calculation here, together with the facts that $a_1 = -.606$ and $a_2 = .606$, shows that one can be reasonably sure the main effect of water quench is negative and the main effect of oil quench is positive. The oil-quench joint strengths are on average larger than the water-quench strengths. But the call is still a relatively "close" one. The $\pm.542$ value is nearly as large as the fitted effects themselves.

Finally, as an illustration of the use of formula (5.22) in the comparison of main effects, consider the estimation of differences in antimony amount main effects, $\beta_j - \beta_{j'}$. Using the last row of Table 5.5, 95% two-sided confidence limits for differences in antimony main effects are

$$b_j - b_{j'} \pm 2.179(1.056)\sqrt{\frac{2}{3(2)}}, \quad \text{that is} \quad b_j - b_{j'} \pm 1.328 \text{ MPa}.$$

Recall that $b_1 = .428$, $b_2 = 1.811$, and $b_3 = -2.239$ and note that while b_1 and b_2 differ by less than 1.328 MPa, b_1 and b_3 differ by substantially more than 1.328, as do b_2 and b_3. This implies that while the evidence of a real difference between average strengths for levels 1 and 2 of antimony is not sufficient to allow one to make definitive statements, both antimony level 1 and antimony level 2 average joint strengths are clearly above that for antimony level 3. This conclusion is in accord with the earlier analysis based entirely on Fig. 5.6. The differences between antimony levels are clearly more marked (and evident above the background/experimental variation) than the cooling method differences.

Example 72 *Computing Factors from Table 5.6.* As a way of illustrating the intended meaning of the components of the formulas in Table 5.6, consider a hypothetical 3×3 factorial where $n_{12} = 1, n_{33} = 1$, and all other n_{ij} are 2. For the (i, j)-pair $(1, 1)$, sums appearing in the table would be

$$\frac{1}{n_{11}} = .5,$$

$$\sum_{j' \neq 1} \frac{1}{n_{1j'}} = \frac{1}{n_{12}} + \frac{1}{n_{13}} = 1.0 + .5 = 1.5,$$

$$\sum_{i' \neq 1} \frac{1}{n_{i'1}} = \frac{1}{n_{21}} + \frac{1}{n_{31}} = .5 + .5 = 1.0,$$

$$\sum_{i' \neq 1, j' \neq 1} \frac{1}{n_{i'j'}} = \frac{1}{n_{22}} + \frac{1}{n_{23}} + \frac{1}{n_{32}} + \frac{1}{n_{33}} = .5 + .5 + .5 + 1.0 = 2.5,$$

$$\sum_{j} \frac{1}{n_{1j}} = \frac{1}{n_{11}} + \frac{1}{n_{12}} + \frac{1}{n_{13}} = .5 + 1.0 + .5 = 2.0,$$

$$\sum_{i} \frac{1}{n_{i1}} = \frac{1}{n_{11}} + \frac{1}{n_{21}} + \frac{1}{n_{31}} = .5 + .5 + .5 = 1.5,$$

$$\sum_{i' \neq 1, j} \frac{1}{n_{i'j}} = \frac{1}{n_{21}} + \frac{1}{n_{22}} + \frac{1}{n_{23}} + \frac{1}{n_{31}} + \frac{1}{n_{32}} + \frac{1}{n_{33}}$$
$$= .5 + .5 + .5 + .5 + .5 + 1.0 = 3.5,$$

and

$$\sum_{i, j' \neq 1} \frac{1}{n_{ij'}} = \frac{1}{n_{12}} + \frac{1}{n_{13}} + \frac{1}{n_{22}} + \frac{1}{n_{23}} + \frac{1}{n_{32}} + \frac{1}{n_{33}}$$
$$= 1.0 + .5 + .5 + .5 + .5 + 1.0 = 4.0.$$

Using sums of these types, intervals for factorial effects can be computed even in this sort of unbalanced data situation.

5.2.3 Fitting and Checking Simplified Models for Balanced Two-Way Factorial Data

The possibility that interactions in a two-way factorial situation are negligible is one that brings important simplification to interpreting how Factors A and B affect the response y. In the absence of important interactions, one may think of A and B acting on y more or less "independently" or "separately." And if, in addition, the A main effects are negligible, one can think about the "two-knob black box system" as having only one knob (the B knob) that really does anything in terms of changing y.

The confidence intervals for two-way factorial effects introduced in this section are important tools for investigating whether some effects can indeed be ignored. A further step in this direction is to derive *fitted values* for y under assumptions that some of the factorial effects are negligible and to use these to compute *residuals*. The idea here is very much like what is done in regression analysis. Faced with a large list of possible predictor variables, one goal of standard regression analysis is to find an equation involving only a few of those predictors that does an adequate job of describing the response variable. In the search for such an equation, y values predicted by a candidate equation are subtracted from observed

Chapter 5. Experimental Design and Analysis for Process Improvement   275

y values to produce residuals, and these are plotted in various ways looking for possible problems with the candidate.

Finding predicted values for y under an assumption that some of the two-way factorial effects are 0 is in general a problem that must be addressed using a regression program and what are known as "dummy variables." And the matter is subtle enough that treating it here is not feasible. The reader is instead referred for the general story to books on regression analysis and intermediate-level books on statistical methods. What *can* be done here is to point out that in the special case that factorial data are balanced, appropriate fitted values can be obtained by simply adding to the grand sample mean fitted effects corresponding to those effects that one does not wish to assume are negligible.

That is, for balanced data, under the assumption that all $\alpha\beta_{ij}$ are 0, an appropriate estimator of the mean response when Factor A is at level i and Factor B is at level j (a fitted value for any y_{ijk}) is

$$\hat{y}_{ijk} = \overline{y}_{..} + a_i + b_j. \tag{5.24}$$

Balanced Data "No-Interaction" Fitted Values

Further, for balanced data, under the assumption that all $\alpha\beta_{ij}$ are 0 and all α_i are also 0, an appropriate estimator of the mean response when Factor B is at level j (a fitted value for any y_{ijk}) is

$$\hat{y}_{ijk} = \overline{y}_{..} + b_j. \tag{5.25}$$

Balanced Data "B Effects Only" Fitted Values

And again for balanced data, under the assumption that all $\alpha\beta_{ij}$ are 0 and all β_j are also 0, an appropriate estimator of the mean response when Factor A is at level i (a fitted value for any y_{ijk}) is

$$\hat{y}_{ijk} = \overline{y}_{..} + a_i. \tag{5.26}$$

Balanced Data "A Effects Only" Fitted Values

Using one of the relationships (5.24) through (5.26), residuals are then defined as differences between observed and fitted values:

$$e_{ijk} = y_{ijk} - \hat{y}_{ijk}. \tag{5.27}$$

Residuals

It is hopefully clear that the residuals defined in Eq. (5.27) are not the same as those defined in Sect. 5.1 and used there to check on the reasonableness of the basic one-way normal model assumptions. The residuals (5.27) are for a more specialized model, one of "no interactions," "B effects only," or "A effects only," depending upon which of Eqs. (5.24) through (5.26) is used to find fitted values.

Once residuals (5.27) have been computed, they can be plotted in the same ways that one plots residuals in regression contexts. If the corresponding simplified model for y is a good one, residuals should "look like noise" and carry no obvious patterns or trends (that indicate that something important has been missed in modeling the response). One hopes to see a fairly linear normal plot of

TABLE 5.7. Fitted values and residuals for the no-interaction model of solder joint strength

i	j	$\hat{y}_{ijk} = \overline{y}_{..} + a_i + b_j$	$e_{ijk} = y_{ijk} - \hat{y}_{ijk}$
1	1	$\hat{y}_{11k} = 19.30\overline{5} + (-.60\overline{5}) + .42\overline{7} = 19.128$	$-.53, .37, -.13$
1	2	$\hat{y}_{12k} = 19.30\overline{5} + (-.60\overline{5}) + 1.81\overline{1} = 20.511$	$1.79, -1.01, -0.01$
1	3	$\hat{y}_{13k} = 19.30\overline{5} + (-.60\overline{5}) + (-2.23\overline{8}) = 16.461$	$-1.26, .64, .14$
2	1	$\hat{y}_{21k} = 19.30\overline{5} + .60\overline{5} + .42\overline{7} = 20.339$	$-.34, .56, .06$
2	2	$\hat{y}_{22k} = 19.30\overline{5} + .60\overline{5} + 1.81\overline{1} = 21.722$	$-.82, 1.18, -1.12$
2	3	$\hat{y}_{23k} = 19.30\overline{5} + .60\overline{5} + (-2.23\overline{8}) = 17.672$	$-1.27, 1.33, .428$

residuals and hopes for "trendless/random scatter with constant spread" plots of residuals against levels of Factors A and B. Departures from these expectations draw into doubt the appropriateness of the reduced or simplified description of y.

Example 73 *(Example 60 continued.) The earlier analysis of the solder joint strength data suggests that a no-interaction description of joint strength might be tenable. To further investigate the plausibility of this, consider the computation and plotting of residuals based on fitted values (5.24). Table 5.7 shows the calculation of the six fitted values (5.24) and lists the 18 corresponding residuals for the no-interaction model (computed from the raw data listed in Table 5.3).*

Figure 5.11 is a normal plot of the 18 residuals listed in Table 5.7, and then Figs. 5.12 and 5.13 are, respectively, plots of residuals versus level of A and then against level of B. The normal plot is reasonably linear (except possibly for its extreme lower end), and the plot of residuals against level of Factor A is completely unremarkable. The plot against level of Factor B draws attention to the fact that the first level of antimony has residuals that seem somewhat smaller than those from the other two levels of antimony. But on the whole, the three plots offer no strong reason to dismiss "normal distributions with constant variance and no interactions between A and B" model of joint strength.

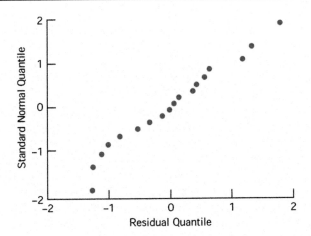

FIGURE 5.11. Normal plot of the residuals for a no-interaction model of solder joint strength

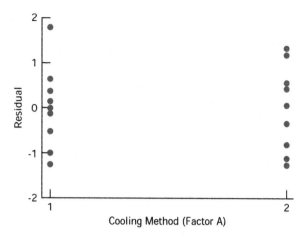

FIGURE 5.12. Plot of no-interaction residuals versus level of Factor A in the soldering study

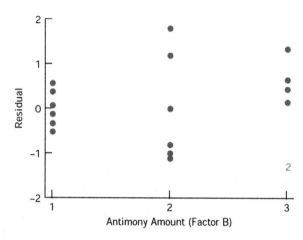

FIGURE 5.13. Plot of no-interaction residuals versus level of Factor B in the soldering study

Section 5.2 Exercises

1. What does the phrase "two-way factorial study" mean? Is each factor restricted to two levels? Briefly, give examples.

278   Chapter 5. Experimental Design and Analysis for Process Improvement

2. In a 3 × 3 two-way factorial study, fitted A main effects are $a_1 = 2$ and $a_2 = 3$, fitted B main effects are $b_1 = -1$ and $b_2 = 2$, and four of the fitted AB interactions are $ab_{11} = 1, ab_{12} = -1, ab_{21} = 2$ and $ab_{22} = -3$. The average of the 9 sample means is $\bar{y} = 10$.

   (a) What was the sample mean response where both A and B were at their third levels?

   (b) Suppose in this problem $m = 2$ and $s_P = 2$. Using a 95 % confidence level, find an appropriate "margin of error" Δ_{ij} for estimating each of the fitted interactions.

   (c) If you were to make an interaction plot (a plot trace of sample means against level of A, one for each level of B), would you see a departure from "parallelism"? Why or why not? (Hint: consider your answer to (b) above.)

3. Suppose in a two-way factorial study, the fitted interactions ab_{ij} are both statistically detectable and large in a practical sense.

   (a) Is it reasonable to think of a Factor A effect without regard to the level of Factor B? Why or why not?

   (b) Is it reasonable to think of a Factor A effect for each level of Factor B separately? Why or why not?

   (c) Can Factor A effects be judged to be essentially the same at each separate level of Factor B? Why or why not?

4. Students Bauer, Brinnk, and Fife studied the performance of 3 different copy machines. $m = 5$ copies of the same CAD drawing were made on each machine for 3 different enlargement settings, and the length of a particular line segment on that drawing was measured. (All the measuring was done by a single student, and as a means of getting a handle on measurement precision, that student measured the line on the original drawing 10 times, obtaining lengths with sample mean 2.0058 in and sample standard deviation .0008 in.) Sample means and standard deviations for the measurements on the copies are below.

		Factor B-enlargement setting		
		100 %	129 %	155 %
Factor A-copier	1	$\bar{y}_{11} = 2.0158$ $s_{11} = .0028$	$\bar{y}_{12} = 2.6058$ $s_{12} = .0024$	$\bar{y}_{13} = 3.1192$ $s_{13} = .0018$
	2	$\bar{y}_{21} = 2.0134$ $s_{21} = .0011$	$\bar{y}_{22} = 2.5996$ $s_{22} = .0009$	$\bar{y}_{23} = 3.1038$ $s_{23} = .0011$
	3	$\bar{y}_{31} = 2.0206$ $s_{31} = .0011$	$\bar{y}_{32} = 2.6098$ $s_{32} = .0022$	$\bar{y}_{33} = 3.1150$ $s_{33} = .0007$

A pooled sample standard deviation computed from the 9 values s_{ij} is $s_P = .0017$.

Chapter 5. Experimental Design and Analysis for Process Improvement 279

(a) Is it a surprise that $s_P = .0017$ is larger than the .0008 in standard deviation obtained in the preliminary measurement study? Why or why not?

(b) The $m = 5$ measurements from copier 1 at the 100 % setting were in fact 2.014, 2.020, 2.017, 2.015, and 2.013. What are the values of the corresponding residuals?

(c) Is it possible from what is given to find *all* the residuals? How might one use *all* the residuals?

(d) Four of the 9 fitted copier $\times$ enlargement setting interactions are $ab_{11} = -.00296, ab_{12} = -.00142, ab_{21} = .00264$, and $ab_{22} = .00038$. Find the other five interactions.

(e) Are ANY of the nine interactions statistically detectable (using, say 95 % two-sided confidence limits for each interaction as a basis of judging this)?

(f) Give 95 % individual two-sided confidence limits for the difference in copier 1 and 2 main effects, $\alpha_1 - \alpha_2$. Is it credible to use your interval for every level of enlargement? Why or why not? (Hint: consider your answer to part (e).)

(g) Make an interaction plot for the set of means. Let the horizontal axis correspond to enlargement setting and the vertical axis correspond to mean line length. You should have one trace for each level of copier. How does your plot support your answer to (e)?

5.3 2^p Factorials

Section 5.2 began discussion of how to profitably conduct and analyze the results of process-improvement experiments involving several factors. The subject there was the case of two factors. We now consider the general case of $p \geq 2$ factors. This will be done primarily for situations where each factor has only two levels, the $2 \times 2 \times \cdots \times 2$ or 2^p factorial studies. This may at first seem like a severe restriction, but in practical terms it is not. As p grows, experimenting at more than two levels of many factors in a full p-way factorial arrangement quickly becomes infeasible because of the large number of combinations involved. And there are some huge advantages associated with the 2^p factorials in terms of ease of data analysis.

The section begins with a general discussion of notation and how effects and fitted effects are defined in a p-way factorial. Then, methods for judging the statistical detectability of effects in the 2^p situation are presented, first for cases where there is some replication and then for cases where there is not. Next, the Yates algorithm for computing fitted effects for 2^p factorials is discussed. Finally, there is a brief discussion of fitting and checking models for balanced 2^p factorials that involve only some of the possible effects.

5.3.1 Notation and Defining Effects in p-Way Factorials

Consider now an instance of the generic process experimentation scenario represented in Fig. 5.1 where there are p knobs under the experimenter's control. Naming the factors involved A, B, C, ... and supposing that they have, respectively, $I, J, K, \ldots$ possible levels, a **full factorial in the p factors** is a study where one has data from all $I \times J \times K \times \cdots$ different possible combinations of levels of these p factors. Figure 5.14 provides a visual representation of these possible combinations laid out in a three-dimensional rectangular array for the case of $p = 3$ factors. Then, Examples 74 and 75 introduce, respectively, $p = 3$ and $p = 4$-way factorial data sets that will be used in this section to illustrate methods of 2^p factorial analysis.

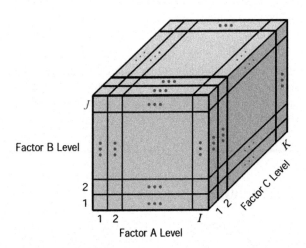

FIGURE 5.14. IJK cells for a three-way factorial

Example 74 *Packing Properties of Crushed T-61 Tabular Alumina Powder. Ceramic Engineering researchers Leigh and Taylor, in their 1990* Ceramic Bulletin *paper "Computer-Generated Experimental Designs," present the results of a 2^3 factorial study on the packing properties of crushed T-61 tabular alumina powder. Densities, y, of the material were determined under several different measurement protocols. Two different "mesh sizes" of particles were employed, full flasks of the material of two different volumes were used, and the flasks were subjected to one of two vibration conditions before calculating densities. This can be*

Chapter 5. Experimental Design and Analysis for Process Improvement   281

thought of as a three-way factorial situation where the factors and their levels are

> Factor A—mesh size   6 mesh vs. 60 mesh
> Factor B—flask   100 cc vs. 500 cc
> Factor C—vibration   none vs. yes

Table 5.8 gives the $m = 5$ densities reported by the researchers for each of the $r = 2 \times 2 \times 2$ measurement protocols and corresponding sample means and standard deviations. The goal of a 2^3 factorial analysis of the data in Table 5.8 will be to identify structure that can be interpreted in terms of the individual and joint effects of the three factors mesh, flask, and vibration.

TABLE 5.8. Crushed T-61 tabular alumina powder densities, sample means, and sample standard deviations for 2^3 different measurement protocols (g/cc)

Mesh	Flask	Vibration	Measured density	$\bar{y}$	s
6	100	None	2.13, 2.15, 2.15, 2.19, 2.20	2.164	.030
60	100	None	1.96, 2.01, 1.91, 1.95, 2.00	1.966	.040
6	500	None	2.23, 2.19, 2.18, 2.21, 2.22	2.206	.021
60	500	None	1.88, 1.90, 1.87, 1.89, 1.89	1.886	.011
6	100	Yes	2.16, 2.31, 2.32, 2.22, 2.35	2.272	.079
60	100	Yes	2.29, 2.29, 2.23, 2.39, 2.18	2.276	.079
6	500	Yes	2.16, 2.39, 2.30, 2.33, 2.43	2.322	.104
60	500	Yes	2.35, 2.38, 2.26, 2.34, 2.34	2.334	.044

Example 75 *Bond Pullouts on Dual In-Line Packages.* The article "An Analysis of Means for Attribute Data Applied to a 2^4 Factorial Design" by R. Zwickl that appeared in the Fall 1985 ASQC Electronics Division Technical Supplement describes a four-way factorial study done to help improve the manufacture of an electronic device called a dual in-line package. Counts were made of the numbers of bonds (out of 96) showing evidence of ceramic pullout (small numbers are desirable) on devices made under all possible combinations of levels of four factors. The factors and their levels used in the study were

> Factor A—ceramic surface   unglazed vs. glazed
> Factor B—metal film thickness   normal vs. 1.5 times normal
> Factor C—annealing time   normal vs. 4 times normal
> Factor D—prebond clean   normal clean vs. no clean

Table 5.9 on page 282 gives Zwickl's data. (We will suppose that the counts recorded in Table 5.9 occurred on one device made under each set of experimental conditions.) Zwickl's data are unreplicated 2^4 factorial data. (In fact, they are attribute or count data. But for present purposes it will suffice to ignore this fact and treat them as if they were measurements obtained from 16 different process *conditions.) The object of a 2^4 factorial analysis will be to find simple structure in the data that can be discussed in terms of the separate and joint effects of the four factors.*

TABLE 5.9. Counts of pullouts on dual in-line packages under 2^4 sets of experimental conditions

A	B	C	D	Pullouts
Unglazed	Normal	Normal	Normal clean	9
Glazed	Normal	Normal	Normal clean	70
Unglazed	1.5×	Normal	Normal clean	8
Glazed	1.5×	Normal	Normal clean	42
Unglazed	Normal	4×	Normal clean	13
Glazed	Normal	4×	Normal clean	55
Unglazed	1.5×	4×	Normal clean	7
Glazed	1.5×	4×	Normal clean	19
Unglazed	Normal	Normal	No clean	3
Glazed	Normal	Normal	No clean	6
Unglazed	1.5×	Normal	No clean	1
Glazed	1.5×	Normal	No clean	7
Unglazed	Normal	4×	No clean	5
Glazed	Normal	4×	No clean	28
Unglazed	1.5×	4×	No clean	3
Glazed	1.5×	4×	No clean	6

It should be obvious from analogy with what was done in the previous section that general notation for p-way factorial analyses will involve at least p subscripts, one for each of the factors. For example, for the case of $p = 3$ factors, one will write

y_{ijkl} = the lth observation at the ith level of A, the jth level of B, and the kth level of C,

μ_{ijk} = the long-run mean system response when A is at level i, B is at level j, and C is at level k,

n_{ijk} = the number of observations at level i of A, level j of B, and the kth level of C,

$\bar{y}_{ijk}$ = the sample mean system response when A is at level i, B is at level j, and C is at level k, and

s_{ijk} = the sample standard deviation when A is at level i, B is at level j, C is at level k, and write the one-way model equation (5.1) in three-way factorial notation as

$$y_{ijkl} = \mu_{ijk} + \epsilon_{ijkl}.$$

Three-Way Model Equation

Further, the obvious "dot subscript" notation can be used to indicate averages of sample or long-run mean responses over the levels of factors "dotted out" of the notation. For example, for the case of $p = 3$ factors, one can write

$$\bar{y}_{.jk} = \frac{1}{I}\sum_{i}\bar{y}_{ijk}, \quad \bar{y}_{.j.} = \frac{1}{IK}\sum_{i,k}\bar{y}_{ijk}, \quad \bar{y}_{...} = \frac{1}{IJK}\sum_{i,j,k}\bar{y}_{ijk}$$

and so on.

The multiple subscript notation is needed to write down technically precise formulas for general p-way factorials. However, it is extremely cumbersome and unpleasant to use. One of the benefits of dealing primarily with 2^p problems is that something more compact and workable can be done when all factors have only two levels. In 2^p contexts it is common to designate (arbitrarily if there is no reason to think of levels of a given factor as ordered) a "first" level of each factor as the "low" level and the "second" as the "high" level. (Often the shorthand "−" is used to designate a low level and the shorthand "+" is used to stand for a high level.) Combinations of levels of the factors can then be named by listing those factors which appear at their second or high levels. Table 5.10 illustrates this naming convention for the 2^3 case.

TABLE 5.10. Naming convention for 2^p factorials

Level of A	i	Level of B	j	Level of C	k	Combination name
−	1	−	1	−	1	(1)
+	2	−	1	−	1	a
−	1	+	2	−	1	b
+	2	+	2	−	1	ab
−	1	−	1	+	2	c
+	2	−	1	+	2	ac
−	1	+	2	+	2	bc
+	2	+	2	+	2	abc

Armed with appropriate notation, one can begin to define effects and their fitted counterparts. The place to start is with the natural analogs of the two-way factorial main effects introduced in Definitions 63 through 66. These were row or column averages of cell means minus a grand average. That is, they were averages of cell means for a level of the factor under discussion minus a grand average. That same thinking can be applied in p-way factorials, provided one realizes that averaging must be done over levels of $(p-1)$ other factors. The corresponding definitions will be given here for $p = 3$ factors with the understanding that the reader should be able to reason by analogy (simply adding some dot subscripts) to make definitions for cases with $p > 3$.

Definition 76 *The (theoretical)* **main effects of factors A, B, and C** *in a complete three-way factorial are*

$$\alpha_i = \mu_{i..} - \mu_{...}, \quad \beta_j = \mu_{.j.} - \mu_{...}, \quad \text{and} \quad \gamma_k = \mu_{..k} - \mu_{...}.$$

Definition 77 *The (estimated or)* **fitted main effects of factors A, B, and C** *in a complete three-way factorial are*

$$a_i = \bar{y}_{i..} - \bar{y}_{...}, \quad b_j = \bar{y}_{.j.} - \bar{y}_{...}, \quad \text{and} \quad c_k = \bar{y}_{..k} - \bar{y}_{...}.$$

It is an algebraic consequence of the form of Definitions 76 and 77 that main effects and fitted main effects sum to zero over the levels of the factor under consideration. That is, for the case of three factors, one has the extension of display (5.20):

$$\sum_i \alpha_i = 0, \sum_j \beta_j = 0, \sum_k \gamma_k = 0, \sum_i a_i = 0, \sum_j b_j = 0, \text{ and } \sum_k c_k = 0.$$

One immediate implication of these relationships is that where factors have only two levels, one need only calculate one of the fitted main effects for a factor. The other is then obtained by a simple sign change.

TABLE 5.11. Alternative notations for the sample mean measured alumina powder densities

Mesh (A)	Flask (B)	Vibration (C)	Sample mean
−	−	−	$\bar{y}_{111} = \bar{y}_{(1)} = 2.164$
+	−	−	$\bar{y}_{211} = \bar{y}_{\mathrm{a}} = 1.966$
−	+	−	$\bar{y}_{121} = \bar{y}_{\mathrm{b}} = 2.206$
+	+	−	$\bar{y}_{221} = \bar{y}_{\mathrm{ab}} = 1.886$
−	−	+	$\bar{y}_{112} = \bar{y}_{\mathrm{c}} = 2.272$
+	−	+	$\bar{y}_{212} = \bar{y}_{\mathrm{ac}} = 2.276$
−	+	+	$\bar{y}_{122} = \bar{y}_{\mathrm{bc}} = 2.322$
+	+	+	$\bar{y}_{222} = \bar{y}_{\mathrm{abc}} = 2.334$

Example 78 (**Example 74 continued.**) *Considering again the density measurements of Leigh and Taylor, one might make the "low" versus "high" level designations as:*

   Factor A—Mesh Size   6 mesh (−) and 60 mesh (+)
   Factor B—Flask   100 cc (−) and 500 cc (+)
   Factor C—Vibration   none (−) and yes (+)

With these conventions, Table 5.11 gives two sets of notation for the sample means listed originally in Table 5.8. Both the triple subscript and the special 2^3 conventions are illustrated.

Chapter 5. Experimental Design and Analysis for Process Improvement   285

It is then the case that

$$\bar{y}_{...} = \frac{1}{8}(2.164 + 1.966 + \cdots + 2.334) = 2.1783,$$

and, for example,

$$\bar{y}_{2..} = \frac{1}{4}(\bar{y}_a + \bar{y}_{ab} + \bar{y}_{ac} + \bar{y}_{abc}) = \frac{1}{4}(1.966 + 1.886 + 2.276 + 2.334) = 2.1155.$$

So using Definition 77

$$a_2 = \bar{y}_{2..} - \bar{y}_{...} = 2.1155 - 2.1783 = -.063.$$

The average of the four "60 mesh" mean densities is .063 g/cc below the overall average of the eight sample means. A simple sign change then says that $a_1 = .063$, and the main effect of mesh size at its low level is positive .063 g/cc.

Similar calculations then show that for the means of Table 5.11

$$b_2 = \bar{y}_{.2.} - \bar{y}_{...} = 2.187 - 2.178 = .009 \quad \text{and}$$
$$c_2 = \bar{y}_{..2} - \bar{y}_{...} = 2.301 - 2.178 = .123.$$

Then switching signs for these two-level factors, one also has $b_1 = -.009$ and $c_1 = -.123$.

*Figure 5.15 on page 286 is a very common and helpful kind of graphic for displaying the 2^3 factorial means sometimes called a **cube plot**. On the plot for this example, the fact that $a_2 = -.063$ g/cc says that the average of the means on the right face of the cube is .063 g/cc below the overall average of the eight sample means pictured. The fact that $b_2 = .009$ says that the average of the means on the top face of the cube is .009 g/cc above the overall average. And the fact that $c_2 = .123$ says that the average of the means on the back face of the cube is .123 g/cc above the overall average.*

Main effects do not completely describe a p-way factorial set of means any more than they completely describe a two-way factorial. There are interactions to consider as well. In a p-way factorial, two-factor interactions are what one would compute as interactions via the methods of the previous section *after averaging out over all levels of all other factors*. For example, in a three-way factorial, two-factor interactions between A and B are what one has for interactions from Sect. 5.2 *after averaging over levels of Factor C*. The precise definitions for the three-factor situation follow (and the reader can reason by analogy and the addition of subscript dots to corresponding definitions for more than three factors).

Definition 79 *The (theoretical) **two-factor interactions** of pairs of factors A, B, and C in a complete three-way factorial are*

$$\alpha\beta_{ij} = \mu_{ij.} - (\mu_{...} + \alpha_i + \beta_j),$$
$$\alpha\gamma_{ik} = \mu_{i.k} - (\mu_{...} + \alpha_i + \gamma_k), \quad \text{and}$$
$$\beta\gamma_{jk} = \mu_{.jk} - (\mu_{...} + \beta_j + \gamma_k).$$

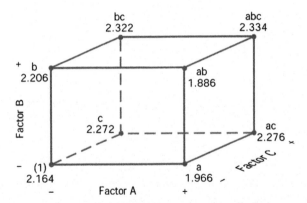

FIGURE 5.15. Cube plot of sample mean measured alumina powder densities (g/cc)

Definition 80 *The (estimated or) fitted two-factor interactions of pairs of factors A, B, and C in a complete three-way factorial are*

$$ab_{ij} = \bar{y}_{ij.} - (\bar{y}_{...} + a_i + b_j),$$
$$ac_{ik} = \bar{y}_{i.k} - (\bar{y}_{...} + a_i + c_k), \quad \text{and}$$
$$bc_{jk} = \bar{y}_{.jk} - (\bar{y}_{...} + b_j + c_k).$$

A main effect is in some sense the difference between what exists (in terms of an average response) and what is explainable in terms of only a grand mean. A two-factor interaction is similarly a difference between what exists (in terms of an average response) and what can be accounted for by considering a grand mean and the factors acting individually.

Just as interactions in two-way factorials sum to zero across rows or columns, it is a consequence of the form of Definitions 79 and 80 that two-factor interactions in p-way factorials also sum to 0 over levels of either factor involved. In symbols

$$\sum_i ab_{ij} = \sum_j ab_{ij} = 0, \quad \sum_i \alpha\beta_{ij} = \sum_j \alpha\beta_{ij} = 0,$$
$$\sum_i ac_{ik} = \sum_k ac_{ik} = 0, \quad \sum_i \alpha\gamma_{ik} = \sum_k \alpha\gamma_{ik} = 0, \quad \text{and}$$
$$\sum_j bc_{jk} = \sum_k bc_{jk} = 0, \quad \text{and} \quad \sum_j \beta\gamma_{jk} = \sum_k \beta\gamma_{jk} = 0.$$

One important consequence of these relationships is that for cases where factors have only two levels, one needs to calculate only one of the four interactions for a given pair of factors. The other three can then be obtained by appropriate changes of sign.

Example 81 *(Examples 74 and 78 continued.) Turning once again to the alumina powder density study, consider the calculation of AB two-factor interactions.*

Averaging front to back on the cube plot of Fig. 5.15 produces the values:

$$\bar{y}_{11.} = 2.218, \quad \bar{y}_{21.} = 2.121, \quad \bar{y}_{12.} = 2.264, \quad \text{and} \quad \bar{y}_{22.} = 2.110.$$

An AB interaction plot of these is shown in Fig. 5.16, and there is some lack of parallelism in evidence. The size of this lack of parallelism can be measured by computing

$$ab_{22} = \bar{y}_{22.} - (\bar{y}_{...} + a_2 + b_2) = 2.110 - (2.178 + (-.063) + .009) = -.014.$$

Then, since $ab_{21} + ab_{22} = 0, ab_{21} = .014$. Since $ab_{12} + ab_{22} = 0, ab_{12} = .014$. And finally, since $ab_{11} + ab_{21} = 0, ab_{11} = -.014$. Similar calculations can be done to find the fitted two-way interactions of A and C and of B and C. The reader should verify that (except possibly for roundoff error)

$$ac_{22} = .067 \quad \text{and} \quad bc_{22} = .018.$$

Others of the AC and BC two-factor interactions in this 2^3 study can be obtained by making appropriate sign changes.

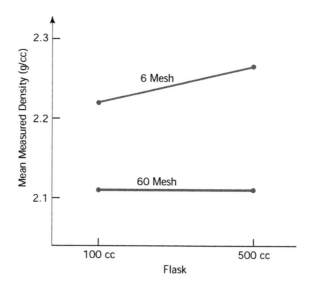

FIGURE 5.16. Interaction plot for alumina powder density after averaging over vibration conditions

Main effects and two-factor interactions do not tell the whole story about a p-way factorial set of means. For example, in a three-factor context, there are many quite different sets of means having a given set of main effects and two-factor interactions. One must go further in defining effects to distinguish between these different possibilities. The next logical step beyond two-factor interactions would seem to be some kind of **three-factor interactions**. To see what these might be, it is helpful to remember that (1) a main effect is the difference between what exists and what is explainable in terms of only a grand mean and (2) a two-factor interaction is the difference between what exists and what can be accounted for by considering a grand mean and main effects. This suggests that one defines a three-factor interaction to be the difference between what exists in terms of an average response and what is explainable in terms of a grand mean, main effects, and two-factor interactions. That is, one has Definitions 82 and 83 for the case of $p = 3$. (The reader can reason by analogy to produce definitions of three-factor interactions in higher way studies by adding some dot subscripts to the three-way factorial expressions.)

Definition 82 *The (theoretical)* **three-factor interactions** *of factors A, B, and C in a complete three-way factorial are*

$$\alpha\beta\gamma_{ijk} = \mu_{ijk} - (\mu_{...} + \alpha_i + \beta_j + \gamma_k + \alpha\beta_{ij} + \alpha\gamma_{ik} + \beta\gamma_{jk}).$$

Definition 83 *The (estimated or)* **fitted three-factor interactions** *of factors A, B, and C in a complete three-way factorial are*

$$abc_{ijk} = \bar{y}_{ijk} - (\bar{y}_{...} + a_i + b_j + c_k + ab_{ij} + ac_{ik} + bc_{jk}).$$

Three-factor interactions sum to 0 over any of their three indices. That means that for the case of $p = 3$ factors

$$\sum_i abc_{ijk} = \sum_j abc_{ijk} = \sum_k abc_{ijk} = 0 \quad \text{and}$$

$$\sum_i \alpha\beta\gamma_{ijk} = \sum_j \alpha\beta\gamma_{ijk} = \sum_k \alpha\beta\gamma_{ijk} = 0.$$

So, in the case of 2^p studies, it again suffices to compute only one fitted interaction for a set of three factors and then obtain all others of that type by appropriate choice of signs.

Example 84 *(Examples 74 through 81 continued.) In the powder density study, the fitted three-factor interaction for the "all-high level" combination is*

$$\begin{aligned} abc_{222} &= \bar{y}_{222} - (\bar{y}_{...} + a_2 + b_2 + c_2 + ab_{22} + ac_{22} + bc_{22}) \\ &= 2.334 - (2.178 + (-.063) + .009 + .123 + (-.014) + .067 + .018) \\ &= .016. \end{aligned}$$

Because the fitted interactions in this 2^3 study must add to 0 over levels of any one of the factors, it is straightforward to see that those abc_{ijk} with an even number of subscripts equal to 1 are .016, while those with an odd number of subscripts equal to 1 are $-.016$.

Hopefully, the pattern of how factorial effects are defined is by now clear. To define an interaction effect involving a particular set of q factors, one first averages means over all levels of all other factors and then takes a difference between such an average mean and the sum of a grand mean, main effects, and interactions of order less than q. The result is a quantity that in some sense measures how much of the system response is explainable only in terms of "what the factors do q at a time." The objective of a p-way factorial analysis is to hopefully identify some few effects that taken together both account for most of the variation in response and also have a simple interpretation. This is not always be possible. But when it is, a factorial analysis can provide important insight into how p factors impact the response.

Objective of a p-Way Analysis

5.3.2 Judging the Detectability of 2^p Factorial Effects in Studies with Replication

Although the examples used in this section have been ones where every factor has only two levels, the definitions of effects have been perfectly general, applicable to any full factorial. But from this point on in this section, the methods introduced are going to be specifically 2^p factorial methods. Of course there are data analysis tools for the more general case (that can be found in intermediate-level statistical method texts). The methods that follow, however, are particularly simple and cover what is with little doubt the most important part of full factorial experimentation for modern process improvement.

We have noted that all effects of a given type in a 2^p factorial differ from each other by at most a sign change. This makes it possible to concentrate on the main effects and interactions for the "all factors at their high levels" treatment combination and still have a complete description of how the factors impact the response. In fact, people sometimes go so far as to call a_2, b_2, ab_{22}, c_2, ac_{22}, bc_{22}, abc_{222}, and so on "the" fitted effects in a 2^p factorial (slurring over the fact that there are effects corresponding to low levels of the factors). This subsection considers the issue of identifying those fitted effects that are big enough to indicate that the corresponding effect is detectable above the baseline experimental variation, under the assumption that there is some replication in the data.

The most effective tool of inference for 2^p factorial effects is the relevant specialization of expression (5.8). As it turns out, every effect in a 2^p factorial is a linear combination of means (an L) with coefficients that are all $\pm 1/2^p$. The corresponding fitted effect is the corresponding linear combination of sample means (the corresponding $\hat{L}$). So under the constant variance normal distribution model assumptions (5.1), if E is a generic 2^p factorial effect and $\hat{E}$ is the corresponding fitted effect, formula (5.8) can be specialized to give confidence limits for E of the form

$$\hat{E} \pm ts_P \frac{1}{2^p}\sqrt{\frac{1}{n_{(1)}} + \frac{1}{n_a} + \frac{1}{n_b} + \frac{1}{n_{ab}} + \cdots}. \quad (5.28)$$

Confidence Limits for a 2^p Factorial Effect

When the plus or minus value prescribed by this formula is larger in magnitude than a fitted effect, the real nature (positive, negative, or 0) of the corresponding effect is in doubt.

Example 85 *(Example 74 through 84 continued.) The fitted effects corresponding to the 60 mesh/500cc/vibrated flask conditions in the alumina powder density study (the "all-high levels" combination) have already been calculated to be*

$$a_2 = -.063, b_2 = .009, ab_{22} = -.014, c_2 = .123,$$
$$ac_{22} = .067, bc_{22} = .018, \text{ and } abc_{222} = .016.$$

Formula (5.28) allows one to address the question of whether any of these empirical values provides clear evidence of the nature of the corresponding long-run effect.

In the first place, using the sample standard deviations given in Table 5.8, one has

$$s_P = \sqrt{\frac{(5-1)(.030)^2 + (5-1)(.040)^2 + \cdots + (5-1)(.044)^2}{(5-1) + (5-1) + \cdots + (5-1)}} = .059 \text{ g/cc}.$$

Actually, before going ahead to use s_P and formula (5.28), one should apply the methods of Sect. 5.1 to check on the plausibility of the basic one-way normal model assumptions. The reader can verify that a normal plot of the residuals is fairly linear. But, in fact, a test like "Bartlett's test" applied to the sample standard deviations in Table 5.8 draws into serious question the appropriateness of the "constant σ" part of the usual model assumptions. For the time being, the fact that there is nearly an order of magnitude difference between the smallest and largest sample standard deviations in Table 5.8 will be ignored. The rationale for doing so is as follows. The t intervals (5.28) are generally thought to be fairly "robust" against moderate departures from the constant σ model assumption (meaning that nominal confidence levels, while not exactly correct, are usually not ridiculously wrong either). So rather than just "give up and do nothing in the way of inference" when it seems there may be a problem with the model assumptions, it is better to go ahead with caution. One should then remember that the confidence levels cannot be trusted completely and agree to avoid making "close calls" of large engineering or financial impact based on the resulting inferences.

Then, assuming for the moment that the one-way model is appropriate, note that s_P has associated with it $\nu = (5-1) + (5-1) + \cdots + (5-1) = 32$ degrees of freedom. So since the .975 quantile of the t_{32} distribution is 2.037, 95% two-sided confidence limits for any one of the 2^3 factorial effects E are

$$\hat{E} \pm 2.037(.059)\frac{1}{2^3}\sqrt{\frac{1}{5} + \frac{1}{5} + \cdots + \frac{1}{5}}, \text{ i.e., } \hat{E} \pm .019 \text{ g/cc}.$$

By this standard, only the A main effects, C main effects, and AC two-factor interactions are clearly detectable. It is comforting here (especially in light of the caution necessitated by the worry over appropriateness of the constant variance assumption) that all of $a_2, c_2,$ and ac_{22} are not just larger than .019 in magnitude but substantially so. It seems pretty safe to conclude that mesh size (Factor A) and vibration condition (Factor C) have important effects on the mean measured density of this powder but that the size of the flask used (Factor B) does not affect mean measured density in any way that can be clearly delineated on the basis of these data. (No main effect or interaction involving B is visible above the experimental variation.)

The fact that the AC interaction is nonnegligible says that one may not think of changing mesh size as doing the same thing to mean measured density when the flask is vibrated as when it is not. Figure 5.17 shows an interaction plot for the average sample means $\bar{y}_{i.k}$ obtained by averaging out over the two flask sizes. 6 mesh material consists of (a mixture of both coarse and finely ground) material that will pass through a fairly coarse screen. 60 mesh material is that (only relatively fine material) that will pass through a fine screen. It is interesting that Leigh and Taylor's original motivation for their experimentation was to determine if their density measurement system was capable of detecting changes in material particle size mix on the basis of measured density. The form of Fig. 5.17 suggests strongly that to detect mix changes on the basis of measured density, their system should be operated with unvibrated samples.

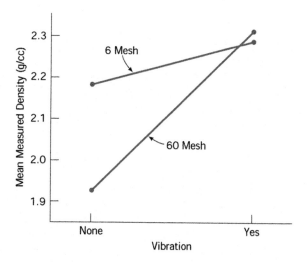

FIGURE 5.17. Interaction plot for alumina powder density after averaging over levels of flask size

5.3.3 Judging the Detectability of 2^p Factorial Effects in Studies Lacking Replication

The use of formula (5.28) to make confidence limits for 2^p factorial effects depends upon the existence of some replication somewhere in a data set, so that s_P can be calculated. When at all possible, experiments need to include some replication. Most fundamentally, replication allows one to verify that experimental results are to some degree repeatable and to establish the limits of that repeatability. Without it, there is no completely honest way to tell whether changing levels of experimental factors is what causes observed changes in y, or if instead changes one observes in y amount only to random variation. But having said all this, one is sometimes forced to make the best of completely unreplicated data. And it is thus appropriate to consider what can be done to analyze data like those in Table 5.9 that include no replication.

The best existing method of detecting factorial effects in unreplicated 2^p data is one suggested by Cuthbert Daniel. His method depends upon the **principle of effect sparsity** and makes use of probability plotting. The principle of effect sparsity is a kind of Pareto principle for experiments. It says that often a relatively few factors account for most of the variation seen in experimental data. And when the principle governs a physical system, the job of detecting real effects amounts only to picking out the largest few. Daniel's logic for identifying those then goes as follows. Any fitted-effect $\hat{E}$ is related to its corresponding effect E as

$$\hat{E} = E + noise.$$

When σ (and therefore the noise level) is large, one might thus expect a normal plot of the $2^p - 1$ fitted effects (for the all-high combination) to be roughly linear. On the other hand, should σ be small enough that one has a chance of seeing the few important effects, they ought to lead to points that plot "off the line" established by the majority (that themselves represent small effects plus noise). These exceptional values thereby identify themselves as more than negligible effects plus experimental variation.

Actually, Daniel's original suggestion for the plotting of the fitted effects was slightly more sophisticated than that just described. There is an element of arbitrariness associated with the exact appearance of a normal plot of a_2, b_2, ab_{22}, c_2, ac_{22}, ab_{22}, abc_{222},.... This is because the signs on these fitted effects depend on the arbitrary designation of the "high" levels of the factors. (And the choice to plot fitted effects associated with the all-high combination is only one of 2^p possible choices.) Daniel reasoned that plotting with the absolute values of fitted effects would remove this arbitrariness.

Now if Z is standard normal and has quantile function $Q_z(p)$, then $|Z|$ has quantile function

$$Q(p) = Q_z\left(\frac{1+p}{2}\right). \quad (5.29)$$

"Half-Normal" Quantile Function

Rather than plotting standard normal quantiles against fitted-effect quantiles as in Sect. 4.1, Daniel's original idea was to plot values of the quantile function $Q(p)$ specified in Eq. (5.29) versus absolute-fitted-effect quantiles, to produce a **half-normal plot**. This text will use both normal plotting of fitted effects and half-normal plotting of absolute fitted effects. The first is slightly easier to describe and usually quite adequate, but as Daniel pointed out, the second is somewhat less arbitrary.

Example 86 *(Example 75 continued.) Return again to Zwickl's data on pullouts on dual in-line packages given in Table 5.9 early in this section. As it turns out, the fitted effects for the counts of Table 5.9 are*

$$a_2 = 11.5, b_2 = -6.0, ab_{22} = -4.6, c_2 = -.6, ac_{22} = -1.5, bc_{22} = -2.3,$$
$$abc_{222} = -1.6, d_2 = -10.3, ad_{22} = -7.1, bd_{22} = 2.9, abd_{222} = 2.5,$$
$$cd_{22} = 3.8, acd_{222} = 3.6, bcd_{222} = -.6, \text{ and } abcd_{2222} = -1.3.$$

Table 5.12 on page 294 then gives the coordinates of the 15 points that one plots to make a normal plot of the fitted effects. And Table 5.13 gives the coordinates of the points that one plots to make a half-normal plot of the absolute fitted effects. The corresponding plots are given in Figs. 5.18 and 5.19 on page 295.

The plots in Figs. 5.18 and 5.19 are not as definitive as one might have hoped. None of the fitted effects or absolute fitted effects stand out as tremendously larger than the others. But it is at least clear from the half-normal plot in Fig. 5.19 that no more than four and probably at most two of the effects should be judged "detectable" on the basis of this data set. There is some indication in these data that the A and D main effects are important in determining bond strength, but the conclusion is unfortunately clouded by the lack of replication.

Example 87 *(Examples 74 through 85 continued.) As a second (and actually more satisfying) application of the notion of probability plotting in the analysis of 2^p data, return again to the powder density data and the issue of the variability in response. We noted earlier that the eight sample standard deviations in Table 5.8 do not really look as if they could all have come from distributions with a common σ. To investigate this matter further, one might look for 2^3 factorial effects on the standard deviation of response. A common way of doing this is to compute and plot fitted effects for the natural logarithm of the sample standard deviation, $\log(s)$. (The logarithm is used because it tends to make the theoretical distribution of the sample standard deviation look more symmetric and Gaussian than it is in*

TABLE 5.12. Coordinates of points for a normal plot of the pullout fitted effects

i	ith smallest $\hat{E}$	$p = \frac{i-.5}{15}$	$Q_z(p)$
1	$d_2 = -10.3$	.033	-1.83
2	$ad_{22} = -7.1$	.100	-1.28
3	$b_2 = -6.0$	.167	$-.97$
4	$ab_{22} = -4.6$	.233	$-.73$
5	$bc_{22} = -2.3$	.300	$-.52$
6	$abc_{222} = -1.6$	.367	$-.34$
7	$ac_{22} = -1.5$	.433	$-.17$
8	$abcd_{2222} = -1.3$	.500	0
9	$bcd_{222} = -.6$	.567	.17
10	$c_2 = -.6$	.633	.34
11	$abd_{222} = 2.5$	.700	.52
12	$bd_{22} = 2.9$	.767	.73
13	$acd_{222} = 3.6$	.833	.97
14	$cd_{22} = 3.8$	.900	1.28
15	$a_2 = 11.5$	.967	1.83

TABLE 5.13. Coordinates of points for a half-normal plot of the pullout absolute fitted effects

i	ith smallest $|\hat{E}|$	$p = \frac{i-.5}{15}$	$Q(p) = Q_z\left(\frac{1+p}{2}\right)$
1	$|bcd_{222}| = .6$	.033	.04
2	$|c_2| = .6$	.100	.13
3	$|abcd_{2222}| = 1.3$	.167	.21
4	$|ac_{22}| = 1.5$	.233	.30
5	$|abc_{222}| = 1.6$	.300	.39
6	$|bc_{22}| = 2.3$	.367	.48
7	$|abd_{222}| = 2.5$	.433	.57
8	$|bd_{22}| = 2.9$	.500	.67
9	$|acd_{222}| = 3.6$	.567	.78
10	$|cd_{22}| = 3.8$	.633	.90
11	$|ab_{22}| = 4.6$	.700	1.04
12	$|b_2| = 6.0$	.767	1.19
13	$|ad_{22}| = 7.1$	.833	1.38
14	$|d_2| = 10.3$	.900	1.65
15	$|a_2| = 11.5$	.967	2.13

its raw form.) The reader is invited to verify that the logarithms of the standard deviations in Table 5.8 produce fitted effects

$$a_2 = -.16, b_2 = -.26, ab_{22} = -.24, c_2 = .57,$$
$$ac_{22} = -.07, bc_{22} = .15, \text{ and } abc_{222} = -.00.$$

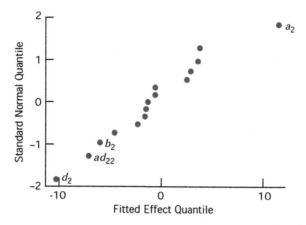

FIGURE 5.18. Normal plot of fitted effects for pullouts on dual in-line packages

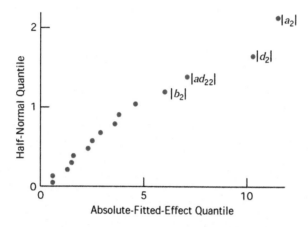

FIGURE 5.19. Half-normal plot of absolute fitted effects for pullouts on dual in-line packages

For example
$$a_2 = \tfrac{1}{4}(\ln(.040) + \ln(.011) + \ln(.079) + \ln(.044))$$
$$- \tfrac{1}{8}(\ln(.030) + \ln(.040) + \cdots + \ln(.104) + \ln(.044)) = -.16.$$

A half-normal plot of the absolute values of these is given in Fig. 5.20 on page 296.

It seems clear from Fig. 5.20 that the fitted main effect of Factor C is more than just experimental noise. Since $c_2 = .57, c_2 - c_1 = 1.14$. One would judge from this analysis that for any mesh size and flask size, the logarithm of the standard deviation of measured density for vibrated flasks is 1.14 more than that for unvibrated flasks. This means that the standard deviation itself is about $\exp(1.14) = 3.13$ times as large when vibration is employed in density determination as when it is not. Not only does the lack of vibration fit best with the researchers' original goal of detecting mix changes via density measurements, but it provides more consis-

tent density measurements than are obtained with vibration.

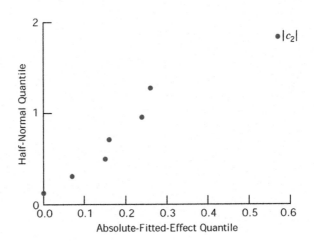

FIGURE 5.20. Half-normal plot of absolute fitted effects on $\log(s)$ in the solder joint strength study

The kind of analysis of sample standard deviations just illustrated in Example 87 is an important rough-and-ready approach to the problem of seeing how several factors affect σ. Strictly speaking, even logarithms of sample standard deviations from normal populations are not really legitimately treated like sample means. But often, at least for balanced 2^p factorial data with m of at least 4 or 5, this kind of crude analysis will draw attention to interpretable structure in observed values of s. And in light of the fact that variation is the enemy of quality of conformance, this can be very important to quality engineering efforts.

5.3.4 The Yates Algorithm for Computing Fitted 2^p Effects

Computing fitted effects directly from definitions like those given in this section is unpleasant. Of course there are computer programs (and even spreadsheets) that will do the work. But it is also very helpful to have an efficient means of computing 2^p fitted effects more or less by hand. Frank Yates discovered such a means many years ago. His algorithm will produce, all at once and with minimum pain, the average sample mean and the fitted effects for the all-high combination.

Yates Algorithm for 2^p Fitted Effects

One sets up the **Yates algorithm** by first listing 2^p sample means in a column in what is called "Yates standard order." This order is easily remembered by beginning with combination (1) (the "all-low" combination) and then combination a, and then "multiplying by b" to get b and ab, then "multiplying by c" to get c, ac, bc, and abc, and so on. One then creates a second column of numbers by adding the numbers in the first column in pairs and then subtracting them in pairs (*the first value in a pair being subtracted from the second*). The additions and subtractions are applied to the new column, and so on, until a total of p successive new columns have been generated from the original list of $\bar{y}$'s. Finally, the entries in the last column generated by additions and subtractions are divided by 2^p to produce the fitted effects (themselves listed in Yates standard order applied to effects rather than combinations).

Example 88 *(Examples 74 through 87 continued.)* *The means listed in Table 5.11 are in fact in Yates standard order. Table 5.14 shows the use of the Yates algorithm to quickly obtain the fitted effects computed earlier by much more laborious means.*

The final column of Table 5.14 is a very compact and helpful summary of what the data of Leigh and Taylor say about mean density measurements. And now that the hard work of defining (and understanding) the fitted effects has been done, the computations needed to produce them turn out to be fairly painless. It should be evident that 2^p factorial analyses are quite doable by hand for as many as four factors.

TABLE 5.14. Use of the Yates algorithm on the alumina powder density means

Combination	$\bar{y}$	Cycle #1	Cycle #2	Cycle #3	Cycle #3 $\div$ 8
(1)	2.164	4.130	8.222	17.426	$2.178 = \bar{y}_{...}$
a	1.966	4.092	9.204	$-.502$	$-.063 = a_2$
b	2.206	4.548	$-.518$	.070	$.009 = b_2$
ab	1.886	4.656	.016	$-.114$	$-.014 = ab_{22}$
c	2.272	$-.198$	$-.038$	.982	$.123 = c_2$
ac	2.276	$-.320$	.108	.534	$.067 = ac_{22}$
bc	2.322	.004	$-.122$	.146	$.018 = bc_{22}$
abc	2.334	.012	.008	.130	$.016 = abc_{222}$

Example 88 and Table 5.14 illustrate the Yates algorithm computations for $p = 3$ factors. There, three cycles of additions and subtractions are done and the final division is by $2^3 = 8$. For the case of $p = 4$, 16 means would be listed, four cycles of additions and subtractions are required, and the final division is by $2^4 = 16$. And so on.

5.3.5 Fitting Simplified Models for Balanced 2^p Data

Just as in the case of two-way factorials, after identifying effects in a 2^p factorial that seem to be detectable above the experimental variation, it is often useful to fit to the data a model that includes only those effects. Residuals can then be computed and examined for indications that something important has been missed in the data analysis. So the question is how one accomplishes the fitting of a simplified model to 2^p data. The general answer involves (as in the case of two-way data) the use of a regression program and dummy variables and cannot be adequately discussed here. What can be done is to consider the case of balanced data.

Fitted Values for Balanced Data

For balanced 2^p data sets, fitted or predicted values for (constant σ, normal) models containing only a few effects can be easily generated by adding to a grand mean only those fitted effects that correspond to effects one wishes to consider. These fitted values, $\hat{y}$, then lead to residuals in the usual way:

Residuals

$$e = y - \hat{y}.$$

Example 89 *(Examples 74 through 88 continued.)* *Return yet again to the alumina powder density data of Leigh and Taylor and consider fitting a model of the form:*

$$y_{ijkl} = \mu + \alpha_i + \gamma_k + \alpha\gamma_{ik} + \epsilon_{ijkl} \qquad (5.30)$$

to their data. This says that mean density depends only on level of Factor A and level of Factor C and (assuming that the ϵ have constant standard deviation) that variation in density measurements is the same for all four AC combinations. The four different corresponding fitted or predicted values are

$$\hat{y}_{1j1l} = \bar{y}_{...} + a_1 + c_1 + ac_{11} = 2.178 + .063 + (-.123) + .067 = 2.185,$$
$$\hat{y}_{1j2l} = \bar{y}_{...} + a_1 + c_2 + ac_{12} = 2.178 + .063 + .123 + (-.067) = 2.297,$$
$$\hat{y}_{2j1l} = \bar{y}_{...} + a_2 + c_1 + ac_{21} = 2.178 + (-.063) + (-.123) + (-.067) = 1.926, \quad \text{and}$$
$$\hat{y}_{2j2l} = \bar{y}_{...} + a_2 + c_2 + ac_{22} = 2.178 + (-.063) + .123 + .067 = 2.305.$$

Residuals for the simple model (5.30) can then be obtained by subtracting these four fitted values from the corresponding data of Table 5.8. And plots of the residuals can again make clear the lack of constancy in the variance of response. For example, Fig. 5.21 is a plot of residuals against level of Factor C and shows the increased variation in response that comes with vibration of the material in the density measurement process.

The brute force additions just used to produce fitted values in Example 89 are effective enough when there are a very few factorial effects not assumed to be 0. But where there are more than four different $\hat{y}$s, such additions become tedious and prone to error. It is thus helpful that a modification of the Yates algorithm can be used to produce 2^p fitted values all at once. The modification is called the **reverse Yates algorithm**.

Reverse Yates Algorithm for Balanced Data Fitted Values

To use the reverse Yates algorithm to produce fitted values, one writes down a column of 2^p effects (including a grand mean) in Yates standard order *from bottom to top*. Then, p normal cycles of the Yates additions and subtractions applied to the column (with no final division) produce the fitted values (listed in reverse Yates order). In setting up the initial column of effects, one sets to 0 all those assumed to be negligible and uses fitted effects for those believed from earlier analysis to be nonnegligible.

Example 90 *(Examples 74 through 89 continued.)* *As an illustration of the reverse Yates calculations, consider again the fitting of model (5.30) to the powder density data. Table 5.15 shows the calculations (to more decimal places that are really justified, in order to avoid roundoff).*

Chapter 5. Experimental Design and Analysis for Process Improvement   299

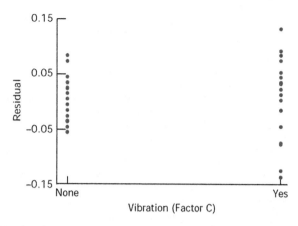

FIGURE 5.21. Plot of residuals from model (5.30) versus level of Factor C in the powder density study

The calculations in Table 5.15 involve $p = 3$ factors and therefore begin with a column of $2^3 = 8$ effects. They involve 3 cycles of the Yates additions and subtractions and end with 8 fitted means. An example involving $p = 4$ factors would start with a column of $2^4 = 16$ effects, involve 4 cycles of the Yates additions and subtractions, and yield 16 fitted means.

As a final point in this chapter, it should be said that while the full factorial analyses presented here are important and useful in their own right, they are rarely the starting point for real engineering experimentation on systems with many potentially important factors. Rather, the fractional factorial methods presented first in the next chapter are often used to "screen" a large number of factors down to what look like they may be the most important few before full factorial experimentation is done. It is, however, necessary to introduce the material in the order that has been employed here, since it is impossible to understand the principles of intelligent fractional factorial experimentation without first understanding complete factorial design and analysis.

TABLE 5.15. Use of the reverse Yates algorithm to fit a model with only A and C effects to the alumina powder density data

Effect	Fitted value	Cycle #1	Cycle #2	Cycle #3
$\alpha\beta\gamma_{222}$	0	0	.1895	$2.305 = \hat{y}_{abc} = \hat{y}_{222l}$
$\beta\gamma_{22}$	0	.1895	2.1155	$2.297 = \hat{y}_{bc} = \hat{y}_{122l}$
$\alpha\gamma_{22}$	.06675	0	.0560	$2.305 = \hat{y}_{ac} = \hat{y}_{212l}$
γ_2	.12275	2.1155	2.2410	$2.297 = \hat{y}_c = \hat{y}_{112l}$
$\alpha\beta_{22}$	0	0	.1895	$1.926 = \hat{y}_{ab} = \hat{y}_{221l}$
β_2	0	.0560	2.1155	$2.185 = \hat{y}_b = \hat{y}_{121l}$
α_2	$-.06275$	0	.0560	$1.926 = \hat{y}_a = \hat{y}_{211l}$
$\mu_{...}$	2.17825	2.2410	2.2410	$2.185 = \hat{y}_{(1)} = \hat{y}_{111l}$

Section 5.3 Exercises

1. Analysts were considering a study involving five factors each having two levels. How many conditions or "treatments" would make up such a study? Express your answer numerically and in the form of a base with an exponent.

2. Cooley, Franklin, and Elrod in a 1999 *Quality Engineering* article describe a 2^3 factorial experiment aimed at understanding the effects of A-hole type (in the fan "spider"), B-barrel type (to which the fan "spider" was attached), and C-assembly method on the torque, y (in ftlbs), required to break some industrial fans (large y is good). $m = 8$ fans of each of the 2^3 types were tested. Below are estimated mean responses for each of the 2^3 combinations. (For all except combination ac, these are sample means, but a peculiarity of the data collection for that combination necessitates the use of something more complicated for ac. (*Here ignore this issue and just treat the values below as if they were sample means of $m = 8$ torques.*)

$$\bar{y}_{(1)} = 53, \bar{y}_a = 105, \bar{y}_b = 44, \bar{y}_{ab} = 50,$$
$$\bar{y}_c = 176, \text{``}\bar{y}_{ac}\text{''} = 197, \bar{y}_{bc} = 154, \text{ and } \bar{y}_{abc} = 166.$$

   (a) Use the Yates algorithm and compute the fitted effects for the "all high" treatment combination.

   (b) Use 7.2 ftlbs as an estimate of the standard deviation of breaking torques for any fixed combination of hole type, barrel type and assembly method. (7.2 is not actually s_P in this problem, but treat it as if it were.) Which of the 2^3 factorial effects do you judge to be "clearly more than noise"? Show appropriate calculations to support your conclusion. Use a 95 % confidence level.

   (c) Suppose that one somehow judges that the only effects of practical importance in this study were A, B, and C main effects and the A × B interactions. What combination of hole type, barrel type, and assembly method has the largest predicted breaking torque, and what is this predicted value, $\hat{y}$?

3. In a particular 2^2 factorial, sample means are $\bar{y}_{(1)} = 8, \bar{y}_a = 10, \bar{y}_b = 6$, and $\bar{y}_{ab} = 16$. Find the fitted main effects of A and B at their "high" levels (a_2 and b_2, respectively).

4. For 2^5 factorial data, the Yates algorithm requires *how many* cycles/columns of additions and subtractions, followed by division by *what*?

5. In a balanced 2^3 factorial study with $m = 4$ observations y per combination, 95 % confidence limits for the factorial effects are of the form $\hat{E} \pm \Delta$. Express Δ as a (numeric) multiple of s_P.

6. In a particular 2^3 factorial study, the only clearly detectable effects are the grand mean and the C main effect. Identify which of the following statements are true:

   (a) Changing from the low to high level of C definitely changes mean response.

   (b) Changing levels of factors A or B has no clearly discernible effect on mean response.

   (c) All of the conditions {(1), a, b, ab} have indistinguishable mean responses, as do all of the conditions {c, ac, bc, abc}.

7. In a particular 2^3 factorial study, the only clearly detectable effects are the grand mean and the AB 2-factor interactions. (In fact $ab_{22} = 3$.) This indicates (select one of the following):

   (a) That for maximization purposes, one wants either {A high and B high} or {A low and B low}.

   (b) That as long as you keep the level of A fixed, changing level of B does not change mean y.

   (c) That since factor C is not involved, the low level of C is best.

   (d) That conditions {(1), a, b, ab} have indistinguishable mean responses.

8. In a 2-way factorial study, if there are strong/important AB interactions, then which of the following are true concerning the quantity $\alpha_1 - \alpha_2$?

   (a) It still measures the difference between average means for levels 1 and 2 of Factor A.

   (b) It can still be estimated (provided there are data from all $I \times J$ combinations of levels of A and B).

   (c) It is of limited practical importance (because "lack of parallelism" says this is not necessarily the change in mean response upon changing from level 1 to level 2 of A for *any* single level of B).

5.4 Chapter Summary

Once a process has been brought to physical stability and is behaving predictably, further improvements typically require fundamental process changes. Intelligent experimentation can be used to guide those changes. This chapter has begun discussion of relevant methods of experimental planning and data analysis. It opened with a presentation of general tools for comparing r experimental conditions that can be used without regard to any special structure in the conditions. Then, statistical methods for two-factor studies with complete factorial structure were considered. Finally, the long final section of the chapter addressed (full factorial) p factor studies, giving primary attention to cases where all p factors each have two levels.

5.5 Chapter 5 Exercises

1. An engineer and a material scientist are interested in a process for making a synthetic material. Contamination of the material during production is thought to be a real possibility. To find manufacturing conditions that produce minimal contamination, they select three different curing times and two different coating conditions (none versus some) for producing the material. Suppose two pieces of the synthetic material are produced for each combination of time and coating condition. The order of production is randomized and the raw material is randomly assigned to combinations of time and coating. The amount of iron (an important contaminant) in the final product is measured.

    (a) What type of (designed) experiment is this?
    (b) How many treatment combinations are there in this experiment?
    (c) To what future production circumstances can conclusions based on this experiment be extended?
    (d) Write a model equation for this experiment. Let "time" be Factor A and "coating condition" be Factor B. Say what each model term means, both in mathematical terms and in the context of the physical experiment.
    (e) What are the degrees of freedom for the estimated variance in response (for pieces made under a given set of process conditions)?
    (f) How many factors are there in this experiment? Identify the number of levels for all factors.
    (g) Write your answer to (b) as $I \times J \times \cdots$, in terms of the number of levels you gave in answer to (f).

2. Refer to problem 1.

    (a) Let i represent the level of Factor A and j the level of Factor B ($j = 1$ indicating no coating). Write out the six equations for treatment means μ_{ij} that follow from the model in problem 1(d).
    (b) For each of the three levels of Factor A (curing time), average your answers to (a) across levels of Factor B, and apply the facts in displays (5.20) and (5.21) to write the averages in terms of $\mu_{..}$ and the curing time main effects (α_is).
    (c) Find the difference between your mean in (a) for the second level of curing time and no coating and your mean in (a) for the first level of curing time and no coating in terms of the appropriate α_is and $\alpha\beta_{ij}$s.

Chapter 5. Experimental Design and Analysis for Process Improvement   303

(d) Find the difference between your mean in (a) for the second level of curing time and some coating and your mean in (a) for the first level of curing time and some coating in terms of the appropriate α_is and $\alpha\beta_{ij}$s.

(e) Find the difference between your answer to (b) for the second level of curing time and your answer to (b) for the first level of curing time in terms of the appropriate α_is .

(f) Reflect on (c), (d), and (e). Would you use your answer in (e) to represent the quantities in (c) or (d)? Why or why not?

(g) What must be true for the two differences identified in (c) and (d) to both be equal to the difference in (e)?

3. Refer to problems 1 and 2.

(a) Average your answers to problem 2(a) across time levels and use the facts in displays (5.20) and (5.21) to express the no coating and coating means in terms of $\mu_{..}$ and the coating main effects (the β_js).

(b) Find the difference between your mean in problem 2(a) for the third level of curing time and no coating and your answer to problem 2(a) for the third level of curing time and some coating in terms of the appropriate β_js and $\alpha\beta_{ij}$s.

(c) Find the difference between your mean in problem 2(a) for the first level of curing time and no coating and your answer to 2(a) for the first level of curing time and some coating.

(d) Reflect on (a), (b), and (c). Would you use the difference of your two answers in (a) to represent the quantities requested in (b) and (c)? Why or why not?

(e) What must be true for the two differences identified in (b) and (c) to both be equal to the difference of your two answers in (a)?

4. Refer to problem 1. Suppose the following table gives mean iron contents (mg) for the set of six different treatment combinations.

	Low time	Medium time	High time
No coating	8	11	7
Coating	2	5	7

(a) Plot the mean responses versus level of time. Use line segments to connect successive points for means corresponding to "no coating." Then separately connect the points for means corresponding to "coating." (You will have then created an interaction plot.)

(b) Find the average (across levels of time) mean for "no coating" pieces. Do the same for "coating" pieces. Find the difference in these two average means (coated minus no coating).

(c) Find the following three differences in means: (low time/coating) minus (low time/no coating), (medium time/coating) minus (medium time/no coating), and (high time/coating) minus (high time/no coating).

(d) Compare the three differences found in (c) to the difference in (b). What is implied about the $\alpha\beta_{ij}$ values? (Hint: use Eq. (5.21) and Definition 68.)

(e) What feature of your graph in (a) reflects your answers in (d)?

(f) If asked to describe the effect of coating on iron content, will your answer depend on the time level? Why or why not?

5. Repeat problem 4 with the table of means below.

	Low time	Medium time	High time
No coating	6	8	7
Coating	1	10	4

6. Repeat problem 4 with the table of means below.

	Low time	Medium time	High time
No coating	7	14	22
Coating	1	5	8

7. **NASA Polymer.** In their article "Statistical Design in Isothermal Aging of Polyimide Resins" that appeared in *Journal of Applied Polymer Science* in 1995, Sutter, Jobe, and Crane reported on their study of polymer resin weight loss at high temperatures. (The work was part of the NASA Lewis Research Center HITEMP program in polymer matrix composites. The focus of the HITEMP program was the development of high-temperature polymers for advanced aircraft engine fan and compressor applications.) The authors' efforts centered on evaluating the thermal oxidative stability of various polymer resins at high temperatures. The larger the weight loss in a polymer resin specimen for a given temperature/time combination, the less attractive (more unstable) the polymer resin. Maximum jet engine temperatures of interest are close to 425° C (700° F), and the experiment designed by the researchers exposed specimens of selected polymer resins to a temperature of 371° C for 400 h. The specimens were initially essentially all the same size. Upon completion of the 400 h exposures, percent weight loss was recorded for each specimen ($m = 4$ specimens per polymer type). The following data are from two of the polymer resins included in the study, Avimid-N, a polymer resin developed by DuPont, and VCAP-75, a polymer resin investigated in earlier engine component development programs.

Avimid-N	VCAP-75
10.6576	25.8805
9.1014	23.5876
9.0909	26.4873
10.2299	30.5398

Chapter 5. Experimental Design and Analysis for Process Improvement   305

   (a) How many experimental factors were there in the study described above? What were the levels of that factor or factors?

   (b) Give a model equation for a "random samples from normal distributions with a common variance" description of this scenario. Say what each term in your model means in the context of the problem. Also, give the numeric range for each subscript.

   (c) How many treatments are there in this study? What are they?

   (d) Find the two estimated treatment means (the two sample means) and the eight residuals.

   (e) Find the estimated (common) standard deviation, s_P.

   (f) Normal plot the eight residuals you found in (d). Does this plot indicate any problems with the "samples from two normal distributions with a common variance" model? Explain.

8. Refer to the **NASA Polymer** case in problem 7. Suppose that the "random samples from two normal distributions with a common variance" model is appropriate for this situation.

   (a) Find simultaneous 95 % two-sided confidence intervals for the two mean percent weight losses using the Bonferroni approach.

   (b) Find a 99 % two-sided confidence interval for the difference in mean percent weight losses for VCAP-75 and Avimid-N (VCAP-75 minus Avimid-N).

   (c) If you were to choose one of the two polymer resins on the basis of the data from problem 7, which one would you choose? Why?

9. Refer to the **NASA Polymer** case in problems 7 and 8.

   (a) Transform each of the percent weight loss responses, y, to $\ln(y)$. Find the sample means for the transformed responses.

   (b) Find the new residuals for the transformed responses.

   (c) Make a normal probability plot of the residuals found in (b). Does normal theory appear appropriate for describing the log of percent weight loss? Does it seem more appropriate for the log of percent weight loss than for percent weight loss itself? Why or why not?

   Henceforth in this problem, assume that usual normal theory is an appropriate description of log percent weight loss.

   (d) Make simultaneous 95 % two-sided confidence intervals for the two average $\ln(y)$s using the Bonferroni approach.

   (e) If $\ln(y)$ is normal, the mean of $\ln(y)$ (say $\mu_{\ln(y)}$) is also the .5 quantile of $\ln(y)$. What quantile of the distribution of y is $\exp(\mu_{\ln(y)})$?

(f) If end points of the intervals produced in (d) are transformed using the exponential function, $\exp(\cdot)$, what parameters of the original y distributions does one hope to bracket?

10. Refer to the **NASA Polymer** case in problem 7. Sutter, Jobe, and Crane experimented with three other polymer resins in addition to the two mentioned in problem 7. ($m = 4$ specimens of all five polymer resins were exposed to a temperature of 371° C for 400 h.) Percent weight losses for specimens of the three additional polymer resins N-CYCAP, PMR-II-50, and AFR700B are given below. All specimens were originally of approximately the same size.

N-CYCAP	PMR-II-50	AFR700B
25.2239	27.5216	28.4327
25.3687	29.1382	28.9548
24.5852	29.8851	24.7863
25.5708	28.5714	24.8217

(a) Use both the data above and the data in problem 7 and answer (a)–(f) of problem 7 for the study including all five polymer resins.

(b) Answer (a) from problem 8 using data from all five polymer resins.

(c) How many different pairs of polymer resins can be made from the five in this study?

(d) Make 95 % individual two-sided confidence intervals for all differences in pairs of weight loss means.

(e) The smaller the weight loss, the better from an engineering/strength perspective. Which polymer resin is best? Why?

11. Refer to the **NASA Polymer** case in problems 7, 8, and 10.

(a) Find the five sample standard deviations (of percent weight loss) for the five different polymer resins (the data are given in problems 7 and 10).

(b) Transform the data for the five polymer resins given in problems 7 and 10 by taking the natural logarithm of each percent weight loss. Then find the five sample standard deviations for the logged values.

(c) Consider the two sets of standard deviations in (a) and (b). Which set is more consistent? What relevance does this comparison have to the use of the methods in Sect. 5.1?

(d) Find the five different sample means of log percent weight loss. Find the five sets of residuals for log percent weight loss (by subtracting sample log means from individual log values).

(e) Plot the residuals found in (d) versus the sample means from (d) (means of the log percent).

(f) Plot the residuals found in part (a) of problem 10 (calculated as in problem 7d) versus the sample means (of percent weight loss). Compare the plot in (e) to this plot.

(g) Which set of data, the original one or the log transformed one, better satisfies the assumption of a common response variance for all five polymer resins?

12. Refer to the **NASA Polymer** case in problems 7, 8, 10, and 11.

    (a) Normal plot the residuals found in part (d) of problem 11. Does the normal theory model of Sect. 5.1 seem appropriate for transformed percent weight loss (for all five resins)? Why or why not?

    (b) Answer (a) from problem 8 using the log transformed data from all five polymer resins. (The requested intervals are for the means of the log transformed percent weight loss.)

    (c) If one exponentiates the end points of the intervals from (b) (plugs the values into the function $\exp(\cdot)$) to produce another set of intervals, what will the new intervals be estimating?

    (d) Make individual 99 % confidence intervals for all differences in pairs of mean log percent weight losses for different polymer resins.

    (e) If one exponentiates the end points of the intervals from (d) (plugs the values into the function $\exp(\cdot)$) to produce another set of intervals, what will the new intervals be estimating? (Hint: $\exp(x - y) = \exp(x)/\exp(y)$.)

13. Refer to the **NASA Polymer** case in problems 7 and 10. The data given in those problems were obtained from one region or position in the oven used to bake the specimens. The following (percent weight loss) data for the same five polymer resins came from a second region or position in the oven. As before, $m = 4$.

Avimid-N	VCAP-75	N-CYCAP	PMR-II-50	AFR700B
9.3103	24.6677	26.3393	25.5882	23.2168
9.6701	23.7261	23.1563	25.0346	24.8968
10.9777	22.1910	25.6449	24.9641	23.8636
9.3076	22.5213	23.5294	25.1797	22.4114

    Suppose the investigators were not only interested in polymer resin effects but also in the possibility of important position effects and position/polymer resin interactions.

    (a) How many experimental factors are there in the complete experiment as represented in problems 7 and 10 and above?

    (b) How many levels does each factor have? Name the levels of each factor.

(c) Give a model equation for a "random samples from normal distributions with a common variance" description of this multifactor study. Say what each term in your model represents in the context of the problem and define each one in terms of the $\mu_{..}, \mu_{i.}, \mu_{.j}$, and μ_{ij}.

(d) Find the complete set of 10 sample average percent weight losses and plot them in an interaction plot format. Plot oven position on the horizontal axis. Use two-sided individual 95 % confidence limits to establish error bars around the sample means on your plot. What is the value of Δ such that $\bar{y}_{ij} \pm \Delta$ represents the 10 intervals? Does your plot suggest that there are strong interactions between position and resin? Why or why not?

(e) Find and plot the 10 sets of residuals. (Make and label 10 tick marks on a horizontal axis corresponding to the 10 combinations of oven position and polymer resin and plot the residuals against treatment combination.)

(f) Find the sample standard deviation for each oven position/polymer resin combination.

(g) Does it appear from (e) and (f) that variability in percent weight loss is consistent from one oven position/polymer resin combination to another? Why or why not?

(h) Make a normal probability plot for the 40 residuals. Does the plot suggest any problems with the "normal distributions with a common variance" model for percent weight loss? Why or why not?

(i) Possibly ignoring your answer to (g), find an estimated standard deviation of percent weight loss (supposedly) common to all oven position/polymer resin combinations, i.e., find s_P. What degrees of freedom are associated with your estimate?

(j) Continuing to possibly ignore your answer to (g), find Δ and $\widehat{ab_{i1}}$ such that $\widehat{ab_{i1}} \pm \Delta$ are individual 99 % two-sided confidence intervals for the five oven position/polymer resin interaction effects for position 1. Should one be 99 % confident that all five intervals are correct? Use the Bonferroni inequality to identify a minimum simultaneous confidence on this whole set of intervals.

(k) In terms of the model from (c), what is the position 1/Avimid-N mean minus the position 1/VCAP-75 mean? Again in terms of the model from (c), what is the difference between mean percent weight loss from Avimid-N averaged over positions 1 and 2 and mean percent weight loss from VCAP-75 averaged over positions 1 and 2? (Use fact (5.21) in answering this second question.) Under what circumstances are the two differences here the same?

14. Refer to the **NASA Polymer** case of problems 7, 10, and 13. Use a "normal distributions with common variance" description of percent weight loss.

Chapter 5. Experimental Design and Analysis for Process Improvement   309

(a) Based on the intervals for $\widehat{ab_{i1}}$ in part (j) of problem 13, is the condition asked for in part (k) of problem 13 satisfied?

(b) Consider the difference in sample mean percent weight losses for position 1/Avimid-N and position 1/VCAP-75. Find the estimated standard deviation of this difference.

(c) Consider the average over positions 1 and 2 of the Avimid-N sample mean percent weight losses minus the average over positions 1 and 2 of the VCAP-75 sample mean weight losses. Find the estimated standard deviation of this difference.

(d) Reflect on your answers to parts (j) and (k) of problem 13 and (a) through (c) of this problem. If you wished to estimate a difference between the mean percent weight losses for two of the polymer resins, would you report two intervals, one for position 1 and one for position 2, or one interval for the difference averaged across positions? Defend your answer. What would be the width (Δ) of your interval(s) (using a 90 % confidence level)? Make the 10 comparisons of pairs of polymer means averaged across position (using individual 90 % confidence levels).

15. Refer to the **NASA Polymer** case in problems 7, 10, and 13. Transform the 40 observed percent weight losses by taking their natural logarithms.

(a) Answer parts (d)–(j) of problem 13 using the transformed data.

(b) Answer parts (a)–(d) of problem 14 using the transformed data.

(c) Should conclusions from the transformed data or original data be presented? Defend your answer.

(d) Consider the final set of intervals made in (b) above (the ones corresponding to part (d) of problem 14). Exponentiate the end points of these intervals (i.e., plug them into the exponential function $\exp(\cdot)$) to get new ones. What do these resulting intervals estimate? (Hint: consider the original percent weight loss distributions and recall that $\exp(x - y) = \exp(x)/\exp(y)$.)

16. **NASA Polymer II.** (See problems 7 and 13 for background.) In a preliminary investigation, Sutter, Jobe, and Crane designed a weight loss study that was to be balanced (with an equal number of specimens per polymer resin/oven position combination). Polymer resin specimens of a standard size were supposed to be randomly allocated (in equal numbers) to each of two positions in the oven. The following percent weight loss values were in fact obtained.

	Position 1	Position 2
Avimid-N	8.9	10.2, 9.1, 9.1, 8.5
PMR-II-50	29.8, 29.8	27.4, 25.5, 25.7

(a) The lack of balance in the data set above resulted from a misunderstanding of how the experiment was to be conducted. How would you respond to a colleague who says of this study "Well, since an equal number of specimens were not measured for each polymer resin/oven position combination, we do not have a balanced experiment so a credible analysis cannot be made"?

(b) How many experimental factors were there in this study? Identify the numbers of levels for each of the factors.

(c) Write a model equation for the "normal distributions with a common variance" description of this study. Give the numeric ranges for all subscripts.

(d) Find the fitted main effect for each level of each factor.

(e) Find the fitted interactions for all combinations of polymer resin and oven position.

(f) Use the model from (c) and find a 99% two-sided confidence interval for the interaction effect for Avimid-N and position 1. Are intervals for the other interaction effects needed? Why or why not?

(g) Use the model from (c) and find 95% two-sided confidence intervals for each of the main effects.

(h) Use the model from (c) and find a 90% two-sided confidence interval for the difference in oven position main effects.

(i) Use the model from (c) and find a 90% confidence interval for the difference in polymer resin main effects.

17. Refer to the **NASA Polymer II** case in problem 16.

    (a) Find the residuals and plot them against the cell means (that serve as fitted values in this context). Does it appear that response variability is consistent from treatment combination to treatment combination? Why or why not?

    (b) Normal plot the residuals found in (a). Does this plot suggest problems with the basic "normal distributions with a common variance" model? Why or why not?

    (c) Transform each response using $y' = \arcsin(y/100)$, where y is a percent (between 0 and 100) as given in problem 16. Compute the four sample means for the transformed values.

    (d) Find residuals for the transformed data from (c). Plot these against the sample means from (c). Does it appear that variability in transformed response is consistent from treatment combination to treatment combination? Normal plot these residuals. Does this plot suggest problems with the basic "normal distributions with a common variance" model for the transformed response? Why or why not?

Chapter 5. Experimental Design and Analysis for Process Improvement   311

(e) Transform each response in problem 16 using $y' = \ln(y)$. Answer (c) and (d) with the newly transformed data.

(f) Which version of the data (the original, the arcsin transformed, or the log transformed) seems best described by the "normal distributions with a common variance" model? Why?

18. Refer to the **NASA Polymer II** case in problems 16 and 17. Consider the original data in problem 16 and use the "normal distributions with a common variance" model.

   (a) Plot the sample means in interaction plot format, placing "oven position" on the horizontal axis and using two-sided individual 90 % confidence limits to establish error bars around each of the sample means. (Note that since the sample sizes vary, the error bars will not be the same length mean to mean.) Does the line segment connecting the two Avimid-N points cross that connecting the two PMR-II-50 points?

   (b) Find a 95 % two-sided confidence interval for the difference in average percent weight losses at position 1 (Avimid-N minus PMR-II-50).

   (c) Find a 95 % two-sided confidence interval for the difference in Avimid-N and PMR-II-50 main effects.

   (d) Which interval, the one from (b) or the one from (c), is better for estimating the difference in mean percent weight losses (Avimid-N minus PMR-II-50) at oven position 1? Why? (See part (f) of problem 16 and (a) above.)

   (e) Find a 90 % two-sided confidence interval for the difference in oven position mean percent weight losses (position 2 minus position 1) for Avimid-N.

   (f) Find a 90 % two-sided confidence interval for the difference in the two oven position main effects (position 2 minus position 1).

   (g) Which interval, the one from (e) or the one from (f), is better for estimating the difference in mean percent weight losses at the two oven positions for Avimid-N? Why? (See part (f) of problem 16 and (a) above.)

19. **NASA Percent Weight Loss**. Sutter, Jobe, and Ortiz designed and conducted an experiment as part of NASA Lewis Research Center's efforts to evaluate the effects of two-level factors kapton, preprocessing time at 700° F, 6 F dianhydride type, and oven position on percent weight loss of PMR-II-50 specimens baked at 600° F for 936 h. Two specimens were produced for each of the possible combinations of kapton, preprocessing time, and dianhydride type. One specimen of each type was randomly selected for baking at position 1 (and the other was assigned to position 2). Exact baking locations of the 8 specimens at each position were randomized

312   Chapter 5. Experimental Design and Analysis for Process Improvement

within that region. The oven was set at 600° F and the specimens were exposed to this temperature for 936 h. All specimens initially had about the same mass. Percent weight losses (y) similar to those in Table 5.16 were observed.

The levels of the factors "kapton," "preprocessing time," and "dianhydride type" were:

Kapton	No kapton (1) vs. With kapton (2)
Preprocessing time	15 min (1) vs. 2 h (2)
Dianhydride type	Polymer grade (1) vs. Electronic grade (2)

(a) How many experimental factors were there in this study? (Include oven position as a factor.) Identify them and say how many levels of each were used.

(b) Describe the factorial structure of the treatment combinations using a "base and exponent" notation.

TABLE 5.16. Data for Problem 19

Oven Position	Kapton	Preprocessing Time	Dianhydride Type	y
1	1	1	1	4.5
2	1	1	1	5.0
1	2	1	1	4.7
2	2	1	1	5.3
1	1	2	1	4.4
2	1	2	1	5.0
1	2	2	1	4.8
2	2	2	1	5.2
1	1	1	2	3.9
2	1	1	2	3.8
1	2	1	2	3.9
2	2	1	2	3.8
1	1	2	2	4.0
2	1	2	2	3.8
1	2	2	2	3.4
2	2	2	2	3.9

(c) How many treatment combinations were there in this experiment? Identify them.

(d) Was there replication in this study? Why or why not?

(e) Give a model equation for the "normal distributions with a common variance" description of percent weight loss that uses "factorial effects" notation.

(f) Find the fitted effects corresponding to the "all-high" treatment combination. Use Yates algorithm.

Chapter 5. Experimental Design and Analysis for Process Improvement   313

(g) Find the $p = (i - .5)/15$ standard normal quantiles (for $i = 1, 2, \ldots, 15$).

(h) Use Eq. (5.29) and find the $p = (i - .5)/15$ half-normal quantiles (for $i = 1, 2, \ldots, 15$).

(i) Using your answers to (f)–(h), make a full normal plot of the fitted effects and a half-normal plot for the absolute fitted effects corresponding to the all-high treatment combination.

(j) Does it appear from your plots in part (i) that there are any statistically detectable effects on mean percent weight loss? Defend your answer.

20. Refer to the **NASA Percent Weight Loss** case in problem 19. Recall that single specimens of all eight different combinations of kapton, preprocessing time, and dianhydride type were randomized within position 1, and another set of eight were randomized within position 2. The factor "position" was not really one of primary interest. Its levels were, however, different, and its systematic contribution to response variability could have been nontrivial. (This type of experimental design is sometimes referred to as a factorial arrangement of treatments (FAT) in a randomized block.) Problem 19 illustrates one extreme of what is possible in the way of analysis of such studies, namely, that where the main effects and all interactions with the "blocking" factor are considered to be potentially important. The other extreme (possibly adopted on the basis of the results of an analysis like that in problem 19) is that where the main effects of the blocking variable and all interactions between it and the other experimental factors are assumed to be negligible. (Intermediate possibilities also exist. For example, one might conduct an analysis assuming that there are possibly main effects of "blocks" but that all interactions involving "blocks" are negligible.)

In this problem consider an analysis of the data from problem 19 that completely ignores the factor "oven position." If one treats the main effects and all interactions with oven position as negligible, the data in problem 19 can be thought of as three-factor factorial data (in the factors kapton, preprocessing time, and dianhydride type) with $m = 2$.

(a) Redo parts (a)–(e) of problem 19 taking this new point of view regarding oven position.

(b) Find the fitted effects for the all-high treatment combination in this three-factor study.

(c) Is normal plotting the only possible means of judging the statistical detectability of the factorial effects you estimated in (b)? Why or why not?

(d) Find the residuals for this study by subtracting sample means from individual percent weight loss measurements.

(e) Plot the residuals versus the sample means. Does it appear the assumption of a common variance across treatment combinations is reasonable? Why or why not?

(f) Find the sample standard deviation, s, for each treatment combination. Let $y^* = \ln(s)$ and find fitted effects for the log sample standard deviations, y^*.

(g) Use Eq. (5.29) and find the $(i - .5)/7$ half-normal quantiles (for $i = 1, 2, \ldots, 7$).

(h) Make a half-normal plot of the absolute values of the fitted effects found in (f) using the quantiles found in (g). Do you see evidence in this plot of any kapton, preprocessing time, or dianhydride type effects on the *variability* of percent weight loss? Why or why not?

21. Refer to the **NASA Percent Weight Loss** case in problems 19 and 20. Continue the analysis begun in problem 20 where the factor "oven position" is ignored.

    (a) How many pairs can be formed from the set of eight different treatment combinations considered by Jobe, Sutter, and Ortiz?

    (b) Find s_P based on the eight "samples" of size $m = 2$. What degrees of freedom are associated with this estimate of σ?

    (c) Find a set of (eight) two-sided interval estimates for the treatment mean weight losses that have associated 95 % individual confidence levels.

    (d) Consider the all-high treatment combination. Find individual 95 % confidence intervals for all three two-factor interactions and the three-factor interaction corresponding to this combination. How "sure" are you that all four intervals bracket their corresponding theoretical interactions? (Use the Bonferroni inequality.)

    (e) Let α_2 correspond to "with kapton," β_2 correspond to "2 h," and γ_2 correspond to "electronic dianhydride." The investigators wanted a good estimate of the effect (averaged over the two time conditions) of kapton on the mean percent weight loss of electronic grade PMR-II-50 polymer resin. Find a 95 % two-sided confidence interval for the mean percent weight loss of electronic grade PMR-II-50 polymer resin made with kapton, minus the mean for specimens made without kapton (both averaged over time), i.e., estimate

    $$\mu_{2.2} - \mu_{1.2} = \frac{1}{2}(\mu_{212} + \mu_{222}) - \frac{1}{2}(\mu_{212} + \mu_{222}).$$

    (Hint: use display (5.8).) Find a 95 % two-sided confidence interval for the difference in the two kapton main effects (with kapton minus without kapton), i.e., estimate $(\alpha_2 - \alpha_1)$. Which interval is better

Chapter 5. Experimental Design and Analysis for Process Improvement   315

for the investigators' purposes? Defend your answer. (Hint: consider your answer to (d) above.)

(f) Let α_2 correspond to "with kapton," β_2 correspond to "2 h," and γ_2 correspond to "electronic dianhydride." The investigators wanted a good estimate of the effect (averaged over the two time conditions) of kapton on the mean percent weight loss of polymer grade PMR-II-50 polymer resin. Find a 95 % two-sided confidence interval for the mean percent weight loss of polymer grade PMR-II-50 polymer resin made with kapton minus the mean for specimens made without kapton (both averaged over time), i.e., estimate

$$\mu_{2.1} - \mu_{1.1} = \frac{1}{2}(\mu_{211} + \mu_{221}) - \frac{1}{2}(\mu_{111} + \mu_{121}).$$

(Hint: use display (5.8).) Find a 95 % two-sided confidence interval for the difference in the two kapton main effects (with kapton minus without kapton), i.e., estimate $(\alpha_2 - \alpha_1)$. Which interval is better for the investigators' purposes? Defend your answer. (Hint: consider your answer to (d) above.)

(g) Let α_2 correspond to "with kapton," β_2 correspond to "2 h," and γ_2 correspond to "electronic dianhydride." The investigators wanted a good estimate of the effect of dianhydride type (grade) (averaged over the two time conditions) on the mean percent weight loss of PMR-II-50 polymer resin with kapton. Find a 95 % confidence interval for the mean percent weight loss of polymer grade PMR-II-50 with kapton minus the mean for specimens of electronic grade PMR-II-50 with kapton (both averaged over time), i.e., estimate

$$\mu_{2.1} - \mu_{2.2} = \frac{1}{2}(\mu_{211} + \mu_{221}) - \frac{1}{2}(\mu_{212} + \mu_{222}).$$

(Hint: use display (5.8).) Find a 95 % interval for the difference in dianhydride type main effects (polymer grade minus electronic grade), i.e., estimate $(\gamma_1 - \gamma_2)$. Which interval is better for the investigators' purposes? Defend your answer. (Hint: consider your answer to (d) above.)

22. **NASA Fe.** Refer to the **NASA Percent Weight Loss** case in problem 19. A portion of each PMR-II-50 specimen represented by the data in problem 19 was not exposed to 600° F temperature for 936 h but instead was analyzed for iron (Fe) content. It is known that electronic grade dianhydride has small amounts of iron and polymer grade dianhydride has larger amounts of iron. NASA researchers Sutter and Ortiz were also aware of the possibility of iron transfer from the pressing mechanism used to form the polymer resin specimens. Thus, the protective kapton coating was used on

316   Chapter 5. Experimental Design and Analysis for Process Improvement

the mechanism in pressing half of the PMR-II-50 polymer resin specimens, and no kapton was used for the others. Data like the sixteen responses (y) (in ppm Fe) in Table 5.17 were obtained.

TABLE 5.17. Data for problem 22

Kapton	Preprocessing Time	Dianhydride Type	y
1	1	1	12.0, 2.5
2	1	1	9.0, 9.1
1	2	1	2.6, 2.5
2	2	1	25.0, 2.5
1	1	2	18.0, 34.0
2	1	2	7.0, 4.5
1	2	2	3.0, 14.0
2	2	2	2.5, 2.6

(a) How many experimental factors were there in this study? Identify them and say how many levels of each were used.

(b) Describe the factorial structure of this arrangement of treatment combinations using a "base and exponent" notation.

(c) How many treatment combinations are there in this experiment? Identify them.

(d) Does "oven position" play any role in the iron content measurements here?

(e) Was there replication in this study? What are the eight sample sizes in this study?

(f) Give a model equation in factorial effects notation for the "normal distributions with a common variance" description of Fe content.

(g) Find the eight sample means.

(h) Find the fitted effects corresponding to the all-high treatment combination using the Yates algorithm. Say what terms in your model from (f) each of these is meant to estimate.

(i) Find the 16 residuals for this data set. Plot the residuals versus the sample means (that can function as fitted or predicted Fe responses). Does it appear that a constant variance model is reasonable? Why or why not?

(j) Find the sample standard deviation of iron content, s, for each of the eight treatment combinations. Let $y^* = \ln(s)$ and find new fitted effects for the log standard deviations, y^*, using Yates algorithm.

(k) Use Eq. (5.29) and find the $(i - .5)/7$ half-normal quantiles (for $i = 1, 2, \ldots, 7$).

(l) Make a half-normal plot of the absolute values of the fitted effects computed in (j). Do you see evidence in this plot of any kapton, pre-processing time, or dianhydride type effects on the *variability* of iron content? Explain.

23. Refer to the **NASA Fe** case in problem 22. Transform each Fe content response by taking its natural logarithm. Answer parts (g)–(l) of problem 22 based on the log iron contents. Should the data in problem 22 be analyzed in terms of the original y or in terms of $\ln(y)$? Why?

24. Refer to **NASA Fe** case in problems 22 and 23. In this problem, use the log transformed responses.

   (a) Find s_P for the log Fe contents. What degrees of freedom are associated with this estimate of σ?

   (b) Make individually 95 % two-sided confidence intervals for the eight mean log iron contents.

   (c) If the end points of each interval in (b) are exponentiated (plugged into the function $\exp(\cdot)$), what parameters of the original iron distributions are estimated?

   (d) Find a value Δ, so that if one attaches an uncertainty of $\pm \Delta$ to any particular difference among the eight sample mean log Fe contents, one may be 95 % confident of that interval (as representing the corresponding difference in theoretical mean log iron contents).

   (e) If the end points of each interval in (d) are exponentiated (plugged into the function $\exp(\cdot)$), what does a given interval estimate in terms of parameters of the original iron distributions?

   (f) Consider the all-high treatment combination. Find individual 99 % confidence intervals for all three two-factor interactions and the three-factor interaction corresponding to this combination. How "sure" are you that all four intervals bracket their corresponding theoretical interactions? (Use the Bonferroni inequality.)

   (g) Let α_2 correspond to "with kapton," β_2 correspond to "2 h," and γ_2 correspond to "electronic dianhydride." The investigators wanted a good estimate of the effect (averaged over the two time conditions) of kapton on the mean log Fe content of electronic grade PMR-II-50 polymer resin. Find a 95 % two-sided confidence interval for the mean log Fe content of electronic grade PMR-II-50 polymer resin made with kapton minus the mean log Fe content for specimens of electronic grade PMR-II-50 polymer resin made without kapton (both averaged over time), i.e., estimate

   $$\mu_{2.2} - \mu_{1.2} = \frac{1}{2}(\mu_{212} + \mu_{222}) - \frac{1}{2}(\mu_{112} + \mu_{122}).$$

(Hint: consider display (5.8).) Find a 95 % two-sided confidence interval for the difference in the two kapton main effects (with kapton minus without kapton), i.e., estimate $(\alpha_2 - \alpha_1)$. Which interval is better for the investigators' purposes? Defend your answer. (Hint: consider your answer to (f) above.)

(h) Let α_2 correspond to "with kapton," β_2 correspond to "2 h," and γ_2 correspond to "electronic dianhydride." The investigators wanted a good estimate of the effect of dianhydride type (grade) (averaged over the two time conditions) on the mean log Fe content of PMR-II-50 polymer resin with kapton. Find a 95 % confidence interval for the mean log Fe content of polymer grade PMR-II-50 with kapton minus the mean for electronic grade PMR-II-50 with kapton (both averaged over time), i.e., estimate

$$\mu_{2.1} - \mu_{2.2} = \frac{1}{2}(\mu_{211} + \mu_{221}) - \frac{1}{2}(\mu_{212} + \mu_{222})$$

(Hint: consider display (5.8).) Find a 95 % confidence interval for the difference in dianhydride type main effects (polymer grade minus electronic grade), i.e. estimate $(\gamma_1 - \gamma_2)$. Which interval is better for the investigators' purposes? Defend your answer. (Hint: consider your answer to (f) above.)

(i) Using the fitted effects computed in part (h) of problem 23, find a set of interval estimates for the factorial effects corresponding to the all-high treatment combination. Make 99 % two-sided individual confidence intervals. What can be said concerning the confidence that each one of these intervals simultaneously contains its corresponding theoretical effect? (Use the Bonferroni inequality.)

(j) If an interval in (i) includes 0, set that fitted effect to 0. Otherwise, use the value from part (h) of problem 23 and find fitted values for all eight treatment combination mean log Fe contents using the reverse Yates algorithm. Show your work in table form.

25. Refer to the **NASA Percent Weight Loss** and **NASA Fe** cases in problems 19 through 24. Reflect on your responses to these problems. Does it appear Fe content is related to percent weight loss? Why or why not? (Hint: consider your conclusions when lnFe content was the response variable and when percent weight loss was the response variable.)

26. Refer to the **Brush Ferrules** case in problem 15 of Chap. 1. Adams, Harrington, Heemstra, and Snyder identified several factors that potentially affected ferrule thickness. Ultimately, they were able to design and conduct an experiment with the two factors "crank position" and "slider position." Two levels (1.625 in and 1.71875 in) were selected for crank position, and two levels (1.75 in and 2.25 in) were selected for slider position. $m = 4$

Chapter 5. Experimental Design and Analysis for Process Improvement 319

new ferrules were produced for every combination of crank position and slider position. The resulting ferrule thicknesses (in inches) are given below.

Crank position	Slider position	Thickness
1.625	1.75	.421, .432, .398, .437
1.71875	1.75	.462, .450, .444, .454
1.625	2.25	.399, .407, .411, .404
1.71875	2.25	.442, .451, .439, .455

(a) Give a model equation for the "normal distributions with a common variance" description of ferrule thickness that uses "factorial effects" notation. Say what each term in your model represents in the context of the problem and define each one in terms of the $\mu_{..}, \mu_{i.}, \mu_{.j}$, and μ_{ij}.

(b) Find the four sample means.

(c) Plot the sample means versus crank position in interaction plot format. (Connect means having the same slider level with line segments.) Does it appear there are strong interaction effects? Why or why not?

(d) Find the residuals. Plot these versus sample means. Does it appear that a common variance (across treatment combinations) assumption is reasonable? Why or why not?

(e) In Example 87, half-normal plotting of absolute fitted effects of two-level factors on logged sample standard deviations is used as a means of looking for factor effects on response *variability*. Why is that method likely to provide little insight here? (Hint: How many points would you end up plotting? How decidedly "nonlinear" could such a plot possibly look?)

(f) Use the Yates algorithm and find fitted effects for the thickness data corresponding to the all-high treatment combination.

(g) Find s_P for the thickness data.

(h) Enhance your plot from (c) with the addition of error bars around the sample means individual 95 % two-sided confidence intervals for the four treatment combination means.

(i) Find a 95 % two-sided confidence interval for the interaction effect for the all-high treatment combination. Interpret this interval and point out in what ways it is consistent with the enhanced plot from (h).

(j) Find a 95 % two-sided confidence interval for the difference in slider position main effects (2.25 in setting minus 1.75 in setting). Would you use this interval to estimate the difference in mean thicknesses for 1.625 in crank position ferrules (2.25 in slider setting minus 1.75 in slider setting)? Why or why not? (Consider your answer to part (i).)

(k) Find a 95 % two-sided confidence interval for the difference in crank position main effects (1.71875 in setting minus 1.625 in setting). Would you use this interval to estimate the difference in mean thicknesses for 2.25 in slider position ferrules (1.71875 in crank setting minus 1.625 in crank setting)? Why or why not? (Consider your answer to part (i).)

27. Refer to the **Brush Ferrules** case in problem 26. The slider controls the first stage of a forming process and the crank controls the second.

    (a) Make individual 95 % two-sided confidence intervals for the two main effects and the two-factor interaction corresponding to the 1.71875 crank position and 2.25 slider position (the all-high) treatment combination. Why is it sufficient to consider only these factorial effects in this 2^2 factorial study? (For instance, why is nothing additional gained by considering the effects corresponding to the "all-low" treatment combination?)

    (b) Replace by 0 any effect whose interval in (a) includes zero and use the reverse Yates algorithm to find fitted means for a model of ferrule thickness that involves only those effects judged (via the intervals from part (a)) to be statistically detectable.

    (c) Plot the fitted thickness values from part (b) versus crank position. Connect the two points for the 1.75 slider position with a line segment and then connect the two points for the 2.25 slider position with another line segment.

    (d) The desired ferrule thickness was .4205 in. Assuming the linearity of your plot in (c) is appropriate, give several (at least two) different combinations of crank and slider positions that might produce the desired ferrule thickness.

28. Refer to the **Cutoff Machine** case in problem 18 of Chap. 1. The focus of project team efforts was improving the tool life of carbide cutting inserts. Carbide cutting inserts are triangular shaped pieces of titanium-coated carbide about 3/16 in thick. All three corners of a given insert can be used to make cuts. A crater, break, and poor quality dimension of a cut part are typical indicators of a "failed" corner, and the objective was to improve (increase) the number of cuts that could be made before failure of an insert.

    Stop delay and feed rate were seen as factors potentially affecting tool life. Stop delay is the time required to insert raw material (during which the insert cools). Feed rate is the rate at which the cutting insert is forced into the tubing being cut. Tubing RPM was also identified as a factor possibly affecting tool life, but because of time constraints, this factor was held fixed during the team's study. The team considered two stop delay settings (low, high) and four different feed rate settings (coded here as 1 through 4). For

Chapter 5. Experimental Design and Analysis for Process Improvement 321

each combination of stop delay and feed rate, a new carbide insert was used to cut 304 stainless steel. The number of tubes cut (until a failure occurred) was recorded for each corner of the insert. Thus, three responses (one from each corner of an insert) were recorded for each combination of stop delay and feed rate. The resulting data are in Table 5.18.

TABLE 5.18. Data for problem 28

Stop delay	Feed rate	Number of tubes cut
1	1	125, 129, 146
1	2	135, 130, 176
1	3	194, 183, 166
1	4	176, 187, 204
2	1	136, 141, 149
2	2	169, 155, 177
2	3	162, 207, 198
2	4	163, 195, 224

(a) How many treatment combinations were there in this study?

(b) Find the eight "sample" means.

(c) Plot the sample means from (b) in interaction plot format, placing feed rate on the horizontal axis. Does there seem to be serious interaction between stop delay and feed rate? Why or why not?

(d) Find all the fitted factorial effects for this study. (Find two stop delay main effects, four feed rate main effects, and eight two-factor interaction effects.)

(e) How many carbide inserts were used in this experiment? Do you think that this experiment was equivalent to one in which only one corner is used from each insert (three inserts per treatment combination)? Would you expect to see more variation or less variation in response than that in the data above, if three inserts (one corner of each) had been used?

(f) Probably the safest analysis of the data above would simply use the averages from (b) as responses, admitting that there was no real replication in the study and that one really has eight samples of size $m = 1$. If this route were taken, could you go beyond the computations in (d) to make confidence intervals for main effects and interactions? Explain.

Henceforth, in this problem, suppose that the data in the table above actually represent total cuts (for all corners) for three different inserts per treatment combination (so that it makes sense to think of the data as eight samples of size $m = 3$).

(g) Give a model equation for the "normal distributions with a common variance" description of number of tubes cut that uses "factorial effects" notation. Say what each term in your model represents in the context of the problem and define each one in terms of the $\mu_{..}, \mu_{i.}, \mu_{.j}$, and μ_{ij}.

(h) Find the 24 residuals by subtracting sample means from observations.

(i) Plot the residuals found in (h) versus the sample means. Does it appear that the constant variance feature of the model in (g) is appropriate for number of tubes cut? Why or why not?

(j) Find s_P, the pooled estimate of the supposedly common standard deviation, σ.

(k) Normal plot the 24 residuals found in (h). What insight does a plot of this type provide?

29. Refer to the **Cutoff Machine** case in problems 28. As in the last half of problem 28, treat the data given in that problem as if they had been obtained using three different inserts per treatment combination. Use the "normal distributions with a common variance" model for number of cuts per insert in the following analysis.

    (a) Make individual 95 % lower confidence bounds for the eight mean numbers of cuts.

    (b) Find a value Δ, so that if one attaches an uncertainty of $\pm \Delta$ to a particular (*pre-chosen*) difference in a pair of sample mean numbers of cuts, one may be 99 % confident the resulting interval (as an estimate of the corresponding difference in long-run mean numbers of cuts).

    (c) There are a total of 28 different comparisons between means that could be made in this context. If one applies the uncertainty from (b) to each, what minimum overall confidence does the Bonferroni inequality guarantee for *all* comparisons? On the basis of your analysis in (b), is there a combination of stop delay and feed rate that is clearly better than the others by the standard of (b)? Why or why not? (Hint: compare the maximum sample average to the minimum sample average.)

    (d) Find a value Δ so that an interval with end points $ab_{ij} \pm \Delta$ can be used as a 95 % two-sided confidence interval for the corresponding interaction $\alpha\beta_{ij}$.

    (e) Make a 99 % two-sided confidence interval for the difference (feed rate 2, stop delay 1) minus (feed rate 2, stop delay 2).

    (f) Make a 99 % two-sided confidence interval for the difference (feed rate 4, stop delay 1) minus (feed rate 3, stop delay 1).

(g) Find a 99% two-sided confidence interval for the difference in stop delay main effects (stop delay 1 minus stop delay 2). Would you use this interval in place of the one from (e)? Why or why not? (Hint: consider (d) above and the estimated interactions from problem 28(d).)

(h) Find a 99% two-sided confidence interval for the difference in feed rate 4 and feed rate 3 main effects. Would you use this interval in place of the one from (f)? Why or why not? (Hint: consider (d) above and the estimated interactions from problem 28(d).)

30. **Tablet Hardness.** Tablet hardness (for medicine) is measured in standard Cobb units (SCU) and specifications for hardness are 17 ± 5 SCU. After analyzing the tablet production process, engineers concluded that tablet press compression level and powder moisture level had large effects on final tablet hardness. Management was using a trial-and-error method to adjust compression and moisture levels. The engineering team decided to adopt a more systematic approach to finding a good combination of compression and moisture. Low and high settings were identified for both factors for purposes of experimentation. The following summaries were obtained from an experiment where the four different combinations of compression and moisture content were used to produce tablets. Given in the table are means and standard deviations of hardness for a number of batches made under each treatment combination. (Batch hardness, y, was defined by testing and averaging test results for ten tablets from the batch. The ns in the following table are numbers of batches. The $\bar{y}$s were obtained by averaging the y's determined from each batch. The s's are the standard deviations of the ys for a given compression/moisture combination.)

Compression	Moisture	n	$\bar{y}$	s
Low	Low	29	17.59	1.22
High	Low	17	16.75	.71
Low	High	27	17.53	.78
High	High	23	17.60	1.02

(a) How many experimental factors were there in this study? Name them.

(b) Give a model equation for the "normal distributions with a common variance" description of batch hardness that uses "factorial effects" notation.

(c) Find the pooled estimate of the (supposedly common) variance of batch hardness under a fixed set of processing conditions. What are the associated degrees of freedom?

(d) Plot the sample means in interaction plot format. Put compression on the horizontal axis. Does it appear that there are strong interactions between compression and moisture? Why or why not?

(e) Find individual 95% two-sided confidence intervals for the four treatment combination means.

(f) Enhance your plot in (d) by drawing error bars around the means based on your confidence limits from (e). What is suggested by these about the statistical detectability of interaction (lack of parallelism) in this study?

(g) Find the fitted effects for each effect in the model given in (b) using Yates algorithm.

(h) Find a 99 % two-sided confidence interval for the two-factor interaction at the high levels of both factors.

(i) Find a 99 % two-sided confidence interval for the high moisture main effect. Find a 99 % two-sided confidence interval for the high compression main effect.

(j) Using the Bonferroni inequality, find a set of 94 % simultaneous confidence intervals for the three effects estimated in (h) and (i).

(k) Based on your intervals from (j) and the enhanced interaction plot from (f) discuss what has been learned about how compression and moisture impact tablet hardness.

(l) Which combination of compression and moisture seems best? Defend your answer. (Hint: recall the ideal hardness value established in the problem description above.)

(m) Using the information given in this problem estimate the fraction of high compression/low moisture batches that have measured "batch hardness" within 1 SCU of the target value of 17 SCU. Adopt a normal model and assume your estimated average for high compression/low moisture is reasonable for the true average and the pooled estimate of variance is reasonable as well. Do you have appropriate information here to estimate the fraction of *individual* tablets made under these same conditions with tested hardness inside the 17 ± 5 SCU specifications (for individuals)? Why or why not?

31. Consider a situation where fatigue life testing of steel bar stock is to be done. Bar stock can be ordered from several different vendors and it can be ordered to several different sets of specifications with regard to each of the factors "dimensions," "hardness," and "chemical composition." There are several different testing machines in a lab and several different technicians could be assigned to do the testing. The response variable will be the number of cycles to failure experienced by a given test specimen. Assume that each factor has two levels.

(a) Describe the three-factor full factorial study that might be carried out in this situation (ignoring testing machine differences, technician differences, etc.). Make out a data table that could be used to record the results. For each "run" of the experiment, specify the levels of each of the 3 factors to be used. What constitutes "replication" in your plan?

(b) Suppose that attention is restricted to steel bar stock from a single vendor, ordered to a single set of specifications, tested on a single machine by a single technician. Suppose further, that either 10 specimens from a single batch or 1 specimen from each of 10 different batches can be used. Under which of the two scenarios would you expect to get the larger variation in observed fatigue life? Under what circumstances would the first test plan be most appropriate? Under what circumstances would the second be most appropriate?

32. **Resistance Measurements.** Anderson, Koppen, Lucas, and Schotter made some resistance measurements on five nominally 1000 Ω resistors. The measurements were made with an analog meter, an old digital meter and a new digital meter. As it turns out, the analog readings and a set of new digital readings were made on one day and the old digital readings and a second set of new digital readings were made on another day. The students' data are given in Table 5.19. We will assume that the readings were all made by the same person, so that the matter of reproducibility is not an issue here.

(a) Discuss why it does not necessarily make sense to treat the two measurements for each resistor made with the new digital meter as a "sample" of size $m = 2$ from a single population, and to then derive a measure of digital meter repeatability from these five samples of size two. If the students had wanted to evaluate measurement repeatability how should they have collected some data?

(b) It is possible to make a judgment as to how important the "day" effect is for the new digital meter by doing the following. For each resistor, subtract the day 1 new digital reading from the day 2 new digital reading to produce five differences d. Then apply the formula (2.10), $\bar{d} \pm t s_d/\sqrt{5}$, to make a two-sided confidence interval for the mean difference, μ_d. Do this using 95 % confidence. Does your interval include 0? Is there a statistically detectable "day" effect in the new digital readings? What does this analysis say about the advisability of ignoring "day," treating the two measurements on each resistor as a sample of size $m = 2$ and attempting to thereby estimate repeatability for the new digital meter?

(c) Differencing the day 1 analog and new digital readings to produce five differences (say analog minus new digital), gives a way of looking for systematic differences between these two meters. One may apply the formula (2.10) to make a confidence interval for the mean difference, μ_d. Do this, making a 95 % two-sided interval. Does your interval include 0? Is there a statistically detectable systematic difference between how these meters read?

(d) Redo (c) for the day 2 old digital and new digital readings.

TABLE 5.19. Data for Problem 32

Resistor	Day	Meter	Measured Resistance	Resistor	Day	Meter	Measured Resistance
1	1	Analog	999	1	2	Old digital	981
1	1	New digital	994	1	2	New digital	993
2	1	Analog	1000	2	2	Old digital	988
2	1	New digital	1001	2	2	New digital	1000
3	1	Analog	999	3	2	Old digital	979
3	1	New digital	992	3	2	New digital	992
4	1	Analog	999	4	2	Old digital	974
4	1	New digital	987	4	2	New digital	987
5	1	Analog	1000	5	2	Old digital	988
5	1	New digital	1001	5	2	New digital	1000

(e) A means of comparing the analog and old digital meters looking for a systematic difference in readings is the following. For each resistor one may compute

$y = (analog - new\ digital\ day\ 1) - (old\ digital - new\ digital\ day\ 2)$

and then apply the formula (2.6), $\bar{y} \pm t s_y / \sqrt{5}$, to make a confidence interval for the mean μ_y. Do this, making a 95 % interval. Does your interval include 0? Is there a statistically detectable systematic difference between how these meters read?

(f) Carefully describe a complete factorial study with the factors "meters," "resistors," and "days" that includes some replication and would allow more straightforward use of the material of this chapter in assessing the effects of these factors. Does your plan allow for the estimation of the repeatability variance component?

33. Consider the situation of Example 60. Suppose a colleague faced with a similar physical problem says "We do not need to change levels of both factors at once. The scientific way to proceed is to experiment one factor at a time. We'll hold the cooling method fixed and change antimony level to see the antimony effect. Then we'll hold antimony level fixed and change cooling method in order to see the cooling method effect." What do you have to say to this person?

34. **Heat Treating Steel**. Bockenstedt, Carrico, and Smith investigated the effects of steel formula (1045 and 1144), austenizing temperature (800° C and 1000° C), and cooling rate (furnace and oil quench) on the hardness of heat-treated steel.

(a) If all possible combinations of levels of the factors mentioned above are included in an experiment, how many treatment combinations total will be studied?

(b) If only a single specimen of each type alluded to in (a) was tested, how would one go about judging the importance of the main effects and interactions of the three factors?

In fact, a 2^3 full factorial with $m = 3$ steel specimens per treatment combination was run and hardness data like those in Table 5.20 were obtained.

(c) Compute the pooled standard deviation here, s_P.

TABLE 5.20. Data for Problem 34

Steel	Temperature	Cooling rate	Hardness, y
1045	800	Furnace cool	186.0, 191.0, 187.0
1144	800	Furnace cool	202.5, 204.0, 202.0
1045	1000	Furnace cool	146.0, 153.0, 147.0
1144	1000	Furnace cool	154.0, 156.0, 156.5
1045	800	Oil quench	222.5, 230.0, 221.5
1144	800	Oil quench	239.5, 248.5, 249.0
1045	1000	Oil quench	268.0, 278.0, 272.5
1144	1000	Oil quench	297.5, 296.0, 299.0

(d) Find the eight sample means and apply the Yates algorithm to find the fitted 2^3 factorial effects corresponding to the all-high (1144/1000/oil quench) treatment combination.

(e) Apply formula (5.28) to make seven individual 95 % two-sided confidence intervals for the main effects, two-factor interactions, and three-factor interaction for the all-high treatment combination here.

(f) The seven intervals from part (e) have at least what level of simultaneous confidence?

(g) Based on the intervals from (e), which effects do you judge to be statistically detectable?

(h) Suppose that maximum hardness is desired, but hardness being equal, the preferable levels of the factors (perhaps for cost reasons) are 1045, 800, and oil quench. In light of your answer to (g), how do you recommend setting levels of these factors? Explain.

35. **Valve Airflow**. In their "Quality Quandaries" article in the 1996, volume 8, number 2 issue of *Quality Progress*, Bisgaard and Fuller further developed an example due originally to Moen, Nolan, and Provost in their *Improving Quality Through Planned Experimentation*. The emphasis of the Bisgaard and Fuller analysis was to consider the effects of four experimental factors on both mean *and standard deviation* of a response variable, y, measuring airflow through a solenoid valve used in an auto air pollution control device. The factors in a 2^4 factorial study with $m = 4$ were length of armature (A) (.595 in vs. .605 in), spring load (B) (70 g vs. 100 g), bobbin depth (C)

328   Chapter 5. Experimental Design and Analysis for Process Improvement

(1.095 in vs. 1.105 in), and tube length (D) (.500 in vs. .510 in). Mean responses and sample standard deviations are given in Table 5.21.

(a) As in Example 87, take the natural logarithms of the sample standard deviations and then use the Yates algorithm and normal plotting (or half-normal plotting) to look for statistically detectable effects on the *variability* of airflow. Which factors seem to have significant effects? How would you recommend setting levels of these factors if the only object were consistency of airflow?

TABLE 5.21. Data for problem 35

A	B	C	D	$\bar{y}$	s	A	B	C	D	$\bar{y}$	s
−	−	−	−	.46	.04	−	−	−	+	.42	.04
+	−	−	−	.42	.16	+	−	−	+	.28	.15
−	+	−	−	.57	.02	−	+	−	+	.60	.07
+	+	−	−	.45	.10	+	+	−	+	.29	.06
−	−	+	−	.73	.02	−	−	+	+	.70	.02
+	−	+	−	.71	.01	+	−	+	+	.71	.02
−	+	+	−	.70	.05	−	+	+	+	.72	.02
+	+	+	−	.70	.01	+	+	+	+	.72	.01

(b) Your analysis in part (a) should suggest some problems with a "constant (across treatment combinations) variance" model for airflow. Nevertheless, at least as a preliminary or rough analysis of *mean* airflow, compute s_P, apply the Yates algorithm to the sample means, and apply the 95 % confidence limits from display (5.28) to judge the detectability of the 2^4 factorial effects on mean airflow. Which variables seem to have the largest influence on this quantity?

36. **Collator Machine Stoppage Rate.** Klocke, Tan, and Chai worked with the ISU Press on a project aimed at reducing jams or stoppages on a large collator machine. They considered the two factors "bar tightness" and "air pressure" in a 2 × 3 factorial study. For each of the six different treatment combinations, they counted numbers of stoppages and recorded machine running time in roughly 5 min of machine operation. (The running times did not include downtime associated with fixing jams. As such, every instant of running time could be thought of as providing opportunity for a new stoppage.) Table 5.22 summarizes their data.

This situation can be thought of as a "mean nonconformities per unit" scenario, where the "unit of product" is a 1-second time interval.

(a) A crude method of investigating whether there are any clear differences in the six different operating conditions is provided by the retrospective u chart material of Sect. 3.3.2. Apply that material to the

Chapter 5. Experimental Design and Analysis for Process Improvement   329

TABLE 5.22. Data for problem 36

Bar tightness	Air pressure	Number of Jams, X	Running Time, k (sec)	$\hat{u}=X/k$
Tight	Low	27	295	.0915
Tight	Medium	21	416	.0505
Tight	High	33	308	.1071
Loose	Low	15	474	.0316
Loose	Medium	6	540	.0111
Loose	High	11	498	.0221

six $\hat{u}$ values given in the table and say whether there is clear evidence of some differences in the operating conditions (in terms of producing stoppages). (Note that a total of 113 stoppages were observed in a total of 2531 seconds of running time, so that a pooled estimate of a supposedly common λ is $113/2531 = .0446$.)

(b) Plot the $\hat{u}$ values in interaction plot format, placing levels of air pressure on the horizontal axis.

(c) As in Sect. 3.3.2, a Poisson model for X (with mean $k\lambda$) produces a standard deviation for $\hat{u}$ of $\sqrt{\lambda/k}$. This in turn suggests estimating the standard deviation of $\hat{u}$ with

$$\hat{\sigma}_{\hat{u}} = \sqrt{\hat{u}/k}.$$

Crude approximate confidence limits for λ might then be made as

$$\hat{u} \pm z\hat{\sigma}_{\hat{u}}$$

(for z a standard normal quantile). For each of the six treatment combinations, make approximate 99% two-sided limits for the corresponding stoppage rates. Use these to place error bars around the rates plotted in (b).

(d) Based on the plot from (b) and enhanced in (c), does it appear that there are detectable bar tightness/air pressure interactions? Does it appear that there are detectable bar tightness or air pressure main effects? Ultimately, how do you recommend running the machine?

(e) The same logic used in (c) says that for r different conditions leading to r values $\hat{u}_1, \ldots, \hat{u}_r$ and r constants $c_1, \ldots, c_r$, the linear combination

$$\hat{L} = c_1\hat{u}_1 + \cdots + c_r\hat{u}_r$$

has a standard error

$$\hat{\sigma}_{\hat{L}} = \sqrt{\sum_{i=1}^{r} \frac{c_i^2 \hat{u}_i}{k_i}},$$

that can be used to make approximate confidence intervals for

$$L = c_1\lambda_1 + \cdots + c_r\lambda_r$$

as

$$\hat{L} \pm z\hat{\sigma}_{\hat{L}}.$$

Use this method and make an approximate 95 % two-sided confidence interval for

$$\alpha_1 - \alpha_2 = \frac{1}{3}(\lambda_{11} + \lambda_{12} + \lambda_{13}) - \frac{1}{3}(\lambda_{21} + \lambda_{22} + \lambda_{23}),$$

the difference in tight and loose bar main effects.

37. Refer to the **Collator Machine Stoppage Rate** case in problem 36. The analysis in problem 36 is somewhat complicated by the fact that the standard deviation of $\hat{u}$ depends not only on k but on λ as well. A way of somewhat simplifying the analysis is to replace $\hat{u}$ with $y = g(\hat{u})$ where g is chosen so that (at least approximately) the variance of y is independent of λ. The Freeman-Tukey suggestion for g is

$$y = g(\hat{u}) = \frac{\sqrt{\hat{u}} + \sqrt{\hat{u} + \frac{1}{k}}}{2},$$

and unless λ is very small

$$\text{Var} y \approx \frac{1}{4k}.$$

This problem considers the application of this idea to simplify the analysis of the collator machine data.

(a) Compute the six y values corresponding to the different observed jam rates, $\hat{u}$, in problem 36.

(b) Plot the y values in interaction plot format, placing levels of air pressure on the horizontal axis.

(c) Approximate confidence limits for the mean of y ($Ey = \mu_y$) are

$$y \pm z\frac{1}{2\sqrt{k}},$$

for z a standard normal quantile. Make individual 99 % two-sided confidence limits for each of the six different means of y. Use these to enhance the plot in (b) with error bars.

(d) Based on the plot from (b) and enhanced in (c), does it appear that there are detectable bar tightness/air pressure interactions? Does it appear that there are detectable bar tightness or air pressure main effects? Ultimately, how do you recommend running the machine?

(e) If r different conditions lead to r different values $y_1, \ldots, y_r$ and one has in mind r constants $c_1, \ldots, c_r$, the linear combination

$$\hat{L} = c_1 y_1 + \cdots + c_r y_r$$

has an approximate standard deviation

$$\sigma_{\hat{L}} \approx \frac{1}{2}\sqrt{\sum_{i=1}^{r} \frac{c_i^2}{k_i}}.$$

The quantity
$$L = c_1 \mathrm{E} y_1 + \cdots + c_r \mathrm{E} y_r$$

then has approximate confidence limits

$$\hat{L} \pm z \sigma_{\hat{L}},$$

and these provide means for judging the statistical detectability of factorial effects (on the mean of the transformed variable). Use this method and make an approximate 95 % two-sided confidence interval for

$$\alpha_1 - \alpha_2 = \frac{1}{3}(\mathrm{E}y_{11} + \mathrm{E}y_{12} + \mathrm{E}y_{13}) - \frac{1}{3}(\mathrm{E}y_{21} + \mathrm{E}y_{22} + \mathrm{E}y_{23}),$$

the difference in tight and loose bar main effects. Does this method show that there is a clear difference between the bar tightness main effects?

CHAPTER **6**

EXPERIMENTAL DESIGN AND ANALYSIS FOR PROCESS IMPROVEMENT PART 2: ADVANCED TOPICS

The basic tools of experimental design and analysis provided in Chap. 5 form a foundation for effective multifactor experimentation. This chapter builds on that and provides some of the superstructure of statistical methods for process-improvement experiments.

Section 6.1 provides an introduction to the important topic of fractional factorial experimentation. The 2^{p-q} designs and analyses presented there give engineers effective means of screening a large number of factors, looking for a few that need more careful subsequent scrutiny. Then Sect. 6.2 is concerned with using regression analysis as a tool in optimizing a process with quantitative inputs. That is, it considers "response surface methods" in systems where the process variables can be changed independently of each other. Finally, the short Sect. 6.3 discusses a number of important qualitative issues in experimentation for process improvement.

6.1 2^{p-q} Fractional Factorials

It is common for engineers engaged in process-improvement activities to be initially faced with many more factors/"process knobs" than can be studied in a practically feasible full factorial experiment. For example, even with just two levels for each of $p = 10$ factors (which is not that many by the standards of real industrial processes), there are already $2^{10} = 1{,}024$ different combinations to be considered. And few real engineering experiments are run in environments where there are both time and resources sufficient to collect 1,000 or more data points.

If one cannot afford a full factorial experiment in many factors, the alternatives are two. One must either hold the levels of some factors fixed (effectively eliminating them from consideration in the experiment) or find some way to vary all of the factors over some appropriate **fraction** of a full factorial (and then make a sensible analysis of the resulting data). This section concerns methods for this second approach. The thinking here is that it is best in early stages of experimentation to run fractional factorial experiments in many factors, letting data (rather than educated guessing alone) help *screen* those down to a smaller number that can subsequently be studied more carefully.

The discussion begins with some additional motivation for the section and some preliminary insights into what can and cannot possibly come out of a fractional factorial study. Then specific methods of design and analysis are provided for half-fractions of 2^p factorials. Finally, the methods for half-fractions are generalized to provide corresponding tools for studies involving only a fraction $1/2^q$ of all possible combinations from a full 2^p factorial.

6.1.1 Motivation and Preliminary Insights

Table 6.1 lists two levels of 15 factors from a real industrial experiment discussed by C. Hendrix in his article, "What Every Technologist Should Know About Experimental Design," which appeared in *Chemtech* in 1979. The object of experimentation was to determine what factors were principal determiners of the cold crack resistance of an industrial product. Now $2^{15} = 32{,}768$, and there is clearly no way that plant experimentation could be carried out in a full 2^{15} factorial fashion in a situation like this. Something else had to be done.

Rather than just guessing at which of the 15 factors represented in Table 6.1 might be most important and varying only their levels in an experiment, Hendrix and his colleagues were able to conduct an effective fractional factorial experiment varying all $p = 15$ factors in only 16 experimental runs. (Only a $1/2048$ fraction of all possible combinations of levels of these factors was investigated!) Methods of experimental design and analysis for problems like this are the subject of this section. But before jumping headlong into technical details, it is best

TABLE 6.1. 15 Process variables and their experimental levels

Factor	Process variable	Levels
A	Coating roll temperature	115° (−) vs. 125° (+)
B	Solvent	Recycled (−) vs. refined (+)
C	Polymer X-12 preheat	No (−) vs. yes (+)
D	Web type	LX-14 (−) vs. LB-17 (+)
E	Coating roll tension	30 (−) vs. 40 (+)
F	Number of chill rolls	1 (−) vs. 2 (+)
G	Drying roll temperature	75° (−) vs. 80° (+)
H	Humidity of air feed to dryer	75 % (−) vs. 90 % (+)
J	Feed air to dryer preheat	Yes (−) vs. No (+)
K	Dibutylfutile in formula	12 % (−) vs. 15 % (+)
L	Surfactant in formula	.5 % (−) vs. 1 % (+)
M	Dispersant in formula	.1 % (−) vs. .2 % (+)
N	Wetting agent in formula	1.5 % (−) vs. 2.5 % (+)
O	Time lapse before coating web	10 min (−) vs. 30 min (+)
P	Mixer agitation speed	100 rpm (−) vs. 250 rpm (+)

to begin with some qualitative/common sense observations about what will ultimately be possible. (And the reader is encouraged to return to this subsection after wrestling with the technical details that will follow, in an effort to avoid missing the forest for the trees.)

To begin, there is no magic by which one can learn from a small fractional factorial experiment all that could be learned from the corresponding full factorial. In the 15-factor situation represented by Table 6.1, there are potentially $2^{15} = 32{,}768$ effects of importance in determining cold crack resistance (from a grand mean and 15 main effects through a 15-way interaction). Data from only 16 different combinations cannot possibly be used to detail all of these. In fact, intuition should say that ys from 16 different conditions ought to let one estimate at most 16 different "things." Anyone who maintains that there is some special system of experimentation by which it is possible to learn all there is to know about a p-variable system from a small fractional factorial study is confused (or is selling snake oil).

In fact, unless the principle of effect sparsity is strongly active, small fractions of large factorials are doomed to provide little useful information for process-improvement efforts. That is, in complicated systems the methods of this section can fail, in spite of the fact that they are indeed the best ones available. So from the outset, the reader must understand that although they are extremely important and can be instrumental in producing spectacular process-improvement results, the methods of this section have unavoidable limitations that are simply inherent in the problem they address.

It is important next to point out that *if* one is to do fractional factorial experimentation, there are more effective and less effective ways of doing it. Take, for example, the (completely unrealistic but instructive) case of two factors A and B, both with two levels, supposing that one can afford to conduct only half of a full 2^2 factorial experiment. In this artificial context, consider the matter of experimental design, the choice of which of the four combinations (1), a, b, and ab to include in one's study. Temporarily supposing that the "all-low" combination, (1), is one of the two included in the experiment, it is obvious that combination ab should be the other. Why? The two combinations a and b can be eliminated from consideration because using one of them together with combination (1) produces an experiment where the level of only one of the two factors is varied. This kind of reasoning shows that the only two sensible choices of half-fractions of the 2^2 factorial are those consisting of "(1) and ab" or of "a and b." And while it is simple enough to reason to these choices of half of a 2^2 factorial, how in general to address the choice of a $1/2^q$ fraction of a 2^p factorial is much less obvious.

The artificial situation of a half-fraction of a 2^2 factorial can be used to make several other points as well. To begin, suppose that one uses either the "(1) and ab" or the "a and b" half-fraction as an experimental plan and sees a huge change in response between the two sets of process conditions. How is that outcome to be interpreted? After all, *both* the A "knob" and the B "knob" are changed as one goes from one set of experimental conditions to the other. Is it the A main effect that causes the change in response? Or is it the B main effect? Or is it perhaps both? There is inevitable ambiguity of interpretation inherent in this example. Happily, in more realistic problems (at least under the assumption of effect sparsity), the prospects of sensibly interpreting the results of a 2^{p-q} experiment are not so bleak as they seem in this artificial example.

Finally, in the case of the half-fraction of the 2^2 factorial, it is useful to consider what information about the 2^2 factorial effects $\mu_{..}$, α_2, β_2, and $\alpha\beta_{22}$ can be carried by, say, $\bar{y}_{(1)}$ and $\bar{y}_{ab}$ alone. Clearly, $\bar{y}_{(1)}$ tells one about $\mu_{(1)}$ and $\bar{y}_{ab}$ tells one about μ_{ab}, but what information about the factorial effects do they provide? As it turns out, the story is this. On the basis of $\bar{y}_{(1)}$ and $\bar{y}_{ab}$, one can estimate two *sums of effects*, namely, $\mu_{..} + \alpha\beta_{22}$ and $\alpha_2 + \beta_2$, but cannot further separate these four effects. The jargon typically used in the world of experimental design is that the A and B main effects are **confounded** or **aliased** (as are the grand mean and the AB interaction). And the fact that such basic quantities as the two main effects are aliased (and in some sense indistinguishable) in even this "best" half-fraction of the 2^2 factorial pretty much lays to final rest the possibility of any practical application of this pedagogical example.

With this motivation and qualitative background, it is hopefully clear that there are three basic issues to be faced in developing tools for 2^{p-q} fractional factorial experimentation. One must know how to

1. choose wisely a $1/2^q$ fraction of all 2^p possible combinations of levels of p two-level factors,

2. determine exactly which effects are aliased with which other effects for a given choice of the fractional factorial, and

3. do intelligent data analysis in light of the alias structure of the experiment.

The following two subsections address these matters, first for the case of half-fractions and then for the general $1/2^q$ fraction situation.

6.1.2 Half-Fractions of 2^p Factorials

Consider now a situation where there are p two-level experimental factors and for some reason one wishes to include only half of the 2^p possible combinations of levels of these factors in an experiment. Standard notation for this kind of circumstance is that a 2^{p-1} **fractional factorial** is contemplated. (The p exponent identifies the number of factors involved, and the -1 exponent indicates that a half-fraction is desired, one that will include 2^{p-1} different treatment combinations.)

The following is an algorithm for identifying a best possible half-fraction of a 2^p factorial:

> Write down for the "first" $p-1$ factors a $2^{p-1} \times (p-1)$ table of plus and minus signs specifying all possible combinations of levels of these, columns giving the levels for particular factors in combinations specified by rows. Then multiply together the "signs" in a given row (treating negative signs as -1s and positive signs as $+1$s) to create an additional column. This new (product) column specifies levels of the "last" factor to be used with the various combinations of levels of the first $p-1$ factors.

Algorithm for Identifying a Best Half-Fraction

Example 91 *A Hypothetical* 2^{4-1} *Design. The preceding prescription can be followed to produce a good choice of 8 out of 16 possible combinations of levels of factors A, B, C, and D. Table 6.2 shows that the combinations following from the algorithm are those that have an even number of factors set at their high levels.*

TABLE 6.2. Construction of a best half-fraction of a 2^4 factorial

A	B	C	(D) Product	Combination
−	−	−	−	(1)
+	−	−	+	ad
−	+	−	+	bd
+	+	−	−	ab
−	−	+	+	cd
+	−	+	−	ac
−	+	+	−	bc
+	+	+	+	abcd

Example 92 *An Unreplicated* 2^{5-1} *Chemical Process-Improvement Study. The article "Experimenting with a Large Number of Variables" by R. Snee, which appeared in the 1985 ASQC Technical Supplement* Experiments in Industry, *describes a $p = 5$ factorial experiment aimed at improving the consistency of the color of a chemical product. The factors studied and their levels were*

Factor A—solvent/reactant   Low (−)   vs.   High (+)
Factor B—catalyst/reactant   .025 (−)   vs.   .035 (+)
Factor C—temperature   150° C (−) vs.   160° C (+)
Factor D—reactant purity   92 % (−)   vs.   96 % (+)
Factor E—pH of reactant   8.0 (−)   vs.   8.7 (+)

Snee's unreplicated 2^{5-1} fractional factorial data are given here in Table 6.3, in a way that makes it clear that the algorithm for producing a best 2^{p-1} fractional factorial was followed. (The reader might also notice that for $p = 5$ factors, the prescription for constructing a good half-fraction picks out those combinations with an odd number of factors set at their high levels.)

TABLE 6.3. Observed color index for 16 combinations of levels of five two-level factors

A	B	C	D	(E) Product	Combination	Color Index, y
−	−	−	−	+	e	−.63
+	−	−	−	−	a	2.51
−	+	−	−	−	b	−2.68
+	+	−	−	+	abe	−1.66
−	−	+	−	−	c	2.06
+	−	+	−	+	ace	1.22
−	+	+	−	+	bce	−2.09
+	+	+	−	−	abc	1.93
−	−	−	+	−	d	6.79
+	−	−	+	+	ade	6.47
−	+	−	+	+	bde	3.45
+	+	−	+	−	abd	5.68
−	−	+	+	+	cde	5.22
+	−	+	+	−	acd	9.38
−	+	+	+	−	bcd	4.30
+	+	+	+	+	abcde	4.05

Having identified which 2^{p-1} combinations of levels of the factors one is going to include in a fractional factorial study, the next issue is understanding the implied **alias structure**, the pattern of what is confounded or aliased with what. As it turns out, 2^p factorial effects in a half-fraction are aliased in 2^{p-1} different *pairs*. One can hope to estimate 2^{p-1} sums of two effects, but cannot further separate the aliased effects. Exactly which pairs are confounded can be identified using a **system of formal multiplication** as follows:

Method for Determining the Alias Structure for a Half-Fraction

One begins by writing down the relationship

$$\text{name of the last factor} \leftrightarrow \text{product of the other } p-1 \text{ factor names} \quad (6.1)$$

called the **generator** of the design, in that it specifies how the column of signs is made up for the last factor. Then one can multiply both sides of the relationship (6.1) by any letter under the rules that

1. any letter times itself produces the letter I, and
2. I times any letter is that letter.

The relationships that then arise identify pairs of effects that are aliased.

For example, it follows from display (6.1) and these rules of multiplication that

$$I \leftrightarrow \text{product of the names of all } p \text{ factors.} \qquad (6.2)$$

Identifying I with the grand mean, this relationship says that the grand mean is aliased with the p factor interaction. Relationship (6.2), having I on the left side, is often called the **defining relation** for the fractional factorial. This is because one can easily multiply through by any string of letters of interest and see what is confounded with the corresponding main effect or interaction. For example, multiplying both sides of a defining relation like (6.2) by A, one sees that the main effect of A is aliased with the $(p-1)$-way interaction of all other factors.

Example 93 (*Example 91 continued.*) *Return to the hypothetical 2^{4-1} situation introduced earlier. In making up the combinations listed in Table 6.2, levels of factor D were chosen using products of signs for factors A, B, and C. Thus the generator for the design in Table 6.2 is*

$$D \leftrightarrow ABC.$$

From this (multiplying through by D and remembering that $D \cdot D$ is I), the defining relation for the design is

$$I \leftrightarrow ABCD,$$

and all aliases can be derived from this relationship. To begin, the grand mean is aliased with the four-factor interaction. That is, based on data from the eight combinations listed in Table 6.2, one can estimate $\mu_{....} + \alpha\beta\gamma\delta_{2222}$ but cannot further separate the summands. Or, multiplying through by A, one has

$$A \leftrightarrow BCD,$$

and the A main effect and BCD three-factor interaction are aliases. One can estimate $\alpha_2 + \beta\gamma\delta_{222}$ but cannot further separate these effects. Or, multiplying both sides of the defining relation by both A and B, one has

$$AB \leftrightarrow CD.$$

The combinations in Table 6.2 produce data leaving the AB two-factor interaction confounded with the CD two-factor interaction. One can estimate $\alpha\beta_{22} + \gamma\delta_{22}$ but cannot further separate these two-factor interactions on the basis of half-fraction data alone.

Example 94 *(Example 92 continued.)* *In Snee's color index study, levels of factor E were set using a column of signs derived as products of signs for factors A, B, C, and D. That means that the generator of the design used in the study is*

$$E \leftrightarrow ABCD,$$

so the defining relation is (upon multiplying through by E)

$$I \leftrightarrow ABCDE.$$

This implies that the grand mean is aliased with the five-factor interaction. Then, for example, multiplying through by A, one has

$$A \leftrightarrow BCDE$$

and the A main effect is confounded with the four-way interaction of B, C, D, and E. On the basis of the data in Table 6.3, one can estimate $\alpha_2 + \beta\gamma\delta\epsilon_{2222}$ but neither of the summands separately.

Examples 93 and 94 are instructive in terms of showing what happens to the alias structure of the best 2^{p-1} designs with increasing p. For the 2^{4-1} case, main effects are aliased with three-factor interactions, while for five factors, main effects are confounded with four-factor interactions. If one expects high-order interactions to typically be negligible, this is a comforting pattern. It says that for moderate to large p, an estimate of a main effect plus the aliased $(p-1)$-way interaction may often be thought of as essentially characterizing the main effect alone. And this kind of thinking suggests what is the standard design doctrine for 2^{p-q} studies. One wants to set things up so that (often important) low-order effects (main effects and low-order interactions) are aliased only with high-order interactions (that in simple systems are small). The virtue of the prescription for half-fractions given in this section is that it produces the best alias structure possible in this regard.

The final issue needing attention in this discussion of half-fractions is the matter of **data analysis**. How does one make sense out of 2^{p-1} fractional factorial data like those given in Table 6.3? This question has a simple answer:

Data Analysis Method for a 2^{p-1} Study

To analyze 2^{p-1} fractional factorial data, one first ignores the existence of the last factor, treating the data as if they were complete factorial data in the first $p-1$ factors. "Fitted effects" are computed and judged exactly as in Sect. 5.3. Then the statistical inferences are interpreted in light of the alias structure, remembering that one actually has estimates of not single effects, but sums of pairs of 2^p effects.

Example 95 *An Artificial 2^{3-1} Data Set.* *For purposes of illustrating the meaning of the instructions for data analysis just given when there is some replication in a 2^{p-1} data set, consider the artificial figures in Table 6.4.*

TABLE 6.4. Some summary statistics from a hypothetical 2^{3-1} study

A	B	C	Combination	n	$\bar{y}$	s^2
−	−	+	c	1	2.6	
+	−	−	a	1	6.4	
−	+	−	b	2	3.4	1.5
+	+	+	abc	3	7.6	1.8

The means in Table 6.4 are listed in Yates standard order as regards the first two factors. The reader should check that (ignoring factor C) application of the Yates algorithm (two cycles and then division by $2^2 = 4$) produces the "fitted effects" (listed in Yates order for A and B) $5.0, 2.0, .5,$ and $.1$.

The pooled sample variance here is

$$s_P^2 = \frac{(2-1)1.5 + (3-1)1.8}{(2-1) + (3-1)} = 1.7,$$

with $n - r = 7 - 4 = 3$ associated degrees of freedom. So then formula (5.28) can be used to make confidence intervals for judging the statistical detectability of the (sums of) effects estimated by the output of the Yates algorithm. The "p" appropriate in formula (5.28) is 2, since one is computing as if the last factor does not exist. Then, since the .975 quantile of the t_3 distribution is 3.182, using individual two-sided 95 % confidence limits, a plus or minus value of

$$3.182\sqrt{1.7}\left(\frac{1}{2^2}\right)\sqrt{\frac{1}{1} + \frac{1}{1} + \frac{1}{2} + \frac{1}{3}} = 1.7$$

should be associated with the values $5.0, 2.0, .5,$ and 1. By this standard, only the first two represent effects visible above the experimental variation.

This statistical analysis has to this point ignored the existence of factor C. But now in interpreting the results, it is time to remember that the experiment was not a 2^2 factorial in factors A and B, but rather a 2^{3-1} fractional factorial in factors A, B, and C. Note that the signs in Table 6.4 show that the generator for this hypothetical study was

$$C \leftrightarrow AB,$$

so that the defining relation is

$$I \leftrightarrow ABC.$$

Now if the means in Table 6.4 were from a 2^2 factorial, the value 5.0 appearing on the first line of the Yates calculations would be an estimate of a grand mean. But in the 2^{3-1} fractional factorial, the grand mean is aliased with the three-factor interaction, that is,

$$5.0 \quad \text{estimates} \quad \mu_{...} + \alpha\beta\gamma_{222}.$$

Similarly, if the means were from a 2^2 factorial, the value 2.0 appearing on the second line of the Yates calculations would be an estimate of the A main effect.

But in the 2^{3-1} fractional factorial, the A main effect is aliased with the BC two-factor interaction, that is,

$$2.0 \quad \text{estimates} \quad \alpha_2 + \beta\gamma_{22}.$$

Of course, the simplest (and possibly quite wrong) interpretation of the fact that both $\mu_{....} + \alpha\beta\gamma_{222}$ and $\alpha_2 + \beta\gamma_{22}$ are statistically detectable would follow from an assumption that the two interactions are negligible and the estimated sums of effects are primarily measuring the overall mean and the main effect of factor A.

TABLE 6.5. The output of the Yates algorithm applied to the 16 color indices

Combination	Color Index, y	(Yates cycle 4) $\div$ 16	Sum estimated
e	$-.63$	2.875	$\mu_{.....} + \alpha\beta\gamma\delta\epsilon_{22222}$
a	2.51	.823	$\alpha_2 + \beta\gamma\delta\epsilon_{2222}$
b	-2.68	-1.253	$\beta_2 + \alpha\gamma\delta\epsilon_{2222}$
abe	-1.66	.055	$\alpha\beta_{22} + \gamma\delta\epsilon_{222}$
c	2.06	.384	$\gamma_2 + \alpha\beta\delta\epsilon_{2222}$
ace	1.22	.064	$\alpha\gamma_{22} + \beta\delta\epsilon_{222}$
bce	-2.09	.041	$\beta\gamma_{22} + \alpha\delta\epsilon_{222}$
abc	1.93	.001	$\alpha\beta\gamma_{222} + \delta\epsilon_{22}$
d	6.79	2.793	$\delta_2 + \alpha\beta\gamma\epsilon_{2222}$
ade	6.47	$-.095$	$\alpha\delta_{22} + \beta\gamma\epsilon_{222}$
bde	3.45	$-.045$	$\beta\delta_{22} + \alpha\gamma\epsilon_{222}$
abd	5.68	$-.288$	$\alpha\beta\delta_{222} + \gamma\epsilon_{22}$
cde	5.22	$-.314$	$\gamma\delta_{22} + \alpha\beta\epsilon_{222}$
acd	9.38	.186	$\alpha\gamma\delta_{222} + \beta\epsilon_{22}$
bcd	4.30	$-.306$	$\beta\gamma\delta_{222} + \alpha\epsilon_{22}$
abcde	4.05	$-.871$	$\alpha\beta\gamma\delta_{2222} + \epsilon_2$

Example 96 *(Examples 92 and 94 continued.)* As an example of how the analysis of a half-fraction proceeds in the absence of replication, consider what can be done with Snee's color index data given in Table 6.3. The observations in Table 6.3 are listed in Yates standard order as regards factors A, B, C, and D. One may apply the (4-cycle, final division by $2^4 = 16$) Yates algorithm to these data and arrive at the fitted sums of effects listed in Table 6.5.

Snee's data include no replication. So the only method for judging the detectability of effects in this study presented in this text is probability plotting. Figures 6.1 and 6.2 (on page 344) show, respectively, a normal plot of the last 15 estimates listed in Table 6.5 and a half-normal plot of their magnitudes. (Since one is probably willing to grant a priori that the mean response is other than 0, there is no reason to include the first estimate listed in Table 6.5 in the plot.) Especially from Fig. 6.2 it is clear that even if the bulk of the estimates in Table 6.5 consists of nothing more than experimental variation, those corresponding to "D+ABCE," "B+ACDE," "ABCD+E," and "A+BCDE" do not. The four sums of effects

$$\delta_2 + \alpha\beta\gamma\epsilon_{2222}, \quad \beta_2 + \alpha\gamma\delta\epsilon_{2222}, \quad \alpha\beta\gamma\delta_{2222} + \epsilon_2, \quad \text{and} \quad \alpha_2 + \beta\gamma\delta\epsilon_{2222}$$

are statistically detectable.

The simplest possible interpretation of the judgment that the four largest (in magnitude) of the last 15 estimates in Table 6.5 correspond to detectable sums of effects is that only the main effects of factors D, B, E, and A are important. This interpretation says

(1) that (in decreasing order of importance) reactant purity, catalyst/reactant ratio, reactant pH, and solvent/reactant ratio affect color index for this product,

(2) that temperature has no appreciable impact on product color, and

(3) that the important factors act separately on the color index.

Since the original motivation for the experimentation was to find a way of improving color consistency, this result would guide engineers to the very careful control of process inputs. Reactant purity would deserve first attention, catalyst/reactant ratio would deserve second attention, and so on.

These tentative conclusions about color index are so clean and intuitively appealing that there would seem to be little reason to doubt that they are the right ones for the color index problem. But it must be kept in mind that they are based on a fractional factorial study and an assumption of simple structure for the chemical system. If, in fact, the chemical system is not simple (and, e.g., there are important four-factor interactions), then they could lead one in wrong directions when looking for a way to improve color consistency.

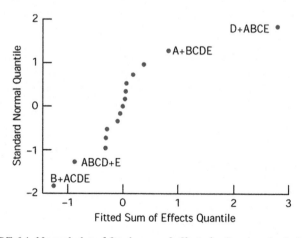

FIGURE 6.1. Normal plot of fitted sums of effects for Snee's color index data

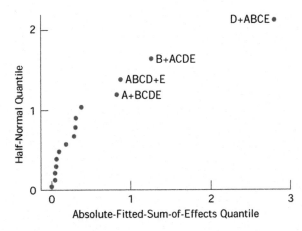

FIGURE 6.2. Half-normal plot of absolute-fitted-sums-of-effects for Snee's color index data

6.1.3 $1/2^q$ Fractions of 2^p Factorials

The tools just presented for half-fractions of 2^p factorials all have their natural extensions to the general case of $1/2^q$ fractions. To begin, the problem of choosing 2^{p-q} out of 2^p possible combinations to include in a fractional factorial study can be addressed using the **product of signs** idea introduced for half-fractions. That is, one may follow this prescription:

Algorithm for Producing a 2^{p-q} Fractional Factorial

Write down for the first $p - q$ factors a $2^{p-q} \times (p - q)$ table of plus and minus signs specifying all possible combinations of levels of these, columns giving the levels for particular factors in combinations specified by rows. Then make up q (different) additional columns of signs as products (a row at a time) of signs in q (different) groups of the first $p - q$ columns. These new product columns specify levels of the last q factors to be used with the various combinations of levels of the first $p - q$ factors.

This set of instructions is somewhat ambiguous in that it does not specify exactly *which* product columns one ought to construct. The fact is that some choices are better than others in terms of the alias structures that they produce. But discussion of this must wait until the matter of actually *finding* an alias structure has been considered.

Example 97 *A 2^{5-2} Catalyst Development Study.* In a paper presented at the 1986 National Meeting of the American Statistical Association, Hanson and Best described an experimental program for the development of an effective catalyst for the production of ethyleneamines by the amination of monoethanolamine. One part of that program involved a quarter-fraction of a 2^5 factorial with the following factors and levels.

Factor A—Ne/Re ratio   2/1 (−) vs. 20/1 (+)
Factor B—precipitant $(NH_4)_2CO_3$ (−) vs. none (+)
Factor C—calcining temperature   300° (−) vs. 500° (+)
Factor D—reduction temperature   300° (−) vs. 500° (+)
Factor E—support used   Alpha-alumina (−) vs. silica-alumina (+)

The response variable of interest was

$$y = \text{the percent water produced.}$$

A quarter-fraction of a full 2^5 factorial involves 8 out of 32 possible combinations of levels of factors A, B, C, D, and E. To follow the prescription just given for choosing such a 2^{5-2} fractional factorial, one begins by writing down an 8×3 table of signs specifying all eight combinations of levels of the factors A, B, and C. Then one must make up $q = 2$ product columns to use in choosing corresponding levels of factors D and E. The particular choice made by Hanson and Best was to use ABC products to choose levels of factor D and BC products to choose levels of E. Table 6.6 shows the construction used, the 2^5 names of the eight combinations selected, the raw data, and some (sample-by-sample) summary statistics. The choice of product columns made by the engineers in this study was by no means the only one possible. Different choices would have led to different confounding patterns (that in other circumstances might have seemed preferable on the basis of engineering considerations).

TABLE 6.6. % H_2O values produced in runs of eight combinations of five factors in the catalyst development study and some summary statistics

A	B	C	(D) ABC Product	(E) BC Product	2^5 Name	%H_2O, y	$\bar{y}$	s^2
−	−	−	−	+	e	8.70, 11.60, 9.00	9.767	2.543
+	−	−	+	+	ade	26.80	26.80	
−	+	−	+	−	bd	24.88	24.88	
+	+	−	−	−	ab	33.15	33.15	
−	−	+	+	−	cd	28.90, 30.98	29.940	2.163
+	−	+	−	−	ac	30.20	30.20	
−	+	+	−	+	bce	8.00, 8.69	8.345	.238
+	+	+	+	+	abcde	29.30	29.30	

The way one finds the alias structure of a general 2^{p-q} fractional factorial is built on the same **formal multiplication** idea used for half-fractions. The only new complication is that one must now work with q generators, and these lead to a defining relation that says that the grand mean is aliased with $2^q - 1$ effects. (So in the end, effects are aliased in 2^{p-q} different groups of 2^q.) To be more precise:

Method for Determining the Alias Structure for a 2^{p-q} Study

One begins by writing down the **q generators** of the design of the form

$$\text{name of one of the last } q \text{ factors} \leftrightarrow$$
$$\text{product of names of some of the first } (p-q) \text{ factors} \quad (6.3)$$

that specify how the columns of signs are made up for choosing levels of the last q factors. Then each of these q relationships (6.3) is multiplied through by the letter on the left to produce a relationship of the form

$$\text{I} \leftrightarrow \text{product}.$$

These are taken individually, multiplied in pairs, multiplied in triples, and so on to produce a **defining relation** of the form

$$\text{I} \leftrightarrow \text{product 1} \leftrightarrow \text{product 2} \leftrightarrow \cdots \leftrightarrow \text{product } (2^q - 1) \quad (6.4)$$

specifying $2^q - 1$ aliases for the grand mean. This defining relation (6.4) and the formal multiplication scheme are then used to find all aliases of any 2^p effect of interest.

A 2^{p-q} fractional factorial of the type described here has q generators. Each of the 2^p factorial effects has $2^p - 1$ aliases. And the defining relation identifies $2^q - 1$ products as equivalent to I.

Example 98 *(Example 97 continued.) Consider again the 2^{5-2} catalyst study of Hanson and Best. Table 6.6 shows the two generators for the eight combinations used in the study to be*

$$\text{D} \leftrightarrow \text{ABC} \quad \text{and} \quad \text{E} \leftrightarrow \text{BC}.$$

Multiplying the first of these through by D and the second by E, one has the two relationships

$$\text{I} \leftrightarrow \text{ABCD} \quad \text{and} \quad \text{I} \leftrightarrow \text{BCE}.$$

But then, multiplying the left sides and the right sides of these two together, one also has

$$\text{I} \cdot \text{I} \leftrightarrow (\text{ABCD}) \cdot (\text{BCE}), \quad \text{that is,} \quad \text{I} \leftrightarrow \text{ADE}.$$

Finally, combining the three "I$\leftrightarrow$product" statements into a single string of aliases, one has the defining relation for this study

$$\text{I} \leftrightarrow \text{ABCD} \leftrightarrow \text{BCE} \leftrightarrow \text{ADE}.$$

This relationship shows immediately that one may estimate the sum of effects $\mu_{....} + \alpha\beta\gamma\delta_{2222} + \beta\gamma\epsilon_{222} + \alpha\delta\epsilon_{222}$ but may not separate the summands. Or, multiplying through the defining relation by A, one has

$$\text{A} \leftrightarrow \text{BCD} \leftrightarrow \text{ABCE} \leftrightarrow \text{DE},$$

Chapter 6. Experimental Design and Analysis for Process Improvement   347

and sees that the A main effect is aliased with the BCD three-factor interaction, the ABCE four-factor interaction, and the DE two-factor interaction. It should be easy for the reader to verify that (as expected) the 2^5 factorial effects are aliased in $8 = 2^{5-2}$ sets of $4 = 2^2$ effects. With data from eight different combinations, one can estimate "eight things," the corresponding eight different sums of four effects.

Example 99 *Defining Relations for Two Different 2^{6-2} Plans.* Consider the choice of generators for a 2^{6-2} plan. Two different possibilities are

$$E \leftrightarrow ABCD \quad \text{and} \quad F \leftrightarrow ABC \qquad (6.5)$$

and

$$E \leftrightarrow BCD \quad \text{and} \quad F \leftrightarrow ABC. \qquad (6.6)$$

The reader should do the work (parallel to that in the previous example) necessary to verify that the defining relation corresponding to the set of generators (6.5) is

$$I \leftrightarrow ABCDE \leftrightarrow ABCF \leftrightarrow DEF.$$

And the defining relation corresponding to the set of generators (6.6) is

$$I \leftrightarrow BCDE \leftrightarrow ABCF \leftrightarrow ADEF.$$

This second defining relation is arguably better than the first. The first shows that the choice of generators (6.5) leaves some main effects aliased with two-factor interactions, fairly low-order effects. In contrast, the choice of generators (6.6) leads to the main effects being aliased with only three-factor (and higher-order) interactions. This example shows that not all choices of q generators in a 2^{p-q} study are going to be equally attractive in terms of the alias structure they produce.

Example 100 *Finding the Defining Relation for a 1/8 Fraction of a 2^6 Study.* As an example of what must be done to find the defining relation for fractions of 2^p factorials smaller than 1/4, consider the set of generators of a 2^{6-3} study:

$$D \leftrightarrow AB, E \leftrightarrow AC, \quad \text{and} \quad F \leftrightarrow BC.$$

These immediately produce

$$I \leftrightarrow ABD, I \leftrightarrow ACE, \quad \text{and} \quad I \leftrightarrow BCF.$$

Then multiplying these in pairs, one has

$$I \cdot I \leftrightarrow (ABD) \cdot (ACE), \text{ i.e., } I \leftrightarrow BCDE,$$

$$I \cdot I \leftrightarrow (ABD) \cdot (BCF), \text{ i.e., } I \leftrightarrow ACDF,$$

and
$$I \cdot I \leftrightarrow (ACE) \cdot (BCF), \text{ i.e., } I \leftrightarrow ABEF.$$

And finally multiplying all three of these together, one has
$$I \cdot I \cdot I \leftrightarrow (ABD) \cdot (ACE) \cdot (BCF), \text{ i.e., } I \leftrightarrow DEF.$$

So then, stringing together all of the aliases of the grand mean, one has the defining relation:
$$I \leftrightarrow ABD \leftrightarrow ACE \leftrightarrow BCF \leftrightarrow BCDE \leftrightarrow ACDF \leftrightarrow ABEF \leftrightarrow DEF,$$

and it is evident that 2^6 effects are aliased in eight groups of eight effects.

Data Analysis Method for a 2^{p-q} Study

Armed with the ability to choose 2^{p-q} plans and find their confounding structures, the only real question remaining is how **data analysis** should proceed. Again, essentially the same method introduced for half-fractions is relevant. That is:

> To analyze 2^{p-q} fractional factorial data, one first ignores the existence of the last q factors, treating the data as if they were complete factorial data in the first $p - q$ factors. "Fitted effects" are computed and judged exactly as in Sect. 5.3. Then the statistical inferences are interpreted in light of the alias structure, remembering that one actually has estimates of not single effects, but sums of 2^p factorial effects.

Example 101 *(Examples 97 and 98 continued.) The sample means in Table 6.6 are listed in Yates standard order as regards factors A, B, and C. The reader may do the arithmetic to verify that the (three-cycle, final division by $2^3 = 8$) Yates algorithm applied directly to these means (as listed in Table 6.6) produces the eight fitted sums of effects listed in Table 6.7.*

There is replication in the data listed in Table 6.6, so one may use the confidence interval approach to judge the detectability of the sums corresponding to the estimates in Table 6.7. First, the pooled estimate of σ must be computed from the sample variances in Table 6.6:

$$\begin{aligned} s_P^2 &= \frac{(3-1)2.543 + (2-1)2.163 + (2-1).238}{(3-1) + (2-1) + (2-1)}, \\ &= 1.872, \end{aligned}$$

so that
$$s_P = \sqrt{1.872} = 1.368 \text{ \% water.}$$

Then the $p = 3$ version of formula (5.28) says that (using individual 95 % two-sided confidence limits) a plus or minus value of

$$2.776(1.368)\frac{1}{2^3}\sqrt{\frac{1}{3} + \frac{1}{1} + \frac{1}{1} + \frac{1}{1} + \frac{1}{2} + \frac{1}{1} + \frac{1}{2} + \frac{1}{1}} = 1.195 \text{ \% water}$$

should be associated with each of the estimates in Table 6.7. By this criterion, the estimates on the first, second, fourth, seventh, and eighth lines of Table 6.7 are big enough to force the conclusion that they represent more than experimental error.

TABLE 6.7. Fitted sums of effects from the catalyst development data

A	B	C	D	E	$\bar{y}$	Estimate (8 Divisor)	Sum estimated
−	−	−	−	+	9.767	24.048	$\mu_{.....}$ + aliases
+	−	−	+	+	26.80	5.815	α_2 + aliases
−	+	−	+	−	24.88	−.129	β_2 + aliases
+	+	−	−	−	33.15	1.492	$\alpha\beta_{22}$ + aliases
−	−	+	+	−	29.94	0.399	γ_2 + aliases
+	−	+	−	−	30.20	−.511	$\alpha\gamma_{22}$ + aliases
−	+	+	−	+	8.345	−5.495	$\beta\gamma_{22}$ + aliases
+	+	+	+	+	29.30	3.682	$\alpha\beta\gamma_{222}$ + aliases

How then does one interpret such a result? Ignoring the sum involving the grand mean, the sums of 2^5 effects corresponding to the four largest (in absolute value) fitted sums are (using the defining relation for this study)

$$\alpha_2 + \beta\gamma\delta_{222} + \alpha\beta\gamma\epsilon_{2222} + \delta\epsilon_{22} \quad \text{estimated as} \quad 5.815,$$
$$\beta\gamma_{22} + \alpha\delta_{22} + \epsilon_2 + \alpha\beta\gamma\delta\epsilon_{22222} \quad \text{estimated as} \quad -5.495,$$
$$\alpha\beta\gamma_{222} + \delta_2 + \alpha\epsilon_{22} + \beta\gamma\delta\epsilon_{2222} \quad \text{estimated as} \quad 3.682, \quad \text{and}$$
$$\alpha\beta_{22} + \gamma\delta_{22} + \alpha\gamma\epsilon_{222} + \beta\delta\epsilon_{222} \quad \text{estimated as} \quad 1.492.$$

Concentrating initially on the first three of these, one can reason to at least four different, relatively simple interpretations. First, if the largest components of these sums are the main effects appearing in them, one might have an "A, E, and D main effects only" explanation of the pattern of responses seen in Table 6.6. But picking out from the first sum the A main effect, from the second the E main effect, and from the third the AE interaction, an explanation involving only factors A and E would be "A and E main effects and interactions only." And that does not end the relatively simple possibilities. One might also contemplate an "A and D main effects and interactions only" or a "D and E main effects and interactions only" description of the data. Which of these possibilities is the most appropriate for the catalyst system cannot be discerned on the basis of the data in Table 6.6 alone. One needs either more data or the guidance of a subject matter expert who might be able to eliminate one or more of these possibilities on the basis of some physical theory. (As it turns out, in the real situation, additional experimentation confirmed the usefulness of an "A, E, and D main effects only" model for the chemical process.)

Exactly what to make of the detectability of the sum corresponding to the 1.492 estimate is not at all clear. Thankfully (in terms of ease of interpretation of the experimental results), its estimate is less than half of any of the others. While it may be statistically detectable, it does not appear to be of a size to rival the other effects in terms of physical importance.

This 2^{5-2} experiment does what any successful fractional factorial does. It gives some directions to go in further experimentation and hints as to which factors may not be so important in determining the response variable. (The factors B and C do not enter into any of the four simple candidates for describing mean response.) But definitive conclusions await either confirmation by actually "trying out" tentative recommendations/interpretations based on fractional factorial results or further experimentation aimed at removing ambiguities and questions left by the study.

TABLE 6.8. 16 Experimental combinations and measured cold crack resistances

Combination	y
eklmnop	14.8
aghjkln	16.3
bfhjkmo	23.5
abefgkp	23.9
cfghlmp	19.6
acefjlo	18.6
bcegjmn	22.3
abchnop	22.2
dfgjnop	17.8
adefhmn	18.9
bdeghlo	23.1
abdjlmp	21.8
cdehjkp	16.6
acdgkmo	16.7
bcdfkln	23.5
abcdefghjklmnop	24.9

Example 102 *Tentative Conclusions in a 2^{15-11} Fractional Factorial Study.* As a final (and fairly extreme) example of what is possible in the way of fractional factorial experimentation, return to the scenario represented by the factors and levels in Table 6.1. The study actually conducted by Hendrix and his associates was a 2^{15-11} study with the 11 generators

$$E \leftrightarrow ABCD, F \leftrightarrow BCD, G \leftrightarrow ACD, H \leftrightarrow ABC, J \leftrightarrow ABD, K \leftrightarrow CD,$$
$$L \leftrightarrow BD, M \leftrightarrow AD, N \leftrightarrow BC, O \leftrightarrow AC, \text{ and } P \leftrightarrow AB.$$

The combinations run and the cold crack resistances observed are given in Table 6.8.

In this scenario it is practically infeasible to write out the whole defining relation. Since only 16 out of the 32,768 possible combinations are involved in this 2^{15-11} study, every 2^{15} factorial effect is aliased with 2,047 other effects! But at least the generators make it clear which main effects are aliased with the effects involving only factors A, B, C, and D.

Chapter 6. Experimental Design and Analysis for Process Improvement   351

The combinations and means in Table 6.8 are listed in Yates standard order as regards the first $15 - 11 = 4$ factors, A, B, C, and D. Applying the (four-cycle and 16 divisors) Yates algorithm to them (in the order listed), one obtains the estimates given in Table 6.9.

Judging the detectability of the sums of 2^{15} factorial effects in this study is complicated by the lack of replication. All that can be done is to probability plot the estimates of Table 6.9. Figures 6.3 and 6.4 on page 352 are, respectively, a normal plot of the last 15 estimates in Table 6.9 and a half-normal plot of the absolute values of these. The plots show clearly that even if all other estimates really represent only experimental variation, the ones corresponding to the B main effect plus its aliases and the F main effect plus its aliases do not. Of course, the simplest possible interpretation of this outcome is that only the factors B and F impact cold crack resistance in any serious way, the factors acting independently on the response and the high levels of both factors leading to the largest values of y. In light of the very small fraction involved here, however, the wise engineer would treat such conclusions as very tentative. They are intriguing and are perhaps even absolutely correct. But it would be foolhardy to conclude such on the basis of 2^{15-11} data. Of course, if one's object is only to find a good combination of levels of these 15 factors, this analysis points out what is in retrospect completely obvious about the data in Table 6.8. It is those four combinations with both of B and F at their high levels that have the largest responses.

TABLE 6.9. Estimates of sums of effects for the cold crack resistance data

Sum of effects estimated	Estimate
Grand mean + aliases	20.28
A main + aliases	.13
B main + aliases	2.87
P main + aliases (including AB)	−.08
C main + aliases	.27
O main + aliases (including AC)	−.08
N main + aliases (including BC)	−.19
H main + aliases (including ABC)	.36
D main + aliases	.13
M main + aliases (including AD)	.03
L main + aliases (including BD)	.04
J main + aliases (including ABD)	−.06
K main + aliases (including CD)	−.26
G main + aliases (including ACD)	.29
F main + aliases (including BCD)	1.06
E main + aliases (including ABCD)	.11

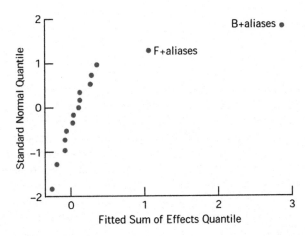

FIGURE 6.3. Normal plot of fitted sums of effects for the cold crack resistance data

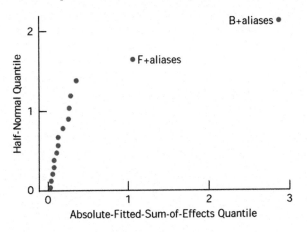

FIGURE 6.4. Half-normal plot of absolute-fitted-sums-of-effects for the cold crack resistance data

Section 6.1 Exercises

1. Consider a 2^{4-1} fractional factorial study.

    (a) How many different combinations of levels of the experimental factors are involved?

    (b) Suppose three combinations were run twice and the other five were run once each. The 95 % confidence limits for sums of pairs of factorial effects (aliased pairs) are $\hat{E} \pm \Delta$. Find Δ as a (numeric) multiple of s_P.

    (c) How many generators are involved in setting up this study?

(d) Suppose as run, D↔ABC and only the "grand mean plus alias," "A main effect plus alias," "BC interaction plus alias," and "ABC interaction plus alias" sums of effects are clearly detectable. Assuming all two-factor and higher-order interactions are negligible, give a simple interpretation of this result.

2. Suppose that in a 2^{4-1} fractional factorial study with D↔ABC, data analysis leads to the conclusion that the detectable sums of effects are "grand mean plus alias," "A main effect plus alias," "B main effect plus alias," "C main effect plus alias," and "ABC 3-factor interaction plus alias." *One plausible interpretation of this is that all of A, B, C, and D have detectable main effects.* Further, suppose that $\alpha_2, \beta_2,$ and γ_2 appear to be positive, while δ_2 appears to be negative. What combination of levels of factors A, B, C, and D then has the largest fitted mean? Is this combination represented in the original data set?

3. **PCB Study.** article "Reduction of Defects in Wave Soldering Process" that appeared in *Quality Engineering* discuss a 2^{9-5} fractional factorial experiment run in an attempt to learn how to reduce defects on printed circuit boards (PCBs). The factors studied and their levels were:

Factor	Process variable	Levels
A	Wave height	11.0 mm (−) vs. 11.5 mm (+)
B	Flux specific gravity	.80 (−) vs. .78 (+)
C	Conveyor speed	1.75 m/mm (−) vs. 1.65 m/mm (+)
D	Preheater temp	80° C (−) vs. 75° C (+)
E	Solder bath temp	225° C (−) vs. 230° C (+)
F	Blower heater temp	215° (−) vs. 210° (+)
G	Foam pressure	.20 kg/cm^2 (−) vs. .15 kg/cm^2 (+)
H	Direction of PCB	Reverse (−) vs. existing (+)
J	Jig height	High (−) vs. low (+)

The response variable was

$y =$ total number of dry solder defects on 20 PCBs.

(a) How many factors that potentially affect the response variable, y, have been considered in this study?

(b) If every combination of levels of the complete list of factors were to be evaluated, how many would there be?

(c) Suppose Chowdhury and Mitra had run a 2^{9-1} half-fraction design, how many combinations would they have considered?

(d) As mentioned above, in fact, Chowdhury and Mitra used a 2^{9-5} fractional factorial design. How many combinations of the factors were included in their study?

(e) What fraction of the number of possible combinations were actually run?

(f) How many generators were needed in choosing the study the authors ran?

(g) For the fractional factorial actually run, how many effects were aliased with the grand mean?

4. Continuing with the preceding **PCB** problem. The generators used for the **PCB** study were

$$E \leftrightarrow BC, F \leftrightarrow BD, G \leftrightarrow ACD, H \leftrightarrow AD, \text{ and } J \leftrightarrow AB.$$

(a) Consider the two experimental runs partially described with the A, B, C, and D levels in the small table below. Give the necessary levels (determined by the generators) for E, F, G, H, and J. (Essentially finish filling in the two rows of the table.)

A	B	C	D	E	F	G	H	J
+	+	−	−					
+	−	−	+					

(b) Name 3 effects aliased with the A main effect in this study.

(c) When the values of y obtained in the study are listed in Yates order for factors A through D and the Yates algorithm is applied, the results (again in Yates order, left to right, top to bottom) are

$$92.0, 3.75, 1.25, -1.25, 0.00, -1.00, -0.75, -3.50, -6.38,$$
$$-10.13, 0.13, -1.88, 5.13, 0.13, -0.38, \text{ and } 1.88.$$

The two largest (in magnitude) of the "fitted effects" computed by the Yates algorithm are -10.13 and -6.38. Give the simplest possible interpretation of these quantities (Hint: assume all 2-factor and higher-order interactions are negligible.)

(d) Suppose the effects you identified in (c) are judged to be the only ones of importance in this study. What settings of factors A through J should one use to produce a minimum y, and what response do you predict if your recommendations are followed? That is, recommend levels of all factors and give a corresponding value of $\hat{y}$.

6.2 Response Surface Studies

When p process variables $x_1, x_2, \ldots, x_p$ are all quantitative, there is the possibility of doing some experimentation and then using the resultant data to produce an equation describing how (at least in approximate terms) a response y depends upon the process variables. This section primarily concerns methods for using such an equation in the exploration of the approximate relationship between y

and $x_1, x_2, \ldots, x_p$. This can then provide direction as one tries to optimize (maximize or minimize) y by choice of settings of the process variables.

Standard methodology for turning n data points $(x_1, x_2, \ldots, x_p, y)$ into an equation for y as a function of the xs is the **multiple regression analysis** that is a main topic in most introductions to engineering statistics. This section does not repeat basic regression material, but instead assumes that the reader is already familiar with the subject and shows how it is useful in process improvement.

The section opens with a discussion of using graphical means to aid understanding of equations fit to $(x_1, x_2, \ldots, x_p, y)$ data. Then the topic of quadratic response functions is introduced, along with the experimental design issues associated with their use. There follows a discussion of analytical tools for interpreting fitted quadratic functions. Finally, the section concludes with a discussion of search strategies that can be used when one's goal is to optimize a response and issues of process modeling are not of particular concern.

6.2.1 Graphics for Understanding Fitted Response Functions

The primary output of a standard regression analysis based on n data points

$$(x_{11}, x_{21}, \ldots, x_{p1}, y_1), (x_{12}, x_{22}, \ldots, x_{p2}, y_2), \ldots, (x_{1n}, x_{2n}, \ldots, x_{pn}, y_n)$$

is an equation that we will temporarily represent in generic terms as

$$\hat{y} = f(x_1, x_2, \ldots, x_p). \tag{6.7}$$

A typical specific version of Eq. (6.7) is, of course,

$$\hat{y} = b_0 + b_1 x_1 + b_2 x_2 + \cdots + b_p x_p, \tag{6.8}$$

an equation linear in all of the process variables. But more complicated equations are possible and, in many cases, are necessary to really adequately describe the relationship of y to the process variables.

Example 103 *A Two-Variable Drilling Experiment.* *The paper "Design of a Metal-Cutting Drilling Experiment: A Discrete Two-Variable Problem" by E. Mielnik appeared in* Quality Engineering *in 1993 and concerns the drilling of 7075–T6 aluminum alloy. The two process variables*

$$x_1 = \textit{feed rate (ipr) and}$$
$$x_2 = \textit{drill diameter (inches)}$$

were varied in (800 rpm) drilling of aluminum specimens and both

$$y_1 = \textit{thrust (lbs) and}$$
$$y_2 = \textit{torque (ft lbs)}$$

were measured. A total of nine different (x_1, x_2) combinations were studied, and 12 data points (x_1, x_2, y) were collected for both thrust and torque. Mielnik's data are given in Table 6.10 on page 356.

Apparently on the basis of established drilling theory, Mielnik found it useful to express all variables on a log scale and fit linear regressions for $y_1' = \ln(y_1)$ and $y_2' = \ln(y_2)$ in terms of the variables $x_1' = \ln(x_1)$ and $x_2' = \ln(x_2)$. Multiple regression analysis then provides fitted equations

$$\widehat{y_1'} = 10.0208 + .6228 x_1' + .9935 x_2'$$

and

$$\widehat{y_2'} = 6.8006 + .8927 x_1' + 1.6545 x_2'.$$

By exponentiating, one gets back to equations for the original responses y_1 and y_2 in terms of x_1 and x_2, namely,

$$\widehat{y_1} = 22{,}489 x_1^{(.6228)} x_2^{(.9935)}$$

and

$$\widehat{y_2} = 898.39 x_1^{(.8927)} x_2^{(1.6545)}.$$

TABLE 6.10. Thrust and torque measurements for nine feed rate/drill diameter combinations

Feed Rate, x_1 (ipr)	Diameter, x_2 (in)	Thrust, y_1 (lbs)	Torque, y_2 (ft lbs)
.006	.250	230	1.0
.006	.406	375	2.1
.013	.406	570	3.8
.013	.250	375	2.1
.009	.225	280	1.0
.005	.318	255	1.1
.009	.450	580	3.8
.017	.318	565	3.4
.009	.318	400, 400, 380, 380	2.2, 2.1, 2.1, 1.9

Example 104 *Lift-to-Drag Ratio for a Three-Surface Configuration.* P. Burris studied the effects of the placement of a canard (a small forward "wing") and a tail, relative to the main wing of a model aircraft. He measured the lift/drag ratio for nine different configurations. With

x_1 = the canard placement in inches above the main wing and
x_2 = the tail placement in inches above the main wing.

part of his data are given in Table 6.11.

Burris' data set has the unfortunate feature that it contains no replication. It is a real weakness of the study that there is no pooled (or "pure error") sample standard deviation against which to judge the appropriateness of a fitted equation. However, making the best of the situation, a multiple regression analysis described on pages 133 through 136 of Vardeman's *Statistics for Engineering Problem Solving* leads to the equation

$$\hat{y} = 3.9833 + .5361x_1 + .3201x_2 - .4843x_1^2 - .5042x_1x_2$$

as a plausible description of lift-to-drag ratio in terms of canard and tail positions. (Equations simpler than this one turn out to have obvious deficiencies when one plots residuals.)

Notice that in Example 103, the equations for log thrust and log torque are linear in the (logs of the) process variables. That is, they are exactly of the form in display (6.8). But the equation for lift-to-drag ratio obtained from Burris' data in Example 104 (while still linear in the fitted parameters) is *not* linear in the variables x_1 and x_2. (It is, in fact, a kind of quadratic equation in the predictor variables x_1 and x_2. Much more will be said about such equations later in this section.)

After one has fit an equation for y in the variables $x_1, x_2, \ldots, x_p$ to data from a process-improvement experiment, the question of how to interpret that equation must be faced. Several possibilities exist for helpfully representing the generic fitted relationship (6.7) in graphical terms. The most obvious is to simply make plots of y against a single process variable, x_i, for various combinations of the remaining variables that are of particular interest. This amounts to viewing **slices** of the fitted response function.

A second possibility is to make **contour plots** of y against a pair of the process variables $(x_i, x_{i'})$, for various combinations of any remaining variables that are of particular interest. Such plots function as "topographic maps" of the response surface. They can be especially helpful when there are several more or less com-

TABLE 6.11. Lift/drag ratio for nine different canard and tail configurations (positions in inches above the main wing)

Canard position, x_1	Tail position, x_2	Lift/Drag ratio, y
−1.2	−1.2	.858
−1.2	0	3.156
−1.2	1.2	3.644
0	−1.2	4.281
0	0	3.481
0	1.2	3.918
1.2	−1.2	4.136
1.2	0	3.364
1.2	1.2	4.018

peting responses and one must find a compromise setting of the process variables that balances off process performance on one response against performance on another.

And a final method of representing a fitted response function is to make use of modern graphics software and produce **surface plots/perspective plots** of y against a pair of the process variables $(x_i, x_{i'})$, again for various combinations of any remaining variables that are of particular interest. These plots attempt to give a "3-D" rendering of the relationship between y and $(x_i, x_{i'})$ (with values of the remaining variables held fixed).

Example 105 *(Example 103 continued.) Figure 6.5 is a plot of the fitted log torque in the drilling study as a function of log feed rate for drills of several diameters. Figure 6.6 is the corresponding plot where both torque and feed rate are in their original units (rather than being portrayed on logarithmic scales).*

On the logarithmic scales of Fig. 6.5, the response changes linearly in the feed rate variable. In addition, the linear traces on Fig. 6.5 are parallel. In the language of factorial analysis, on the logarithmic scales there are no interactions between feed rate and drill diameter. On the logarithmic scales, the fitted relationship between feed rate, diameter, and torque is a very simple one. The relationship on the original scales of measurement represented by Fig. 6.6 is not impossibly complicated, but neither is it as simple as the one in Fig. 6.5.

Figures 6.7 and 6.8 are, respectively, a contour plot and a surface plot of the fitted relationship between log torque and the logarithms of feed rate and drill diameter. They illustrate clearly that on the logarithmic scales, the fitted equation for torque defines a plane in three space.

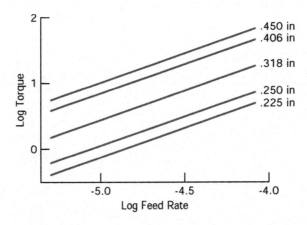

FIGURE 6.5. Fitted log torque as a function of log feed rate for five different drill diameters

Chapter 6. Experimental Design and Analysis for Process Improvement 359

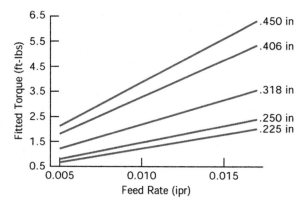

FIGURE 6.6. Fitted torque as a function of feed rate for five different drill diameters

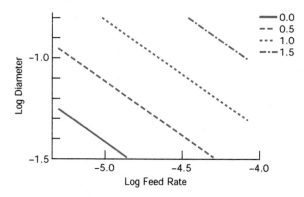

FIGURE 6.7. Contour plot of fitted log torque as a function of log feed rate and log drill diameter

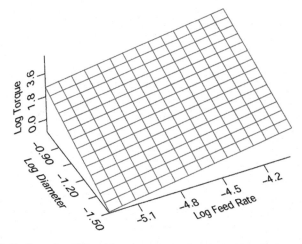

FIGURE 6.8. Surface plot of fitted log torque as a function of log feed rate and log drill diameter

Example 106 *(Example 104 continued.)* *Figures 6.9 and 6.10 are, respectively, a contour plot and surface plot for the equation for lift-to-drag ratio fit to the data of Table 6.11. They show a geometry that is substantially more complicated than the planar geometry in Figs. 6.7 and 6.8. It is evident from either examining the fitted equation itself or viewing the surface plot that, for a fixed tail position (x_2), the fitted lift-to-drag ratio is quadratic in the canard position (x_1). And even though for fixed canard position (x_1) the fitted lift-to-drag ratio is linear in tail position (x_2), the relevant slope depends upon the canard position. That is, there are canard positions by tail position interactions.*

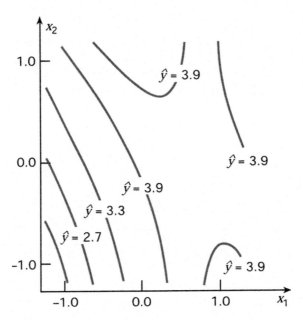

FIGURE 6.9. Contour plot for fitted lift/drag ratio as a function of canard and tail positions

Having raised the issue of interactions in the examples, it is worth pointing out how one can tell from the form of a fitted equation whether or not it implies the existence of interactions between the process variables. The general story is this. If the function $f(x_1, x_2, \ldots, x_p)$ in Eq. (6.7) can be written as a *sum* of two functions, the first of which has as its arguments $x_1, x_2, \ldots, x_l$ and the second of which has as its arguments $x_{l+1}, x_{l+2}, \ldots, x_p$, then the fitted equation implies that there are no interactions between any variables in the first set of arguments and any variables in the second set of arguments. Otherwise there *are* interactions. For example, in the case of the lifting surface study, the existence of the cross product term $-.5042x_1x_2$ in the fitted equation makes it impossible to write it in terms of a sum of a function of x_1 and a function of x_2. So the interactions noted on Fig. 6.9 are to be expected.

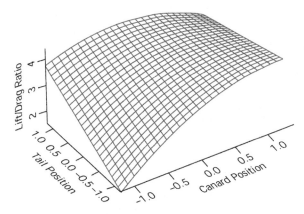

FIGURE 6.10. Surface plot for fitted lift/drag ratio as a function of canard and tail positions

TABLE 6.12. Yields and filtration times for nine combinations of condensation temperature and amount of B

Temperature, x_1 (° C)	Amount of B, x_2 (cc)	Yield, y_1 (g)	Time, y_2 (sec)
90	24.4	21.1	150
90	29.3	23.7	10
90	34.2	20.7	8
100	24.4	21.1	35
100	29.3	24.1	8
100	34.2	22.2	7
110	24.4	18.4	18
110	29.3	23.4	8
110	34.2	21.9	10

Example 107 *Yield and Cost Associated with a Chemical Process.* *The data in Table 6.12 are from the paper "More on Planning Experiments to Increase Research Efficiency" by Hill and Demler, which appeared in* Industrial and Engineering Chemistry *in 1970. The responses are a yield (y_1) and a filtration time (y_2) for a chemical process run under nine different combinations of the process variable condensation temperature (x_1) and amount of boron (x_2). Notice that the filtration time is a cost factor, and ideally one would like large yield and small filtration time. But chances are that these two responses more or less work against each other, and some compromise is necessary.*

The study represented in Table 6.12 suffers from a lack of replication, making it impossible to do a completely satisfactory job of judging the appropriateness of regressions of the responses on the process variables. But as far as is possible to tell given the inherent limitations of the data, the fitted equations (derived using multiple regression analysis)

$$\widehat{y_1} \approx -113.2 + 1.254x_1 + 5.068x_2 - .009333x_1^2 - .1180x_2^2 + .01990x_1x_2,$$

and

$$\widehat{\ln(y_2)} \approx 99.69 - .8869x_1 - 3.348x_2 + .002506x_1^2 + .03375x_2^2 + .01196x_1x_2$$

do a reasonable job of summarizing the data of Table 6.12. Figure 6.11 goes a long way toward both making the nature of the fitted response surfaces clear and providing guidance in how compromises might be made in the pursuit of good process performance. The figure shows contour plots for yield and log filtration time overlaid on a single set of (x_1, x_2)-axes.

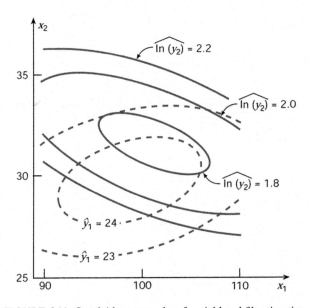

FIGURE 6.11. Overlaid contour plots for yield and filtration time

The three real applications used thus far in this section (begun in Examples 103, 104, and 107) are all ones involving only two process variables, x_1 and x_2. As such, the figures shown here provide reasonably straightforward and complete pictures of the fitted response functions. When more than two variables are involved, it becomes harder to make helpful graphics. In the case of three process variables, one can make plots like those shown here involving x_1 and x_2 for several different values of x_3. And in the case of four process variables, one can make plots involving x_1 and x_2 for several different (x_3, x_4) combinations. But unless the fitted equation is such that the function $f(x_1, x_2, \ldots, x_p)$ in Eq. (6.7) can be written as a sum of several functions of small and disjoint sets of process variables (that can be individually represented using the kind of plots illustrated in the examples), the plots become less and less helpful as p increases.

6.2.2 Using Quadratic Response Functions

Equation (6.8) represents the simplest kind of response function that can be easily fit to $(x_1, x_2, \ldots, x_p, y)$ data. As illustrated in Example 103, this kind of equation represents a p-dimensional plane in $(p + 1)$-dimensional space. As useful as it is, it does not allow for any **curvature** in the response. This means, for example, that when used to describe how y varies over some region in $(x_1, x_2, \ldots, x_p)$-space, an equation of the type in display (6.8) will always predict that optimum y occurs on the boundary of the region. And that is not appropriate for many situations where one is *a priori* fairly certain that optimum settings for the process variables occur in the interior of a region of experimentation. But to allow for this kind of circumstance, one needs alternatives to relationship (6.8) that provide for curvature.

The simplest convenient alternative to the linear relationship (6.8) with the ability to portray curvature is the **general quadratic relationship** between process variables $x_1, x_2, \ldots, x_p$ and a response y. This involves a constant term, p linear terms in the process variables, p pure quadratic terms in the process variables, and $\binom{p}{2}$ cross product terms in the process variables. In the simple case of $p = 2$, this means fitting the approximate relationship

$$y \approx \beta_0 + \beta_1 x_1 + \beta_2 x_2 + \beta_3 x_1^2 + \beta_4 x_2^2 + \beta_5 x_1 x_2 \qquad (6.9)$$

General Quadratic Relationship Between Two Process Variables and a Response

via least squares to obtain the fitted equation

$$\hat{y} = b_0 + b_1 x_1 + b_2 x_2 + b_3 x_1^2 + b_4 x_2^2 + b_5 x_1 x_2.$$

The use of quadratic equations like that in display (6.9) can be motivated on a number of grounds. For one, such equations turn out to be convenient to fit to data (multiple linear regression programs can still be used). For another, they can provide a variety of shapes (depending on the values of the coefficients β) and thus can be thought of as a kind of "mathematical French curve" for summarizing empirical data. And they can also be thought of as natural approximations to more complicated theoretical relationships between process variables and a response. That is, if there is some general relationship of the form

$$y \approx h(x_1, x_2, \ldots, x_p)$$

at work, making a second-order Taylor approximation of h about any relevant base point (i.e., finding a function with the same first- and second-order partial derivatives as h at the point of interest, but all partials of higher order equal to 0 at that point) will produce a relationship like that in Eq. (6.9).

Quadratic response functions have already proved helpful in Examples 104 and 107, and once one realizes that (taking the coefficients on the pure quadratic and cross product terms to be 0) linear equations of the form (6.8) are just special quadratics, it is evident that even Example 103 can be thought of as using them. The potential usefulness of quadratic equations then prompts the question of what **data requirements** are in order to be able to fit them. Not every set of n data points $(x_1, x_2, \ldots, x_p, y)$ will be adequate to allow the fitting of a quadratic response function. At a minimum, one needs at least as many different $(x_1, x_2, \ldots, x_p)$ combinations as coefficients β. So one must have $n \geq 1 + 2p + \binom{p}{2}$. But in addition, the set of $(x_1, x_2, \ldots, x_p)$ combinations must be "rich" enough, sufficiently spread out in p-space.

One kind of pattern in the $(x_1, x_2, \ldots, x_p)$ combinations that is sufficient to support the fitting of a quadratic response function is a full 3^p factorial design. For example, the data sets in Examples 104 and 107 are (unreplicated) 3^2 factorial data sets. But as p increases, 3^p grows very fast. For example, for $p = 4$, a full 3^4 factorial requires a minimum of 81 observations. And it turns out that for $p > 2$, a full 3^p factorial is really "overkill" in terms of what is needed. On the other hand, a 2^p factorial *will not* support the fitting of a quadratic response function.

There are two directions to go. In the first place, one might try to "pare down" a 3^p factorial to some useful (small) fraction that will still allow the use of a full quadratic response function. One popular kind of fraction of a 3^p of this type is due to Box and Behnken and is discussed, for example, in Section 15.4 of *Empirical Model-Building and Response Surfaces* by Box and Draper.

A second route to finding experimental designs that allow the economic (and precise) fitting of second-order surfaces is to begin with a 2^p factorial in the p process variables and then augment it until it provides enough information for the effective fitting of a quadratic. This route has the virtue that it suggests how the designs can be used in practice in a **sequential** manner. One can begin with a 2^p factorial experiment, fitting a linear equation of form (6.8), and augment it only when there is evidence that a response function allowing for curvature is really needed to adequately describe a response. This kind of experimental strategy is important enough to deserve a thorough exposition.

Sequential Experimental Strategy

So suppose that in a study involving the process variables $(x_1, x_2, \ldots, x_p)$, one desires to center experimentation about the values $x_1^*, x_2^*, \ldots, x_p^*$. Then for $\Delta_1, \Delta_2, \ldots, \Delta_p$ positive constants, one can let the low level of the ith process variable be $x_i^* - \Delta_i$ and the high level be $x_i^* + \Delta_i$ and begin experimentation with a 2^p factorial. It is a good idea to initially also collect a few responses at the center point of experimentation $(x_1^*, x_2^*, \ldots, x_p^*)$. Data from such a **$2^p$ plus repeated center point** experimental design provide a good basis for checking the adequacy of a linear equation like the one in display (6.8). If (applying regression techniques) one finds important lack of fit to the linear equation and/or (applying the Yates algorithm to the 2^p part of the data) important interactions among the process variables, the need for more data collection is indicated.

Then, a clever way of choosing additional points at which to collect data is to augment the 2^p with **star or axial points**. For some constant α (usually taken to be at least 1), these are $2p$ points of the form

$$(x_1^*, x_2^*, \ldots, x_{l-1}^*, x_l^* \pm \alpha \Delta_l, x_{l+1}^*, \ldots, x_p^*)$$

for $l = 1, 2, \ldots, p$. Geometrically, if one thinks of setting up a coordinate system with origin at the center of the 2^p factorial part of the experimental design, these are points on the axes of the system, α times as far from the center as the "faces" of the 2^p design. Figure 6.12 shows $p = 2$ and $p = 3$ versions of this "2^p factorial plus center points plus star points" arrangement of combinations of process variables. In practice, it is also wise to make some additional observations at the

Chapter 6. Experimental Design and Analysis for Process Improvement   365

center point in the second round of experimentation. (These repeats of a condition from the first round of experimentation allow one to investigate whether experimental conditions appear to be different in the second round than they were in the first.)

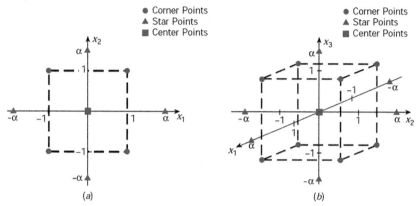

FIGURE 6.12. (a) 2^2 factorial plus center points and star points and (b) 2^3 factorial plus center points and star points

When the whole "2^p plus center points plus star points" experimental program is completed, the jargon typically employed is that one has used a **central composite** experimental design. Allowing for the fact that only certain settings of the drilling machine were possible and only certain drill diameters were available, the data set in Table 6.10 came approximately from a $p = 2$ central composite design. The next example involves a $p = 3$ central composite study.

TABLE 6.13. Seal strengths under 15 different combinations of $p = 3$ process variables

x_1	x_2	x_3	y
225	46	.5	6.6
285	46	.5	6.9
225	64	.5	7.9
285	64	.5	6.1
225	46	1.7	9.2
285	46	1.7	6.8
225	64	1.7	10.4
285	64	1.7	7.3
255	55	1.1	10.1, 9.9, 12.2, 9.7, 9.7, 9.6
204.54	55	1.1	9.8
305.46	55	1.1	5.0
255	39.862	1.1	6.9
255	70.138	1.1	6.3
255	55	.0908	4.0
255	55	2.1092	8.6

Example 108 *A Seal Strength Optimization Study. The article "Sealing Strength of Wax-Polyethylene Blends" by Brown, Turner, and Smith appeared in* Tappi *in 1958 and contains an early and instructive application of a central composite design and a quadratic response function. Table 6.13 on page 365 contains the data from the article. Three process variables,*

$$x_1 = \text{seal temperature (°F)},$$
$$x_2 = \text{cooling bar temperature (°F)}, \text{and}$$
$$x_3 = \text{polyethylene content (\%)}$$

were considered regarding their effects on

$$y = \text{the seal strength of a paper bread wrapper stock (g/in)}.$$

The data in Table 6.13 comprise a central composite design with center point $(255, 55, 1.1)$, $\Delta_1 = 30$, $\Delta_2 = 9$, $\Delta_3 = .6$, *and* $\alpha = 1.682$. *The first eight rows of Table 6.13 represent a* 2^3 *design in the three process variables. The ninth row represents six observations taken at the center point of the experimental region. And the last six rows represent the axial or star points of the experimental design. It is important to note that the repeated center point provides an honest estimate of experimental variability. The sample standard deviation of the six ys at the center of the design is* $s_P = 1.00$ g/in *to two decimal places.*

It is instructive to consider the analysis of the data in Table 6.13 in two stages, corresponding to what could have first been known from the "2^3 *factorial plus repeated center point" part of the data and then from the whole data set. To begin, the reader can verify that fitting a linear regression*

$$y \approx \beta_0 + \beta_1 x_1 + \beta_2 x_2 + \beta_3 x_3.$$

to the data on the first nine lines of Table 6.13 produces the prediction equation

$$\hat{y} = 13.0790 - .0292 x_1 + .0306 x_2 + 1.2917 x_3,$$

$R^2 = .273$, $s_{SF} = 1.75$ *a regression-based (or "surface fitting") estimate of* σ, *and residuals that when plotted against any one of the process variables have a clear "up then back down again pattern." There are many indications that this fitted equation is not a good one. The* R^2 *value is small by engineering standards, the regression-based estimate of* σ *shows signs of being inflated (when compared to* $s_P = 1.00$) *by lack of fit (in fact, a formal test of lack of fit has a p-value of .048), and the fitted equation underpredicts the response at the center of the experimental region, while overpredicting at all but one of the corners of the* 2^3 *design. (This last fact is what produces the clear patterns on plots of residuals against* x_1, x_2, *and* x_3.)

A way of formally showing the discrepancy between the mean response at the center of the experimental region and what an equation linear in the process variables implies is based on the fact that if a mean response is linear in the process variables, the average of the means at the 2^p corner points is the same as the mean at the center point. This implies that the average of the ys on the first eight lines of Table 6.13 should be about the same as the average of the ys on the ninth line. But this is clearly not the case. Using the obvious notation for corners and center points,

$$\bar{y}_{\text{center}} - \bar{y}_{\text{corners}} = 10.20 - 7.65 = 2.55 \text{ g/in}.$$

Then formula (5.8) can be applied to make 95% two-sided confidence limits (based on the 5-degree-of-freedom s_P computed from the repeated center point) for the linear combination of nine μs corresponding to this combination of $\bar{y}$s. Doing this, an uncertainty of only

$$\pm 2.571(1.00)\sqrt{\frac{(1)^2}{6} + \frac{8\left(-\frac{1}{8}\right)^2}{1}} = \pm 1.39 \text{ g/in}$$

can be associated with the 2.55 g/in figure. That is, it is clear that the 2.55 figure is more than just noise. A linear equation simply does not adequately describe the first part of the data set. There is some curvature evident in how seal strength changes with the process variables.

So based on the first part of the data set, one would have good reason to collect data at the star points and fit a quadratic model to the whole data set. The reader is invited to verify that using multiple regression to fit a quadratic to all 20 observations represented in Table 6.13 produces

$$\begin{aligned}\hat{y} &= -104.82 + .49552x_1 + 1.72530x_2 + 14.27454x_3 - .00084x_1^2 - .01287x_2^2 \\ &\quad -3.19013x_3^2 - .00130x_1x_2 - .02778x_1x_3 + .02778x_2x_3,\end{aligned}$$

$R^2 = .856$, $s_{\text{SF}} = 1.09$, and residuals that look much better than those for the linear fit to the first part of the data. This fitted equation is admittedly more complicated than the linear equation, but it is also a much better empirical description of how seal strength depends upon the process variables.

As a means of aiding understanding of the nature of the quadratic response function, Fig. 6.13 on page 368 shows a series of contour plots of this function versus x_1 and x_2, for $x_3 = 0, .5, 1.0, 1.5$, and 2.0. This series of plots suggests that optimum (maximum) seal strength may be achieved for x_1 near 225°F, x_2 near 57°F, and x_3 near 1.5%, and that a mean seal strength exceeding 11 g/in may be possible by proper choice of values for the process variables.

It is worth noting that on the basis of some theoretical optimality arguments, it is common to recommend values of α for constructing central composite designs that are larger than 1.0 (thereby placing the star points outside of the region in $(x_1, x_2, \ldots, x_p)$-space with corners at the 2^p design points). There are cases, however, where other considerations may come into play and suggest smaller choices for α. For example, there are experimental scenarios where one really

368 Chapter 6. Experimental Design and Analysis for Process Improvement

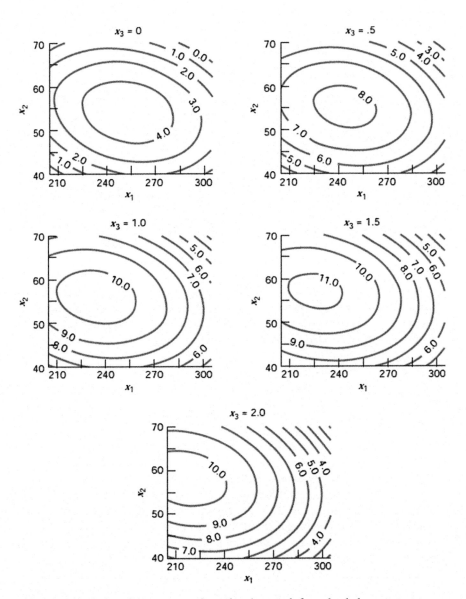

FIGURE 6.13. Series of five contour plots of seal strength for polyethylene contents x_3 between 0 and 2.0 (%)

wants to minimize the number of different levels of a given factor that one uses. And the choice $\alpha = 1$ places the star points on the "faces" of the 2^p design and makes the resulting central composite a fraction of the 3^p design (so that only three different levels of each factor are used, rather than five).

6.2.3 Analytical Interpretation of Quadratic Response Functions

For p much larger than 2 or 3, graphical interpretation of a fitted quadratic response function becomes difficult at best and impossible at worst. Happily there are some analytic tools that can help. Those tools are the subject of this subsection. Readers without a background in matrix algebra can skim this material (reading for main points and not technical detail) without loss of continuity.

For the quadratic function of $p = 1$ variable,

$$y \approx \beta_0 + \beta_1 x + \beta_2 x^2,$$

it is the coefficient β_2 that governs the basic nature of the relationship between x and y. For $\beta_2 > 0$, the equation graphs as a parabola opening up, and y has a minimum at $x = -\beta_1/2\beta_2$. For $\beta_2 < 0$, the equation graphs as a parabola opening down, and y has a maximum at $x = -\beta_1/2\beta_2$. And for $\beta_2 = 0$, the function is actually linear in x and (if β_1 is not 0) has neither a maximum nor a minimum value when x is allowed to vary over all real numbers. Something similar to this story is true for p larger than 1.

It is the coefficients of the pure quadratic and mixed terms of a multivariate quadratic relationship like that in display (6.9) which govern the nature of the response function. In order to detail the situation, some additional notation is required. Suppose that based on n data points $(x_1, x_2, \ldots, x_p, y)$, one arrives at a quadratic regression equation with fitted coefficients

$$\begin{aligned} b_i &= \text{the fitted coefficient of the linear term } x_i, \\ b_{ii} &= \text{the fitted coefficient of the pure quadratic term } x_i^2, \text{ and} \\ b_{ii'} &= b_{i'i} = \text{the fitted coefficient of the mixed term } x_i x_{i'}. \end{aligned}$$

Then using these, define the $p \times 1$ vector b and $p \times p$ matrix B by

$$b = \begin{bmatrix} b_1 \\ b_2 \\ \vdots \\ b_p \end{bmatrix} \quad \text{and} \quad B = \begin{bmatrix} b_{11} & \frac{1}{2}b_{12} & \cdots & \frac{1}{2}b_{1p} \\ \frac{1}{2}b_{21} & b_{22} & \cdots & \frac{1}{2}b_{2p} \\ \vdots & \vdots & & \vdots \\ \frac{1}{2}b_{p1} & \frac{1}{2}b_{p2} & \cdots & b_{pp} \end{bmatrix}. \quad (6.10)$$

Vector and Matrix of Coefficients for a p Variable Quadratic

It is the **eigenvalues** of the matrix B that govern the shape of the fitted quadratic, and b and B together determine where (if at all) the quadratic has a minimum or maximum.

370 Chapter 6. Experimental Design and Analysis for Process Improvement

The eigenvalues of the matrix B are the p solutions λ to the equation

---
Equation Solved by Eigenvalues

$$\det(B - \lambda I) = 0. \tag{6.11}$$

The fitted surface has a stationary point (i.e., a point where all first partial derivatives with respect to the process variables are 0) at the point

---
Location of the Stationary Point for a p Variable Quadratic

$$x = \begin{bmatrix} x_1 \\ x_2 \\ \vdots \\ x_p \end{bmatrix} = -\frac{1}{2} B^{-1} b. \tag{6.12}$$

When all p eigenvalues are positive, the fitted quadratic has a **minimum** at the point defined by relationship (6.12). When all p eigenvalues are negative, the fitted quadratic has a **maximum** at the point defined in display (6.12). When some eigenvalues are positive and the rest are negative, the fitted response function has a **saddle** geometry. (Moving away from the point defined in display (6.12) in some directions causes an increase in fitted y, while moving away from the point in other directions produces a decrease in fitted y.) And when some eigenvalues are 0 (or in practice, nearly 0), the fitted quadratic has a **ridge** geometry.

Example 109 *(Example 107 continued.) The fitted equation for yield in the Hill and Demler chemical process study is a quadratic in $p = 2$ variables, and the corresponding vector and matrix defined in display (6.10) are*

$$b = \begin{bmatrix} 1.254 \\ 5.068 \end{bmatrix} \quad \text{and} \quad B = \begin{bmatrix} -.009333 & \frac{1}{2}(.01990) \\ \frac{1}{2}(.01990) & -.1180 \end{bmatrix}.$$

Figure 6.11 indicates that the fitted yield surface has a mound-shaped geometry, with maximum somewhere near $x_1 = 99$ and $x_2 = 30$. This can be confirmed analytically by using Eq. (6.12) to find the stationary point and examining the eigenvalues defined in Eq. (6.11).

To begin,

$$-\frac{1}{2} B^{-1} b = -\frac{1}{2} \begin{bmatrix} -.009333 & \frac{1}{2}(.01990) \\ \frac{1}{2}(.01990) & -.1180 \end{bmatrix}^{-1} \begin{bmatrix} 1.254 \\ 5.068 \end{bmatrix} = \begin{bmatrix} 99.09 \\ 29.79 \end{bmatrix},$$

so that the stationary point has $x_1 = 99.09$ and $x_2 = 29.79$. Then Eq. (6.11) for the eigenvalues is

$$0 = \det\left(\begin{bmatrix} -.009333 & \frac{1}{2}(.01990) \\ \frac{1}{2}(.01990) & -.1180 \end{bmatrix} - \lambda \begin{bmatrix} 1 & 0 \\ 0 & 1 \end{bmatrix} \right),$$

that is,

$$0 = (-.009333 - \lambda)(-.1180 - \lambda) - \frac{1}{4}(.01990)^2.$$

This is a quadratic equation in the variable λ, and the quadratic formula can be used to find its roots

$$\lambda = -.0086 \quad \text{and} \quad \lambda = -.1187.$$

The fact that these are both negative is an analytical confirmation that the stationary point provides a maximum of the fitted yield surface.

Example 110 *(Examples 104 and 106 continued.)* Figures 6.9 and 6.10 portraying the fitted equation for Burris's lift/drag ratio data indicate the saddle surface nature of the fitted response. This can be confirmed analytically by noting that $\boldsymbol{B}$ corresponding to the fitted equation is

$$\boldsymbol{B} = \begin{bmatrix} -.4843 & \frac{1}{2}(-.5042) \\ \frac{1}{2}(-.5042) & 0 \end{bmatrix},$$

and that in this instance Eq. (6.11) for the eigenvalues,

$$0 = \det(\boldsymbol{B} - \lambda \boldsymbol{I}) = (-.4843 - \lambda)(-\lambda) - \frac{1}{4}(-.5042)^2,$$

has roots

$$\lambda = -.592 \quad \text{and} \quad \lambda = .107,$$

one of which is positive and one of which is negative.

For p larger than two, the calculations indicated in Eqs. (6.11) and (6.12) are reasonably done only using some kind of statistical or mathematical software that supports matrix algebra. Table 6.14 on page 372 holds code and output from an R session for finding the eigenvalues of $\boldsymbol{B}$ and the stationary point identified in display (6.12) for the quadratic response function fit to the seal strength data of Table 6.13. It illustrates that as suggested by Fig. 6.13, the quadratic fit to the data has a maximum when $x_1 = 226°\text{F}$, $x_2 = 57.2°\text{F}$, and $x_3 = 1.50\,\%$.

A final caution needs to be sounded before leaving this discussion of interpreting fitted response surfaces. This concerns the very real possibility of overinterpreting a fitted relationship between p process variables and y. One needs to be sure that a surface is really fairly well identified by data in hand before making conclusions based on the kind of calculations illustrated here (or, for that matter, based on the graphical tools of this section).

One useful rule of thumb for judging whether a surface is well enough determined to justify its use is due to Box, Hunter, and Hunter (see their *Statistics for Experimenters*) and goes as follows. If a response function involving l coefficients b (including a constant term where relevant) is fit to n data points via multiple regression, producing n fitted values $\hat{y}$ and an estimate of σ (based on surface fitting) s_{SF}, then one checks to see whether

$$\max \hat{y} - \min \hat{y} > 4\sqrt{\frac{l s_{\text{SF}}^2}{n}} \quad (6.13)$$

Criterion for Judging Whether a Response is Adequately Determined

TABLE 6.14. R code and output for analysis of the quadratic fit to bread wrapper seal strength

```
> B<-matrix(c(-.00084,-.00065,-.01389,-.00065,
              -.01287,.01389, + -.01389, .01389,
              -3.19013),nrow=3,byrow=T)
> b<-c(.49552,1.72530,14.27454)
> eigen(B)
$values
[1] -0.0007376919 -0.0128511209 -3.1902511872

$vectors
              [,1]          [,2]          [,3]
[1,]   0.998261929 -0.05877213 -0.004354062
[2,]  -0.058753124 -0.99826298  0.004370551
[3,]  -0.004603366 -0.00410714 -0.999980970

> x<-(-.5)*solve(B)%*%b
> x
              [,1]
[1,]   225.793876
[2,]    57.246829
[3,]     1.503435
```

before interpreting the surface. The difference on the left of inequality (6.13) is a measure of the movement of the surface over the region of experimentation. The fraction under the root on the right is an estimate of average variance of the n values $\hat{y}$. The check is meant to warn its user if the shape of the fitted surface is really potentially attributable completely to random variation.

6.2.4 Response Optimization Strategies

The tools of this section are primarily tactical devices, useful for understanding the "local terrain" of a response surface. What remains largely unaddressed in this section is the broader, more strategic issue of how one finds a region in $(x_1, x_2, \ldots, x_p)$-space deserving careful study, particularly where the ultimate objective is to optimize one or more responses y. Many sensible strategies are possible and this subsection discusses two. The first is something called evolutionary operation (or EVOP for short), and the second is an empirical optimization strategy that uses the linear and quadratic response surface tools just discussed.

EVOP is a strategy for conservative ongoing experimentation on a working production process that aims to simultaneously produce good product and also provide information for the continual improvement (the "evolution") of the process. The notion was first formally discussed by Box and Wilson and is thoroughly detailed in the book *Evolutionary Operation* by Box and Draper. Various authors have suggested different particulars for implementation of the general EVOP strategy, including a popular "simplex" empirical hill-climbing algorithm put forth by Spendley, Hext, and Himsworth. As the present treatment must be brief, only the simplest (and original) factorial implementation will be discussed and that without many details.

So consider a situation where p process variables $x_1, x_2, \ldots, x_p$ are thought to affect l responses (or quality characteristics) $y_1, y_2, \ldots, y_l$. If $x_1^*, x_2^*, \ldots, x_p^*$ are current standard operating values for the process variables, an EVOP program operates in the following set of steps:

Outline of an EVOP Strategy

1. An EVOP Committee (consisting of a broad group of experts with various kinds of process knowledge and interests in process performance) chooses a few (two or three) of the process variables as most likely to provide improvements in the response variables. For the sake of concreteness, suppose that x_1 and x_2 are the variables selected. A "two-level factorial plus center point" experimental design is set up in the selected variables with (x_1^*, x_2^*) as the center point. The high and low levels of the variables are chosen to be close enough to x_1^* and x_2^* so that any change in any mean response across the set of experimental conditions is expected to be small.

2. Holding variables $x_3, x_4, \ldots, x_p$ at their standard values $x_3^*, x_4^*, \ldots, x_p^*$, in normal process operation, (x_1, x_2) is cycled through the experimental design identified in step 1, and values for all of $y_1, y_2, \ldots, y_l$ are recorded. This continues until for every response, enough data have been collected so that the changes in mean as one moves from the center point to corner points are estimated with good precision (relative to the sizes of the changes).

3. If, in light of the estimated changes in the means for all of $y_1, y_2, \ldots, y_l$, the EVOP Committee finds no corner point of the design to be preferable to the center point, the program returns to step 1 and, a different set of process variables is chosen for experimentation.

4. If there is a corner of the two-level factorial design that the EVOP Committee finds preferable to the center point, *new* standard values of x_1 and x_2, x_1^* and x_2^*, are established between the previous ones and the values for the superior corner point. A new two-level factorial plus center point experimental design is set up in variables x_1 and x_2 with the new (x_1^*, x_2^*) as the center point. The high and low levels of the variables are chosen to be close enough to the new x_1^* and x_2^* so that any change in any mean response across the experimental conditions is expected to be small. The EVOP program then returns to step 2.

Evolutionary Operation is intended to be a relatively cautious program that succeeds in process improvement because of its persistence. Only a few variables are changed at once, and only small moves are made in the standard operating conditions. Nothing is done in an "automatic" mode. Instead the EVOP Committee considers potential impact on all important responses before authorizing movement. The caution exercised in the ideal EVOP program is in keeping with the fact that the intention is to run essentially no risk of noticeably degrading process performance in any of the experimental cycles.

A Second Response Optimization Strategy

A more aggressive posture can be adopted when experimentation on a process can be done in a "get in, make the improvement, and get out" mode, and there is no serious pressure to assure that every experimental run produces acceptable product. The following steps outline a common kind of strategy that makes use of the linear and quadratic response function ideas of this section in the optimization of a single response y, where p process variables are at work and initial standard operating values are $x_1^*, x_2^*, \ldots, x_p^*$:

1. A "2^p plus center points" experimental design is run (with center at the point $(x_1^*, x_2^*, \ldots, x_p^*)$), and a linear response function for y is fit to the resulting data and examined for its adequacy using the tools of regressions analysis.

2. In the event that the linear equation fit to the 2^p factorial plus center points data is adequate, a sequence of observations is made along a ray in $(x_1, x_2, \ldots, x_p)$-space beginning at the center point of the design and proceeding in the direction of the steepest ascent (or the steepest descent depending upon whether the object is to maximize or to minimize y). That is, if the fitted values of the parameters of the linear response function are $b_0, b_1, b_2, \ldots, b_p$, observations are made at points of the form

$$(x_1^* + ub_1, x_2^* + ub_2, \ldots, x_p^* + ub_p)$$

for positive values of u if the object is to maximize y (or for negative values of u if the object is to minimize y). If the object is to maximize y, the magnitude of u is increased until the response seems to cease to increase (or seems to cease to decrease if the object is to minimize u). Polynomial regression of y on u can be helpful in seeing the pattern of response to these changes in u. The point of optimum response (or optimum fitted response if one smooths the y values using regression on u) along the ray becomes a new point $(x_1^*, x_2^*, \ldots, x_p^*)$, and the algorithm returns to step 1.

3. If the linear surface fit in step 1 is not an adequate description of the 2^p plus center point data, star points are added, and a quadratic surface is fit. The quadratic is examined for adequacy as a local description of mean response, and the location of the best fitted mean within the experimental region is identified. This point becomes a new center of experimental effort $(x_1^*, x_2^*, \ldots, x_p^*)$, and the algorithm returns to step 1.

Fairly obviously, at some point this algorithm typically "stalls out" and ceases to provide improvements in mean y. At that point, engineering attention can be turned to some other process or project.

Section 6.2 Exercises

1. **Polymer Density.** In a 1985 *Journal of Quality Technology* article, R. Snee discussed a study of the effects of annealing time and temperature on polymer density. Data were collected according to a central composite design, except that no observation was possible at the highest level of temperature (because the polymer melted at that temperature), and an additional run was made at 170°C and 30 min. Snee's data are given below:

Temperature, x_1 (°C)	Time, x_2 (min)	Density, y
140	30	70
155	10	70
155	50	72
170	30	91
190	60	101
190	30	98
190	0	70
225	10	83
225	50	101

(a) Plot the nine design points (x_1, x_2) in two space, labeling each with the corresponding observed density, y.

(b) Use a multiple regression program and fit the equation

$$y \approx \beta_0 + \beta_1 x_1 + \beta_2 x_2$$

to these data.

(c) Make a contour plot for the fitted response from part (a) over the "region of experimentation." (This region is discernible from the plot in (a) by enclosing the nine design points with the smallest possible polygon.) Where in the experimental region does one have the largest predicted density? Where is the smallest predicted density?

(d) Compute residuals from the fitted equation in (b), and plot them against both x_1 and x_2 and against $\hat{y}$. Do these plots suggest that an equation that allows for curvature of response might better describe these data? Why or why not?

(e) Use a multiple regression program to fit the equation

$$y \approx \beta_0 + \beta_1 x_1 + \beta_2 x_2 + \beta_3 x_1^2 + \beta_4 x_2^2 + \beta_5 x_1 x_2$$

to these data.

(f) Make a contour plot for the fitted response from (e) over the experimental region. Use this plot to locate a part of the experimental region that has the largest predicted density.

(g) Compute residuals from the fitted equation in (e), and plot them against both x_1 and x_2 and against $\hat{y}$. Do these plots look "better" than the ones in (d)? Why or why not? Does it seem that the quadratic equation is a better description of density than the linear one?

2. **Tar Yield.** In the article "Improving a Chemical Process Through Use of a Designed Experiment" that appeared in *Quality Engineering*, 1990–1991, J.S. Lawson discussed an experiment intended to find a set of production conditions that would produce minimum tars in a process that can be described in rough terms as

$$A + B \xrightarrow[\text{CATALYST}]{\text{SOLVENT}} \text{PRODUCT} + \text{TARS} .$$

(In the process studied, side reactions create tars that lower product quality. As more tars are produced, yields decrease, and additional blending of finished product is needed to satisfy customers. Lower yields and increased blending raise the average production cost.)

High, medium, and low levels of the three process variables

$$\begin{aligned} x_1 &= \text{temperature,} \\ x_2 &= \text{catalyst concentration, and} \\ x_3 &= \text{excess reagent B} \end{aligned}$$

were identified, and 15 experimental runs were made with high reagent A purity and low solvent stream purity. Then 15 additional runs were made with low reagent A purity and low solvent stream purity. In both sets of experimental runs, the response variable of interest was a measure of tars produced, y. Lawson's data are below in separate tables for the two different reagent A purity/solvent stream purity conditions. (The process variables have been coded by subtracting the plant standard operating values and then dividing by the amounts that the "high" and "low" values differ from the standard values.)

Chapter 6. Experimental Design and Analysis for Process Improvement 377

"High" reagent A purity and "low" solvent stream purity data				"Low" reagent a purity and "low" solvent stream purity data			
x_1	x_2	x_3	y	x_1	x_2	x_3	y
−1	−1	0	57.81	−1	−1	0	37.29
1	−1	0	24.89	1	−1	0	4.35
−1	1	0	13.21	−1	1	0	9.51
1	1	0	13.39	1	1	0	9.15
−1	0	−1	27.71	−1	0	−1	20.24
1	0	−1	11.40	1	0	−1	4.48
−1	0	1	30.65	−1	0	1	18.40
1	0	1	14.94	1	0	1	2.29
0	−1	−1	42.68	0	−1	−1	22.42
0	1	−1	13.56	0	1	−1	10.08
0	−1	1	50.60	0	−1	1	13.19
0	1	1	15.21	0	1	1	7.44
0	0	0	19.62	0	0	0	12.29
0	0	0	20.60	0	0	0	11.49
0	0	0	20.15	0	0	0	12.20

(a) Make a "three-dimensional" sketch of the cube-shaped region with $-1 \leq x_1 \leq 1$, $-1 \leq x_2 \leq 1$, and $-1 \leq x_3 \leq 1$. Locate the 13 different (x_1, x_2, x_3) design points employed in Lawson's experiment on that sketch.

(b) Do the 13 design points in Lawson's study constitute a central composite design? Explain.

For (c) through (g), consider only the first set of 15 data points, the ones for high reagent A purity.

(c) Was there any replication in this study? Explain.

(d) Use a multiple regression program to fit the linear equation

$$y \approx \beta_0 + \beta_1 x_1 + \beta_2 x_2 + \beta_3 x_3$$

to the high reagent A purity data.

(e) Use a multiple regression program to fit a full quadratic function in x_1, x_2, and x_3 to the high reagent A purity data. (You will need a constant term, three linear terms, three pure quadratic terms, and three cross product terms.)

(f) Is the quadratic equation in (e) a better description of the relationship between the process variables and tar production than the equation fit in (d)? Explain.

(g) Find an "honest" estimate of σ, the experimental variability in y, for a fixed set of process conditions. (Consider the repeated center point of the design.)

6.3 Qualitative Considerations in Experimenting for Quality Improvement

The discussion of experimental design and analysis in Chap. 5 and thus far in Chap. 6 has concentrated primarily on technical statistical tools. These are essential. But there are also a number of more qualitative matters that must be handled intelligently if experimentation for process improvement is to be successful. This section discusses some of these. It begins by considering the implications of some "classical" experimental design issues for process-improvement studies. Then some matters given emphasis by followers of Genichi Taguchi are discussed.

6.3.1 "Classical" Issues

It is a truism that the world is highly multivariate. Accordingly, almost any process response variable that a person would care to name is potentially affected by *many, many* factors. And successful quality-improvement experimentation requires that those myriad factors be handled in intelligent ways. Section 1.3 has already presented some simple aids for the identification/naming of the most important of those factors. But there is also the matter of how to treat them during an experiment.

There are several possibilities in this regard. For factors of primary interest, it is obvious that one will want to vary their levels during experimentation, so as to learn how they impact the response variable. That is, one will want to manipulate them, treating them as **experimental variables.**

For factors of little interest (usually because outside of the experimental environment, they are not under the direct influence of those running a process) but that nevertheless may impact the response, one approach is to **control** them in the sense of holding their levels fixed. This is the laboratory/pilot plant approach, where one tries to cut down the background noise and concentrate on seeing the effects of a few experimental variables. Scientists doing basic research often think of this approach as "the" scientific method of experimentation. But for technologists, it has its limitations. Results produced in carefully controlled laboratory/pilot plant environments are notoriously hard to reproduce in full-scale facilities where many extraneous variables are not even close to constant. That is, when a variable is controlled during experimentation, there is in general no guarantee that the kind of responses one sees at the single level of that variable will carry over to other levels met in later practice. Controlling a factor can increase experimental sensitivity to the effects of primary experimental variables, *but it also limits the scope of application of experimental results.*

Another approach to the handling of extraneous factors (those not of primary interest) is to include them as experimental factors, purposefully varying them in spite of the fact that in later operation one will have little or no say in their values.

If a full-scale chemical process is going to be run in a building where humidity and temperature change dramatically over the course of a year, technological experimentation with the basic process chemistry might well be carried out in a variety of temperatures and humidities. Or if a metal-cutting process is affected by the hardness of steel bar stock that itself cannot be guaranteed to be completely constant in plant operations, it may be wise to purposely vary the hardness in a process-improvement experiment. This is in spite of the fact that what may be of primary interest are the effects of tooling and cutting methods. This notion of possibly including as experimental variables some whose precise effects are of at most secondary interest is related to two concepts that need further discussion. The first is the notion of **blocking** from classical experimental design, and the second is Taguchi's notion of **noise variables** that will be discussed later in the section.

There are situations in which experimental units or times of experimentation can be segmented into groups that are likely to be relatively homogeneous in terms of their impact on a response variable, but when looked at as a whole are quite heterogeneous. In such cases, it is common (and useful) to break the experimental units or times up into homogeneous **blocks** and to essentially conduct an independent experiment in each of these blocks. If one has in mind two types of tooling and two cutting methods, it makes sense to conduct a 2^2 factorial experiment on a block of "soft" steel bar stock specimens and another one on a block of "hard" steel bar stock specimens. Of course, a way to then think about the study as a whole is to recognize that one has done a complete 2^3 factorial, one factor of which is "blocks." Blocking has the effect of allowing one to account for variation in response attributable to differences between the groups (blocks) rather than, so to speak, lumping it into an omnibus measure of experimental error. It amounts to a sort of local application of the notion of control and provides several relatively homogeneous environments in which to view the effects of primary experimental variables.

Of course there are limits to the size of real-world experiments, and it is not possible to purposely vary *every* variable that could potentially affect a response. (Indeed, one can rarely even be aware of all factors that could possibly affect it!) So there need to be means of protecting the integrity of experimental results against effects of variables not included as experimental factors (and perhaps not even explicitly recognized). Where one is aware of a variable that will not be controlled or purposely varied during data collection and yet might influence experiment results, it makes sense to at least record values for that **concomitant variable**. It is then often possible to account for the effects of such variables during data analysis by, for example, treating them as additional explanatory variables in regression analyses.

In addition, whether or not one is going to try to account for effects of non-experimental variables in data analysis, it is a good idea to take steps to try and balance their effects between the various experimental conditions using the notion of **randomization.** Randomization is the use of a table of random digits (or other randomizing devices) to make those choices of experimental protocol not already specified by other considerations. For example, if one is doing a machining study with two types of tooling and two cutting methods and is planning to apply each of the 2^2 combinations to three steel specimens, one might initially divide 12 specimens up into four groups of three using a randomizing device. (The hope is that the randomization will treat all four combinations "fairly" and, to the extent possible, average between them the effects of all named and unnamed extraneous factors like hardness, surface roughness, microstructure, and so on. This is, of course, a different approach to handling an extraneous factor than treating it as a blocking variable.)

And one might well also randomize the order of cutting in such a study, particularly if there is some concern that ambient conditions might change over the course of experimentation and impact a response of interest. If, for example, one were to make all experimental runs involving tooling type 1 before making the runs for tooling type 2, any factor that changed over time and had an impact on the response would contribute to what the experimenter perceives as the effects of tool type. This is a very unhappy possibility. In terms of understanding what is really affecting a response, it would be far better to either completely randomize the order of the 12 experimental runs or to create three (time) blocks of size 4 and run a full 2^2 factorial within each of the blocks of experimental runs.

This matter of randomizing order in an engineering experiment brings up two additional qualitative issues. The first is that engineers frequently argue that there are cases where some experimental factors have levels that cannot be easily changed, and in those cases it is often much more economical to make runs in an order that minimizes the number of changes of levels of those factors. It is impossible to say that this argument should never be allowed to stand. Physical and economic realities in engineering experimentation are just that, and the object is not to stick to inflexible rules (like "you must always randomize") but rather to artfully make the best of a given set of circumstances *and recognize the implications of the choices one makes* in experimental design. Where it is absolutely necessary to make all experimental runs with level 1 of Factor A before those with level 2 of factor A, so be it. But the wise analyst will recognize that what looks like an effect of factor A is really the effect of A *plus* the effects of any other important but unnamed factors whose levels change over time.

The second matter is the whole issue of what one really means by the word **replication.** Not all methods of obtaining multiple observations associated with a given set of process conditions are equal. Consider, for example, an injection molding process where one is studying the effects of two raw material mixes, two shot sizes, and two temperatures on some property of parts produced by the process. To set up a machine for a given material mix, shot size, and temperature combination and to make and measure five consecutive parts are not at all the same as making that setup and measuring one part, going to some other combination(s) and then returning later to make the second part at that setup, and so on. Five consecutive parts could well look much more alike than five manufactured with intervening time lapses and changes in setup. And it is probably a variation of the second (typically larger) magnitude that should be viewed as the baseline against which one correctly judges the effects of material mix, shot size, and temperature. A way of describing this concern is to contrast **repetition** or **remeasurement** with "true" **replication**. With the former, one learns a lot about a particular setup, but not nearly so much about all setups of that type as with the latter.

This issue is also sometimes called the **unit of analysis problem**. If one wishes to compare physical properties for two steel formulas, the logical unit of analysis is a heat of metal. Making one heat of the new formula and one heat from the old, pouring these into ingots, and measuring some property of each of 10 ingots produce 20 measurements. But in effect one has two (highly multivariate) samples of size *1*, not two samples of size 10. Why? The object is almost surely to compare the formulas, not the two specific heats. And that being the case, the unit of analysis is a heat, and there is but one of these for each formula. Replication means repeating the unit of analysis, not just remeasuring a single one. Engineers who are constrained by economics to build and tweak a single prototype of a new product need to be very conscious that they have but a single unit of analysis, whose behavior will almost certainly exhibit much less variation than would several "similar" prototypes.

A final matter that should be mentioned in this discussion of classical issues in experimental design concerns resource allocation. It is implicit in much of what has been said in this chapter, but needs to be clearly enunciated, that experimentation for process improvement is almost always a **sequential** business. One typically collects some data, rethinks one's view of process behavior, goes back for a second round of experimentation, and so on. This being true, it is important to spend resources wisely and not all on round one. A popular rule of thumb (traceable to Box, Hunter, and Hunter) is that not more than 25 % of an initial budget for process experimentation ought to go into a first round of data collection.

6.3.2 "Taguchi" Emphases

Genichi Taguchi was a Japanese statistician and quality consultant whose ideas about "offline quality control" have been extremely popular in some segments of US manufacturing. Some of the "Taguchi methods" are repackaged versions of well-established classical statistical tools like those presented in this text. Others are, in fact, less than reliable and should be avoided. But the general philosophical outlook and emphases brought by Taguchi and his followers are important and deserve some discussion. The reader who finds his or her interest piqued by this subsection is referred to the article "Taguchi's Parameter Design: A Panel Discussion," edited by V. Nair, which appeared in *Technometrics* in 1992, for more details and an entry into the statistical literature on the subject.

Taguchi's offline quality control ideas have to do with engineering experimentation for the improvement of products and processes. Important points that have been emphasized by him and his followers are

1. small variation in product and process performance is important,

2. the world of engineered products and processes is not a constant variance world,

382   Chapter 6. Experimental Design and Analysis for Process Improvement

3. not all means of producing small variation are equal ... some are cheaper and more effective than others, and

4. it is experimental design and analysis that will enable one to discover effective means of producing products and processes with consistent performance.

The goal of consistent product and process performance in a variety of environments has become known as the goal of **robust** (product and process) **design**.

As an artificial but instructive example, consider a situation where, unbeknownst to design engineers, some system response variable y is related to three other variables x_1, x_2, and x_3 via

$$y = 2x_1 + (x_2 - 5)x_3 + \epsilon, \tag{6.14}$$

where x_1 and x_2 are system variables that one can pretty much set at will (and expect to remain where set), x_3 is a random input to the system that has mean 0 and standard deviation η, and ϵ has mean 0 and standard deviation σ and is independent of x_3. Suppose further that it is important that y have a distribution tightly packed about the number 7.0 and that if necessary η can be made small, but only at huge expense.

Having available the model (6.14), it is pretty clear what one should do in order to achieve desired results for y. Thinking of x_1 and x_2 as fixed and x_3 and ϵ as independent random variables,

$$E y = 2x_1 \quad \text{and} \quad \text{Var } y = \sqrt{(x_2 - 5)^2 \eta^2 + \sigma^2}.$$

So taking $x_2 = 5$, one achieves minimum variance for y without the huge expense of making η small, and then choosing $x_1 = 3.5$, one gets an ideal mean for y. The object of Taguchi's notions of experimentation is to enable an empirical version of this kind of optimization, applicable where one has no model like that in display (6.14) but does have the ability to collect and analyze data.

Notice, by the way, what it is about the model (6.14) that allows one to meet the goal of robust design. In the first place, the variable x_2 interacts with the system input x_3, and it is possible to find a setting of x_2 that makes the response "flat" as a function of x_3. Then, x_1 interacts with neither of x_2 nor x_3 and is such that it can be used to put the mean of y on target (without changing the character of the response as a function of x_2 and x_3).

How common it is for real products and processes to have a structure like this hypothetical example just waiting to be discovered and exploited is quite unknown. But the Taguchi motivation to pay attention to response variance as well as response mean is clearly sound and in line with classical quality control philosophy (which, e.g., has always advised "get the R chart under control before worrying about the $\bar{x}$ chart"). Statistical methods for plotting residuals after fitting "mean-oriented" models can be used to look for indications of change in response variance. And the analysis of Example 87 presented in this book shows explicitly (using a logged standard deviation as a response variable) what can be done in the way of applying methods for response mean to the analysis of patterns in response variance. The genuinely important reminder provided by Taguchi is that nonconstant variance is not only to be discovered and noted, but is to be *exploited* in product and process design.

There is some special terminology and additional experimental philosophy used by Taguchi that needs to be mentioned. In a product or process design experiment, those factors whose levels are under the designer's control both in the experiment and in later use are termed **control factors**. The designer gets to choose levels of those variables as parameters of the product or process. Other factors which may affect performance but will not be under the designer's influence in normal use are then called **noise factors**. And Taguchi emphasized the importance of including *both* control variables and noise variables as factors in product- and process-development experiments, which he termed **parameter design experiments.**

In the synthetic example represented by display (6.14), Taguchi would have called the variables x_1 and x_2 control variables, and the variable x_3 would be termed a noise variable. The objects of experimentation on a system like that modeled by Eq. (6.14) then become finding settings of some control variables that minimize sensitivity of the response to changes in the noise variables and finding other control variables that have no interactions with the noise variables and that can be used to bring the mean response on target.

There has been a fair amount of discussion in the statistical literature about exactly how one should treat noise factors in product- and process-development experimentation. The "Taguchi" approach has been to develop separate experimental designs involving first the control factors and then the noise factors. These are sometimes referred to as, respectively, the **inner array** and the **outer array.** Then each combination of control factors is run with every combination of the noise factors, and summary measures (like a mean and "standard deviation" taken over all combinations of the noise factors) of performance are developed for each combination of control factors. These summary measures are (for better or worse) sometimes called **signal-to-noise ratios** and serve as responses in analyses of experimental results in terms of the "effects" of only the control variables.

This approach to handling noise variables has received serious criticism on a number of practical grounds. In the first place, there is substantial evidence that this Taguchi **product array** approach ultimately leads to very large experiments (much larger than are really needed to understand the relationships between experimental factors and the response). (For more on this point, the reader is referred to the panel discussion mentioned at the beginning of this section and to the paper "Are Large Taguchi-Style Experiments Necessary? A Reanalysis of Gear and Pinion Data" by Miller, Sitter, Wu, and Long, which appeared in *Quality Engineering* in 1993.) Further, there seems to be little to guarantee that any pattern of combinations of noise variables set up for purposes of experimentation will necessarily mimic how those variables will fluctuate in later product or process use. And the whole notion of "summarizing out" the influence of noise variables

384   Chapter 6. Experimental Design and Analysis for Process Improvement

before beginning data analysis in earnest seems misguided. It appears to preclude the possibility of completely understanding the original response in terms of main effects and interactions of all experimental variables or in terms of the kind of response function ideas discussed in Sect. 6.2. The statistical literature indicates that it is both more economical and more effective to simply treat control variables and noise variables on equal footing in setting up a single **combined array** experimental design, applying the methods of design and analysis already discussed in Chap. 5 and the present chapter of this book.

If nothing else is obvious on a first reading of this section, hopefully it is clear that the tools of experimental design and analysis presented in this book are just that, tools. They are not substitutes for consistent, careful, and clear thinking. They can be combined in many different and helpful ways, but no amount of cleverness in their use can salvage a study whose design has failed to take into account one or more of the important qualitative issues discussed here. Experimentation for process improvement cannot be reduced to something that can be looked up in a cookbook. Instead, it is a subtle but also tremendously interesting and rewarding enterprise that can repay an engineer's best efforts with order of magnitude quality improvements.

6.4 Chapter Summary

Chapter 5 introduced the basics of experimental design and analysis for process improvement, covering full factorial studies. This chapter has filled in the picture of process experimentation sketched in Chap. 5. It opened with a discussion of the design and analysis of fractional factorial studies, which are widely used to efficiently screen a large number of factors looking for plausible descriptions of process behavior in terms of a few factorial effects. The second section considered the use of regression analysis and response surface methods in problems where all experimental factors are quantitative. Finally, the last section discussed a number of qualitative issues in experimental planning/design that must be thoughtfully addressed if one hopes to achieve process improvement.

6.5 Chapter 6 Exercises

1. **Tile Manufacturing.** Taguchi and Wu (*Introduction to OffLine Quality Control,* 1980) discussed a tile manufacturing experiment. The experiment involved seven factors, each having two levels. A current operating condition and a newly suggested value were used as levels for each of the factors. A set of 100 tiles was produced for each treatment combination included in the study, and the number of nonconforming tiles from each set of 100 was recorded:

(a) If all possible treatment combinations had been included in the study, how many different experimental runs would have been required (at a minimum)? (A single experimental run consists of all items produced under a single set of process conditions.) How many tiles would that have required?

(b) In fact, a $1/16$ fraction of the full factorial set of possible treatments was actually employed. How many different treatment combinations were studied? How many total tiles were involved?

Suppose the seven factors considered in the experiment were A, B, C, D, E, F, and G, where the current operating conditions are (arbitrarily) designated as the "high" or "+" levels of the factors.

(c) How many generators are needed for specifying which combinations are to be run in this 2^{7-4} study?

(d) Using the multiplication-of-signs convention introduced in Sect. 6.1, suppose

   i. levels of factor D are chosen by making products of signs for levels of factors A and B,
   ii. levels of factor E are chosen by making products of signs for levels of factors A and C,
   iii. levels of factor F are chosen by making products of signs for levels of factors B and C, and
   iv. levels of factor G are chosen by making products of signs for levels of factors A, B, and C.

   Make the table of signs that indicates which eight treatment combinations will be run in the experiment. (List the combinations in Yates standard order as regards factors A, B, and C.)

(e) The grand mean will be aliased with how many other 2^7 factorial effects? Give the defining relation for this experiment.

2. Refer to the **Tile Manufacturing** case in problem 1. The data actually collected are given in Table 6.15 in a format close to that used by the original authors. (Note that the real experiment was *not* run according to the hypothetical set of generators used for exercise in problem 1d).) The response variable, y, is the fraction of tiles that were nonconforming.

   (a) Suppose that levels of factors C, E, F, and G were determined by multiplication of signs for some combination of the factors A, B, and D. (The table below is in reverse Yates order for the three factors D, B, and A—from bottom to top instead of top to bottom.) What four generators were then used? (Hint: Find the four possible "product" columns, and identify which combination of factors (A, B, and D) multiplied together produces the given "sign" for the identified column.)

TABLE 6.15. Data for problem 2

Run	A	B	C	D	E	F	G	y
1	+	+	+	+	+	+	+	.16
2	+	+	+	−	−	−	−	.17
3	+	−	−	+	+	−	−	.12
4	+	−	−	−	−	+	+	.06
5	−	+	−	+	−	+	−	.06
6	−	+	−	−	+	−	+	.68
7	−	−	+	+	−	−	+	.42
8	−	−	+	−	+	+	−	.26

(b) Find the defining relation from the four generators in (a).

(c) Find the 15 effects that are aliased with the A main effect.

(d) Rearrange the rows of the table (the bottom row should be the top row, the second from the bottom row becomes the second row..., the top row becomes the bottom row) to produce the Yates standard order for factors A, B, and D.

(e) Apply the Yates algorithm to the rearranged list of y's, and find the eight fitted sums of effects.

(f) Normal plot the last seven fitted sums from (e).

(g) Make a half-normal plot of the absolute values of the last seven fitted values from (e).

(h) What conclusions do you draw from the plots in (f) and (g) regarding statistically significant effects on the fraction nonconforming?

3. Refer to the **Tile Manufacturing** case in problems 1 and 2. The analysis of the data suggested in problem 2 ignores the fact that the responses "y" were really fractions nonconforming "$\hat{p}$" and that if p is the long-run fraction nonconforming for a given set of conditions, reasoning as in Sect. 3.3.1, Var $\hat{p} = p(1-p)/n$. This in turn suggests that here (where each sample size is $m = 100$ tiles) for $\hat{E}$ a fitted sum of effects, one might estimate the standard deviation for $\hat{E}$ as

$$\hat{\sigma}_{\hat{E}} = \frac{1}{2^3 \sqrt{100}} \sqrt{\sum \hat{p}(1-\hat{p})},$$

where the sum is over the 2^3 treatment combinations included in the study.

Then, based on such a standard error for fitted sums of effects, it is further possible to make crude approximate confidence intervals for the corresponding sum of effects, E, using the end points

$$\hat{E} \pm z\hat{\sigma}_{\hat{E}}.$$

Chapter 6. Experimental Design and Analysis for Process Improvement   387

(In this equation, z is standing for an appropriate standard normal quantile. For example, a two-sided interval made with $z = 1.96$ would have associated approximate confidence of 95 %.) Make intervals such as this (with approximate individual confidence levels of 95 %) for the eight sums of effects based on the data of problem 2. Are your intervals consistent with your conclusions in problem 2? Explain.

4. **Nickel-Cadmium Cells**. In a 1988 *Journal of Quality Technology* paper, Ophir, El-Gad, and Snyder reported results of quality-improvement efforts focusing on finding optimal process conditions for the making of nickel-cadmium cells. (Their study resulted in a process producing almost no defects, annual monetary savings of many thousands of dollars, and an improved workplace atmosphere.) In the production of nickel-cadmium batteries, sometimes contact between the two electrodes occurs. This causes an internal short and the shorted cell must be rejected.

The authors used Pareto and Ishikawa diagrams in their efforts to identify factors possibly influencing battery shorts. Seven factors, each at two levels, were selected as most likely to control the production of shorts. A full factorial design would have involved 128 "treatments." Instead, the levels of the four factors "rolling order," "rolling direction," "nylon sleeve on edge," and "side margin of the plate" were held constant at plant standard levels. The three experimental factors considered were:

A—Method of sintering   Old (−) vs. new (+)
B—Separator   Thin (−) vs. thick (+)
C—Rolling pin   Thin (−) vs. thick (+)

Data like the ones in Table 6.16 resulted.

TABLE 6.16. Data for problem 4

Run	Sintering method	Separator	Rolling pin	Number short	Number tested
1	New	Thick	Thick	0	50
2	New	Thick	Thin	0	50
3	New	Thin	Thick	1	50
4	New	Thin	Thin	1	50
5	Old	Thick	Thick	1	50
6	Old	Thick	Thin	1	50
7	Old	Thin	Thick	1	50
8	Old	Thin	Thin	2	41

(a) Reorder the rows in the table to put it in Yates standard order for the factors A, B, and C. (Hint: Current row 8 becomes row 1.) Compute the sample fraction nonconforming, $\hat{p}$, for each treatment combination.

(b) The most natural way to think of this study is as a full factorial in the three factors A, B, and C. Is it possible to think of it as a fractional factorial in the original seven factors? Why or why not? What can potentially be learned from the data above about the effects of the four factors "rolling order," "rolling direction," "nylon sleeve on edge," and "side margin of the plate"?

(c) Find the fitted effects of the three factors corresponding to the all-high treatment combination by applying the Yates algorithm to the sample fractions $\hat{p}$ calculated in (a).

(d) Make a normal probability plot for the fitted effects found in (c).

(e) Make a half-normal plot for the absolute values of the fitted effects found in (c).

(f) Are there factors that seem to be important? Why? (Notice that your conclusions here apply to operation at the standard levels of the factors "rolling order," "rolling direction," "nylon sleeve on edge," and "side margin of the plate.")

(g) A rough-and-ready method of investigating whether there is evidence of *any* real differences in the fractions nonconforming across the eight treatment combinations is to apply the retrospective p chart ideas from Sect. 3.3.1. Do that here (temporarily entertaining the possibility that one has samples from eight equivalent sets of process conditions). Is there clear evidence of a difference somewhere in this factorial set of treatment combinations?

5. Refer to the **Nickel-Cadmium Cells** case in problem 4 and the discussion in problem 3. Since the sample sizes in problem 4 were not all the same, the estimated standard deviation of a fitted effect $\hat{E}$ is a bit messier than that suggested in problem 3. In place of the $m = 100$ formula given in problem 3, one has

$$\hat{\sigma}_{\hat{E}} = \frac{1}{2^3}\sqrt{\sum \frac{\hat{p}(1-\hat{p})}{n}},$$

where again the sum is over the 2^3 treatment combinations included in the study. Use the formula

$$\hat{E} \pm z\hat{\sigma}_{\hat{E}}$$

and make approximate 90 % two-sided individual confidence intervals for the 2^3 factorial effects on fraction of cells nonconforming. What do these say about the importance of the factors in determining fraction nonconforming? How do your conclusions here compare to those in problem 4?

6. Refer to the **Nickel-Cadmium Cells** case in problems 4 and 5. The investigators decided that the new sintering methodology combined with a thick separator produced a minimum of shorts and that rolling pin thickness had no appreciable influence on this quality measure. Using the new sintering

Chapter 6. Experimental Design and Analysis for Process Improvement   389

method, a thick separator, and a thin rolling pin thickness, half of a full 2^4 factorial experiment in the factors originally held constant was designed and carried out. Data like those in Table 6.17 were obtained.

TABLE 6.17. Data for problem 6

A Nylon sleeve	B Rolling direction	C Rolling order	D Margin	Number of Shorts	n
No	Lower edge first	Negative first	Narrow	1	80
No	Lower edge first	Positive first	Wide	8	88
No	Upper edge first	Negative first	Wide	0	90
No	Upper edge first	Positive first	Narrow	2	100
Yes	Lower edge first	Negative first	Wide	0	90
Yes	Lower edge first	Positive first	Narrow	1	90
Yes	Upper edge first	Negative first	Narrow	0	90
Yes	Upper edge first	Positive first	Wide	0	90

(a) Designate "yes," "upper edge first," "positive first," and "wide" as the high levels of factors A through D. Rearrange the rows of the table above to place the combinations in Yates standard order as regards factors A, B, and C.

(b) Is this design a fractional factorial of the type discussed in Sect. 6.1? Why or why not? If your answer is yes, describe the structure in terms of 2 raised to some "$p - q$" power.

(c) Find the defining relation for the design.

(d) What (single) 2^4 factorial effects are aliased with each of the main effects in this study?

(e) Using the Yates algorithm, find the estimated sums of effects on the long-run fractions nonconforming.

(f) Normal plot the last seven estimates from (e).

(g) Make a half-normal plot of the magnitudes (absolute values) of the last seven estimates from (e).

(h) Based on your plots from (f) and (g), do any of the sums of effects appear to be significantly different from 0? Why or why not?

(i) A rough-and-ready method of investigating whether there is evidence of any real differences in the fractions nonconforming across the eight treatment combinations is to apply the retrospective p chart ideas from Sect. 3.3.1. Do that here (temporarily entertaining the possibility that one has samples from eight equivalent sets of process conditions). Is there clear evidence of a difference somewhere in this fractional factorial set of treatment combinations?

7. Refer to the **Nickel-Cadmium Cells** case in problem 6 and the discussion in problem 5. Since there are $2^{4-1} = 2^3 = 8$ treatment combinations represented in the study in problem 6, the formulas in problem 5 can be used to make rough judgments of statistical detectability of sums of 2^4 effects on the fraction of cells with shorts.

   Use the formulas in problem 5, and make an approximate 95 % two-sided individual confidence interval for each of the eight different sums of the 2^4 effects that can be estimated based on the data given in problem 6. What do these say about the importance of the factors in determining fraction nonconforming? How do your conclusions here compare to those in problem 6?

8. **Paint Coat Thickness**. In an article that appeared in the *Journal of Quality Technology* in 1992, Eibl, Kess, and Pukelsheim reported on experimentation with a painting process. Prior to the experimentation, observed paint thickness varied between 2 mm and 2.5 mm, clearly exceeding the target value of .8 mm. The team's goal was to find levels of important process factors that would yield the desired target value without substantially increasing the cost of production. Preexperimental discussions produced the following six candidate factors and corresponding experimental levels:

A—Belt speed	Low (−) vs. high (+)
B—Tube width	Narrow (−) vs. wide (+)
C—Pump pressure	Low (−) vs. high (+)
D—Paint viscosity	Low (−) vs. high (+)
E—Tube height	Low (−) vs. high (+)
F—Heating temperature	Low (−) vs. high (+)

   The (−) or low levels of each of the experimental factors were in fact the same number of units below standard operating conditions as the (+) or high levels were above standard operating conditions. For some purposes, it is then useful to think of variables $x_A, x_B, \ldots, x_F$ giving (coded) values of the factors in the range −1 to 1, −1 corresponding to the (−) level of the factor, 1 corresponding to the (+) level, and 0 corresponding to a level halfway between the experimental ones (and corresponding to standard plant conditions).

   Since an experiment including all possible combinations of even two levels of all six factors was judged to be infeasible, $1/8$ of the possible treatment combinations were each applied to $m = 4$ different work pieces. The resulting paint thicknesses, y, were measured and are given in Table 6.18 (in millimeters).

   (a) Is this experiment a fractional factorial of the type discussed in Sect. 6.1? Why or why not?

   (b) Describe the experiment in terms of an appropriate base and exponent. Say what the "base" and "exponent" correspond to in the context of the problem.

TABLE 6.18. Data for problem 8

Combination	A	B	C	D	E	F	y
1	+	−	+	−	−	−	1.09, 1.12, .83, .88
2	−	−	+	−	+	+	1.62, 1.49, 1.40, 1.59
3	+	+	−	−	−	+	.88, 1.29, 1.04, 1.31
4	−	+	−	−	+	−	1.83, 1.65, 1.70, 1.76
5	−	−	−	+	−	+	1.46, 1.51, 1.59, 1.40
6	+	−	−	+	+	−	.74, .98, .79, .83
7	−	+	+	+	−	−	2.05, 2.17, 2.36, 2.12
8	+	+	+	+	+	+	1.51, 1.46, 1.42, 1.40

(c) Rearrange the rows of the data table to put the treatment combinations into Yates standard order as regards factors A, B, and C.

(d) Find the sample averages and standard deviations for each row.

(e) Find the 3 generators for this study. Find the defining relation for this study.

(f) Name the seven 2^6 factorial effects aliased with the A main effect, α_2. Do the same for the B main effect, β_2.

(g) Use the Yates algorithm, and find the 7 estimated sums of effects corresponding to the all-high treatment combinations and the estimated grand mean plus aliases.

(h) Use Eq. (5.28) and (g) above to construct the 90% individual two-sided confidence intervals for each of the 8 sums of effects on mean paint thickness. Which of these intervals fail to include 0 and thus indicate statistically detectable sums? What is the simplest (assuming all 2-factor and higher interactions are 0) possible interpretation of these results (in terms of 2^6 factorial effects)?

(i) Suppose that only main effects of the factors A through F (and no interactions) are important and that any main effect not found to be detectable by the method of part (h) is judged to be ignorable. Based only on estimated main effects and the grand mean estimate, find a predicted (or estimated mean) coating thickness for the all-high treatment combination. Based only on estimated main effects and the grand mean estimate, find a predicted coating thickness for the "all-low" treatment combination. (Notice that this is *not* one included in the original study.)

(j) Use a multiple regression program to fit the equation

$$y \approx \beta_0 + \beta_A x_A + \beta_B x_B + \beta_C x_C + \beta_D x_D + \beta_E x_E + \beta_F x_F$$

to the data ("+" means 1 and a "−" means −1). (You will have $n = 32$ data points $(x_A, x_B, x_C, x_D, x_E, x_F, y)$ to work with.) How do the estimates of the coefficients β compare to the results from part (g)?

(k) Drop from your fitted equation in (j) all terms x with corresponding fitted coefficients that are not "statistically significant." (Drop all terms x corresponding to β's whose 90% two-sided confidence intervals include 0.) Use the reduced equation to predict coating thickness when all x's are 1. Then use the reduced equation to predict coating thickness when all x's are -1. How do these predictions compare to the ones from (i)?

(l) The factorial-type analysis in parts (g) through (i) only presumes to infer mean paint thickness at combinations of $(-)$ and $(+)$ levels of the factors. On the other hand, the linear regression analysis in parts (j) and (k) could be used (with caution) for extrapolation or interpolation to other levels of the factors. Using your fitted equation from part (k), give values for x_A, x_B, x_C, and x_D (either -1 or $+1$) such that the predicted paint thickness is about .846. Your set of x_A, x_B, x_C, and x_D values should satisfy the defining relation for x_D.

9. **Speedometer Cable.** In a 1993 *Journal of Quality Technology* paper, Schneider, Kasperski, and Weissfeld discussed and reanalyzed an experiment conducted by Quinlan and published in the *American Supplier Institute News* in 1985. The objective of the experiment was to reduce post-extrusion shrinkage of speedometer casing. Fifteen factors each at two levels were considered in 16 experimental trials. Four measurements were made on pieces of a single (3000 ft) cable segment for each different treatment combination. (A different cable segment was made using each of the different combinations of process conditions.) The fifteen factors were A (liner tension), B (liner line speed), C (liner die), D (liner outside diameter (OD)), E (melt temperature), F (coating material), G (liner temperature), H (braiding tension), J (wire braid type), K (liner material), L (cooling method), M (screen pack), N (coating die type), O (wire diameter), and P (line speed). The log ($\ln$) transformation was applied to each of the 4×16 shrinkage responses, and "sample" means for the 16 combinations included in the study were computed. Estimated sums of effects (on the log scale) for the all-high treatment combination were as in Table 6.19.

(a) If one objective of the experiment was to say what would happen for different 3,000 ft segments of cable made under a given set of process conditions, why would it be a bad idea to treat the four measurements made for a given treatment combination as "replicates" and base inferences on formulas for situations where a common sample size is $m = 4$? What is the real "sample size" in this study?

(b) Normal plot the last 15 estimated sums of effects listed in Table 6.19.

(c) Make a half-normal plot of the absolute values of the last 15 estimated sums of effects.

(d) Based on your plots from (b) and (c), which (if any) sums appear to be clearly larger than background noise?

TABLE 6.19. Estimates for problem 9

Sum of effects estimated	Estimate
Grand mean + aliases	−1.430
A main + aliases	.168
B main + aliases	.239
C main + aliases	−.028
D main + aliases	.222
E main + aliases	−.119
F main + aliases	.046
G main + aliases	−.084
H main + aliases	.212
J main + aliases	−.882
K main + aliases	−.317
L main + aliases	−.102
M main + aliases	−.020
N main + aliases	.309
O main + aliases	−.604
P main + aliases	−.025

(e) If one adopts the tentative conclusion that any sums of effects judged in (d) to be important consist primarily of main effects, it might make sense to follow up the initial 2^{15-11} fractional factorial study with a replicated full factorial in the seemingly important factors (holding the other, apparently unimportant, factors fixed at some standard levels). Describe such a 2^p study with, say, $m = 3$.

(f) What will constitute "replication" in your new study from (e)?

10. **Bond Strength.** Grego (in a 1993 *Journal of Quality Technology* paper) and Lochner and Matar (in the 1990 book *Designing for Quality*) analyzed the effects of four two-level factors on the bond strength of an integrated circuit mounted on a metalized glass substrate. The four factors identified by engineers as potentially important determiners of bond strength were:

A—Adhesive type	D2A (−) vs. H-1-E (+)
B—Conductor material	Copper (Cu) (−) vs. nickel (Ni) (+)
C—Cure time at 90°C	90 min (−) vs. 120 min (+
D—Deposition material	Tin (−) vs. silver (Ag) (+)

Half of all 2^4 possible combinations were included in an experiment. $m = 5$ observations were recorded for each treatment combination. Summary statistics from the experiment are given in Table 6.20 on page 394.

(a) Describe the structure of this study in terms of a base raised to some power $p - q$. (Give numerical values for the base, p and q.)

TABLE 6.20. Summary statistics for problem 10

A	B	C	D	$\bar{y}$	s^2
D2A	Cu	90	tin	73.48	2.452
D2A	Cu	120	Ag	87.06	.503
D2A	Ni	90	Ag	81.58	.647
D2A	Ni	120	tin	79.38	1.982
H-1-E	Cu	90	Ag	83.88	4.233
H-1-E	Cu	120	tin	79.54	8.562
H-1-E	Ni	90	tin	75.60	26.711
H-1-E	Ni	120	Ag	90.32	3.977

(b) The treatment combinations in the table are presently arranged in Yates standard order as regards factors C, B, and A. Rearrange the rows so that the table is in Yates standard order as regards factors A, B, and C.

(c) Find the defining relation for this study.

(d) For each factor A through D, find the effect aliased with that factor's main effect. (Use notation like A↔BCD.)

(e) Write out (in terms of subscripted individual lowercase Greek letters and products of the same, like $\alpha_2 + \beta\gamma\delta_{222}$) the eight sums of 2^4 factorial effects that can be estimated based on the data from this study.

(f) Use the Yates algorithm to find estimates of the eight sums identified in (e).

(g) Find 95 % two-sided individual confidence intervals for the 8 sums identified in (e). (Use display (5.28).)

(h) Use your intervals from (g) to identify statistically detectable sums of effects. What is it about an interval in (g) that says the associated sum of effects is statistically significant?

11. Refer to the **Bond Strength** case in problem 10. Factor screening is one motivation for using a fractional factorial experiment. Providing at least tentative suggestions as to which treatment combination(s) might maximize or minimize a response is another (regardless of whether such a combination was part of an original fractional factorial data set).

(a) Reflect on the results of part (h) of problem 10. What is the simplest possible interpretation (assuming all 2-factor and higher order interactions are 0) of your results in terms of the four original factors? How many of the original four factors are involved in this interpretation? Which factors are not involved and might possibly be "inert," not affecting the response in any way?

(b) Set up a "design" table for a full 2^p factorial in any factors that from part (a) you suspect of being "active," that is, affecting the response. Say which levels you will use for any factors in (a) that you suspect to be "inert."

(c) Again, consider the results of part (h) of problem 10. What combination (or combinations) of levels of factors A through D do you suspect might maximize mean bond strength? (Assume all 2-factor and higher order interactions are 0.) Is such a combination in the original data set? (In general, it need not be.)

(d) For the combination(s) in (c), what is the predicted bond strength?

12. **Solder Thickness.** Bisgaard, in a 1994 *Journal of Quality Technology* paper, discussed an experiment designed to improve solder layer mean thickness and thickness uniformity on printed circuit boards. (A uniform solder layer of a desired thickness provides good electrical contacts.) To stay competitive, it was important for a manufacturer to solve the problem of uneven solder layers. A team of engineers focused on the operation of a hot air solder leveler (HASL) machine. A 16-run screening experiment involving six two-level factors was run. A measure of solder layer uniformity was obtained from each of the 16 runs (y is a sample variance thickness based on 24 thickness measurements). The design generators were E↔ABC and F↔BCD. The 16 combinations of levels of factors A through D and corresponding values of the response variable are given in Table 6.21.

TABLE 6.21. Data for problem 12

A	B	C	D	y
−	−	−	−	32.49
+	+	+	−	46.65
+	−	−	+	8.07
−	+	+	+	6.61
−	+	−	−	25.70
+	−	+	−	16.89
+	+	−	+	29.27
−	−	+	+	42.64
+	+	−	−	31.92
−	−	+	−	49.28
−	+	−	+	11.83
+	−	+	+	18.92
+	−	−	−	35.52
−	+	+	−	24.30
−	−	−	+	32.95
+	+	+	+	40.70

(a) Rearrange the rows in the table to produce Yates standard order for factors A, B, C, and D. Then add two columns to the table giving the levels of factors E and F that were used.

(b) Discuss why, even though there were 24 thickness measurements (probably from a single PC board) involved in the computation of "y" for a given treatment combination, this is really an experiment with no replication ($m = 1$).

(c) Find the defining relation for this 2^{6-2} fractional factorial experiment.

(d) The effects are aliased in 16 groups of four effects each. For each of the six main effects (α_2, β_2, γ_2, δ_2, ϵ_2, and ϕ_2), find the sums involving those which can be estimated using data from this experiment. (Use notation like $\alpha_2 + \beta\gamma\epsilon_{222} + \alpha\beta\gamma\delta\phi_{22222} + \delta\epsilon\phi_{222}$.) (Hint: Consider your answer in (c).)

(e) Transform the responses y by taking natural logarithms, $y' = \ln(y)$. Then use the Yates algorithm to find the estimated sums of effects on y'.

(f) Make a normal probability plot for the last 15 estimated sums of effects in (e). Then make a half-normal plot for the absolute values of these estimates.

(g) Based on your plots in (f), do you detect any important effects on solder layer uniformity? Why or why not? What is the simplest possible interpretation of the importance of these? Based on your results from the full normal plot and using the simplest interpretation (assume all interaction effects are 0), what combination or combinations of levels of the six factors do you predict will have the best uniformity of solder layer? Why? (Hint: the best uniformity corresponds to a small y', i.e., a sample standard deviation that is small.)

(h) Would you be willing to recommend adoption of your projected best treatment combination(s) from (g) with no further experimentation? Explain.

13. A 1/2 replicate of a full 2^4 factorial experiment is to be conducted. Unfortunately, only four experimental runs can be made on a given day, and it is feared that important environmental variables may change from day to day and impact experimental results.

(a) Make an intelligent recommendation of which ABCD combinations to run on each of 2 consecutive days. Defend your answer. (You should probably think of "day" as a fifth factor and at least initially set things up as a 1/4 replicate of a full 2^5 factorial. Let the two generators be D↔ABC and E↔BC where E is (+) for day 2 and (−) for day 1.)

(b) Assuming that environmental changes do not interact with the experimental factors A through D, list the eight sets of aliases associated with your plan from (a). Your sets must include the main effects and interactions of A through D and the main effect (only) of E (day). (Hint: Consider what the sets of aliases would have been with your setup in (a). Drop any of the alias terms including E interacting with one or more terms.)

(c) Discuss how data analysis would proceed for your study.

14. Return to the **Polymer Density** problem 1 in the Sect. 6.2 Exercises.

    (a) Use Eq. (6.12), and find the stationary point for the fitted quadratic equation from part (e) of problem 1 in the Sect. 6.2 Exercises. Is the stationary point the location of a maximum, a minimum, or a saddle point? Why?

    (b) Is the stationary point determined in (a) above located inside the experimental region? Why might the answer to this question be important to the company producing the polymer?

    (c) The experiment was originally planned as a perfect central composite design, but when it became clear that no data collection would be possible at the set of conditions $x_1 = 240$ and $x_2 = 30$, the "extra" data point was added in its place. Snee suggests that when one suspects that some planned experimental conditions may be infeasible, those combinations should be run early in an experimental program. Why does this suggestion make good sense?

15. **Surface Finish in Machining.** The article "Including Residual Analysis in Designed Experiments: Case Studies" by Collins and Collins that appeared in *Quality Engineering* in 1994 contains discussions of several machining experiments aimed at improving surface finish for some torsion bars. The following is a description of part of those investigations. Engineers selected surface finish as a quality characteristic of interest because it is directly related to part strength and product safety and because it seemed possible that its variation and production cost could simultaneously be reduced. Surface roughness was measured using a gauge that records total vertical displacement of a probe as it is moved a standard distance across a specimen. (The same 1-in section was gauged on each torsion bar included in the experiment.)

    Speed rate (rate of spin during machining, in rpm) and feed rate of the machining (rate at which the cutting tool was moved across a bar, in inches per revolution) were the two factors studied in the experiment. Three levels of speed rate and three levels of feed rate were selected for study. $m = 2$ bars were machined at each combination of speed and feed. Speed rate levels were 2,500, 3,500, and 4,500 rpm. Feed rate levels were .001, .005,

Chapter 6. Experimental Design and Analysis for Process Improvement

and .009 in per revolution. In Table 6.22, the speeds x_1 have been coded to produce values x_1' via

$$x_1' = \frac{x_1 - 3{,}500}{1{,}000}$$

and the feed rates x_2 have similarly been coded to produce values x_2' using

$$x_2' = \frac{x_2 - .005}{.004}.$$

(Note that with this coding, low, medium, and high levels of both variables are respectively $-1, 0,$ and 1.)

(a) Describe the factorial experimental design employed here in terms of a base and an exponent.

(b) How many different (x_1, x_2) pairs were there in this study? Is this set sufficient to fit an equation like that given in display (6.8) and like that in display (6.9)? Why or why not?

(c) Use a multiple regression program to fit an equation for y that (like that in display (6.8)) is linear in x_1' and x_2'. (Note that there are 18 data points (x_1', x_2', y) indicated in the table below.)

TABLE 6.22. Data for problem 15

x_1'	x_2'	y
-1	-1	7, 9
-1	0	77, 77
-1	1	193, 190
0	-1	7, 9
0	0	75, 85
0	1	191, 191
1	-1	9, 18
1	0	79, 80
1	1	192, 190

(d) For the equation from (c), plot contours in the (x_1', x_2')-plane corresponding to fitted values, $\hat{y}$, of 5, 55, 105, and 155.

(e) Where in the experimental region (the square specified by $-1 \leq x_1' \leq 1$ and $-1 \leq x_2' \leq 1$) is $\hat{y}$ optimum (minimum)? What is the optimum predicted value? Do you find this value reasonable? (Can y be negative?)

(f) Find the residuals for the equation fit in (c). Plot these against each of $x_1', x_2',$ and $\hat{y}$.

(g) Find R^2 for the equation fit in (c).

(h) Based on (f) and (g), does it appear the fitted model in (c) fits the data well? Why or why not?

(i) Fit a full quadratic equation in x'_1, x'_2 to the surface roughness data. (That is, fit an equation like (6.9) to the data.)

(j) For the equation from (i), plot contours in the (x'_1, x'_2)-plane corresponding to fitted values, $\hat{y}$, of $5, 55, 105$, and 155.

(k) Based on the plot from (j), where in the experimental region does it seem that the predicted roughness is optimum? What is the optimum predicted value? Do you find this value reasonable? (Can y be negative?)

16. Refer to the **Surface Finish** case in problem 15. In problems like the surface finish case, where responses at different conditions vary over an order of magnitude, linear and quadratic fitted response functions often have difficulty fitting the data. It is often helpful to instead try to model the logarithm of the original response variable.

    (a) Redo the analysis of problem 15 using $y' = \ln(y)$ as a response variable. (Notice, for one thing, that modeling the logarithm of surface roughness deals with the embarrassment of possible negative predicted values of y.) Do the methods of Sect. 6.2 seem better suited to describing y or to describing y'?

    (b) For your quadratic description of y', find the stationary point of the fitted surface in both coded and raw units. Is the stationary point inside the experimental region? Is it a maximum, a minimum, or a saddle point? Why?

    (c) Use the rule of thumb summarized in display (6.13) to judge whether first the linear and then the quadratic fitted equations for y' are clearly "tracking more than experimental noise."

17. Refer to the **Surface Finish** case in problem 15. Suppose that in a similar situation, experimental resources are such that only 10 torsion bars can be machined and tested. Suppose further that the lowest possible machining speed is 2,200 rpm and the highest possible speed is 4,800 rpm. Further, suppose that the smallest feed rate of interest is .0002 in/rev and the maximum one of interest is .0098 in/rev.

    (a) Set up a central composite experimental plan for this situation. Use the four "corner points" of the design in problem 15 as "corner points" here. (These are the points with $x'_1 = \pm 1$ and $x'_2 = \pm 1$.) Base the "star points" on the minimum and maximum values of the process variables suggested above, and allow for replication of the "center point." (Hint: choose the minimum of the different αs.) Make out a data collection form that could be used to record the 10 measured roughness values next to the values of the process variables that produce them.

(b) What are α, Δ_1, and Δ_2 for your plan in (a)? (Answer this for both the raw (x) and coded (x') representations of your data collection plan.)

(c) In what order would you suggest collecting the 10 data points specified in part (a)? Why is this a good order?

18. Revisit the **Tar Yield** case in problem 2 of the Sect. 6.2 Exercises. Consider the second set of 15 data points, the ones for low reagent A purity. Redo parts (c) through (g) for this second situation.

19. Refer to the **Tar Yield** case in problem 2 of the Sect. 6.2 Exercises and problem 18.

    (a) Consider the high reagent A purity data. Use the fitted quadratic equation from part (e) of problem 2 of the Sect. 6.2 Exercises. Set x_1 equal to 1 to produce a quadratic for y in terms of x_2 and x_3. Make a contour plot and find the associated (x_2, x_3) stationary point. Is the point a maxima, minima, or saddle point? Is the stationary point within the experimental region (i.e., does it satisfy $-1 \le x_2 \le 1$ and $-1 \le x_3 \le 1$)?

    (b) Consider the low reagent A purity data. Use the fitted quadratic equation from problem 18. Again, set x_1 equal to 1 to produce a quadratic for y in terms of x_2 and x_3. Make a contour plot and find the associated (x_2, x_3) stationary point. Is the point a maxima, minima, or saddle point? Is the stationary point within the experimental region (i.e., does it satisfy $-1 \le x_2 \le 1$ and $-1 \le x_3 \le 1$)?

    (c) Find the stationary point for the quadratic equation in three process variables found in part (e) of problem 2 of the Sect. 6.2 Exercises—high reagent A purity data. Is this a minimum, a maximum, or a saddle point? Explain. Is the point inside the experimental region defined by $-1 \le x_1 \le 1$, $-1 \le x_2 \le 1$, and $-1 \le x_3 \le 1$?

    (d) Find the stationary point for the quadratic equation in three process variables found in part (e) of problem 18—low reagent A purity data. Is this a minimum, a maximum, or a saddle point? Explain. Is the point inside the experimental region defined by $-1 \le x_1 \le 1$, $-1 \le x_2 \le 1$, and $-1 \le x_3 \le 1$?

    (e) Reflect on your answers to problem 18 and (a), (b), (c), and (d) above. What combination of temperature, catalyst concentration, excess reagent B, reagent A purity, and solvent stream purity (inside the experimental region) seems to produce the minimum tar? Defend your answer.

Chapter 6. Experimental Design and Analysis for Process Improvement

20. **Chemical Process Yield.** The text, *Response Surface Methodology*, by Raymond Myers contains the results of a four-variable central composite experiment. The conversion of 1,2-propanediol to 2,5-dimethyl-piperazine is affected by

$$NH_3 = \text{amount of ammonia (g)},$$
$$T = \text{temperature } (^\circ C),$$
$$H_2O = \text{amount of water (g), and}$$
$$P = \text{hydrogen pressure (psi)}.$$

The response variable of interest was a measure of yield, y. For purposes of defining a central composite experimental plan, it was convenient to define the coded process variables

$$x_1 = (NH_3 - 102)/51,$$
$$x_2 = (T - 250)/20,$$
$$x_3 = (H_2O - 300)/200, \text{ and}$$
$$x_4 = (P - 850)/350.$$

(The 2^4 factorial points of the design had $x_1 = \pm 1, x_2 = \pm 1, x_3 = \pm 1,$ and $x_4 = \pm 1$.)

(a) Find the raw (uncoded) high and low levels of each process variable in the 2^4 factorial part of the design.

(b) α for this study was 1.4. Find the uncoded (NH_3, T, H_2O, P) coordinates of the "star points" in this study. What was the (uncoded) center point of this design?

(c) How many design points total were there in this study (including the 2^4 factorial, the star points, and the center point)?

(d) The quadratic equation fit to the data set was

$$\hat{y} = 40.198 - 1.511x_1 + 1.284x_2 - 8.739x_3 + 4.995x_4$$
$$-6.332x_1^2 - 4.292x_2^2 + .020x_3^2 - 2.506x_4^2$$
$$+2.194x_1x_2 - .144x_1x_3 + 1.581x_1x_4$$
$$+8.006x_2x_3 + 2.806x_2x_4 + .294x_3x_4.$$

Find the stationary point for the fitted response surface. Is the stationary point a minimum, maximum, or saddle point? Is the stationary point within the experimental region? Why or why not?

(e) Consider the following two different (x_3, x_4) ordered pairs

$$(-1.4, -1.4) \text{ and } (-1.4, 0).$$

Substitute each of these pairs into the equation in (d), and produce a quadratic equation for y in terms of x_1 and x_2. Find the stationary points for these two different equations, and say whether they locate minima, maxima, or saddle points. What predicted responses are associated with these points?

(f) Based on your analysis in this problem, recommend a point (x_1, x_2, x_3, x_4) that (within the experimental region) maximizes the fitted conversion yield. Explain your choice, and translate the recommendation to raw (uncoded) values of process variables.

21. **Turning and Surface Roughness.** The article "Variation Reduction by Use of Designed Experiments" by Sirvanci and Durmaz appeared in *Quality Engineering* in 1993 and discusses a fractional factorial study on a turning operation and the effects of several factors on a surface roughness measurement, y. Below are the factors and levels studied in the experiment:

A—Insert	#5023 ($-$) vs. #5074 ($+$)
B—Speed	800 rpm ($-$) vs. 1000 rpm ($+$)
C—Feed rate	50 mm/min ($-$) vs. 80 mm/min ($+$)
D—Material	Import ($-$) vs. Domestic ($+$)
E—Depth of cut	.35 mm ($-$) vs. .55 mm ($+$)

Only $2^{5-2} = 8$ of the possible combinations of levels of the factors were considered in the study. These eight combinations were derived using the generators D↔AB and E↔AC. For each of these eight combinations, $m = 5$ different overhead cam block auxiliary drive shafts were machined, and surface roughness measurements, y (in μ inches), were obtained.

(a) Finish Table 6.23 specifying which eight combinations of levels of the five factors were studied.

TABLE 6.23. Treatment combinations for problem 21

A	B	C	D	E	Combination name
−	−	−			
+	−	−			
−	+	−			
+	+	−			
−	−	+			
+	−	+			
−	+	+			
+	+	+			

(b) Use the two generators D↔AB and E↔AC and find the entire defining relation for this experiment. Based on that defining relation, determine which effects are aliased with the A main effect.

Chapter 6. Experimental Design and Analysis for Process Improvement   403

(c) The experimenters made and measured roughness on a total of 40 drive shafts. If, in fact, the total number of shafts (and not anything connected with *which kinds of shafts*) was the primary budget constraint in this experiment, suggest an alternative way to "spend" 40 observations that might be preferable to the one the experimenters tried. (Describe an alternative and possibly better experimental design using 40 shafts.)

(d) In Table 6.24 are the eight sample means ($\bar{y}$) and standard deviations (s) that were obtained in the experiment, listed in Yates standard order as regards factors A, B, and C, along with results of applying the Yates algorithm to the means.

TABLE 6.24. Summary statistics and estimates for problem 21

$\bar{y}$	s	Estimate
74.96	36.84	65.42
57.92	3.72	−3.01
50.44	8.27	−6.80
49.16	4.19	2.27
80.04	8.25	7.30
75.96	3.69	1.57
68.28	5.82	1.52
66.60	6.32	−1.67

In the list of estimates, there are five that correspond to main effects and their aliases. Give the values of these.

(e) The pooled sample standard deviation here is $s_P = 14.20$. For purposes of judging the statistical significance of the estimated sums of effects, one might make individual 95 % two-sided confidence limits of the form $\hat{E} \pm \Delta$. Find Δ. See display (5.28).

(f) Based on your answers to parts (d) and (e), does it seem that the D main effect here might be tentatively judged to be statistically detectable (assuming all interactions are 0)? Explain.

(g) What about the values of s listed in the table calls into question the appropriateness of the confidence interval analysis outlined in parts (e) and (f)? Explain.

22. Professor George Box was famous for saying that to find out what happens to a system when you interfere with it, you have to interfere with it (not just passively observe it). Reflect on the roles (in modern quality improvement) of (1) control charting/process monitoring and (2) experimental design and analysis in the light of this maxim. What in the maxim corresponds to experimental design and analysis? What corresponds to control charting?

23. Consider a hypothetical 2^{3-1} experiment with generator $C \leftrightarrow AB$ and involving some replication that produces sample means $\bar{y}_c = 1$, $\bar{y}_a = 13$, $\bar{y}_b = 15$, and $\bar{y}_{abc} = 11$ and s_P small enough that all four of the sums of effects that can be estimated are judged to be statistically detectable. One possible simple interpretation of this outcome is that the grand mean $\mu_{...}$ and all of the main effects α_2, β_2, and γ_2 (and no interactions) are important.

(a) If one adopts the above simple interpretation of the experimental results, what combination of levels of factors A, B, and C would you recommend as a candidate for producing maximum response? What mean response would you project?

(b) Is the combination from (a) in the original experiment? Why would you be wise to recommend a "confirmatory run" for your combination in (a) before ordering a large capital expenditure to permanently implement your candidate from (a)?

24. Problems 3 and 5 offer one means of dealing with the fact that sample fractions nonconforming $\hat{p}$ have variances that depend not only on n, but on p as well, so that somewhat complicated formulas are needed for standard errors of estimated effects based on them. Another approach is to (for analysis purposes) transform $\hat{p}$ values to $y = g(\hat{p})$ where y has a variance that is nearly independent of p. The Freeman-Tukey transformation is

$$y = g(\hat{p}) = \frac{\arcsin\sqrt{\frac{n\hat{p}}{(n+1)}} + \arcsin\sqrt{\frac{n\hat{p}+1}{(n+1)}}}{2}.$$

Using this transformation, as long as p is not extremely small or extremely large,

$$\mathrm{Var}\, y \approx \frac{1}{4n}.$$

That means that if one computes fitted effects $\hat{E}$ based on the transformed values (using the k-cycle Yates algorithm), then standard errors can be computed as

$$\hat{\sigma}_{\hat{E}} = \frac{1}{2^{k+1}}\sqrt{\sum \frac{1}{n}},$$

where the sum of reciprocal sample sizes is over the 2^k treatment combinations in the study. (Approximate confidence limits are $\hat{E} \pm z\hat{\sigma}_{\hat{E}}$ as in problems 3 and 5.)

(a) Redo the analysis in part (e) through (h) of problem 2 using transformed values y.

(b) Redo the analysis in problem 3 using the transformed values y and $\hat{\sigma}_{\hat{E}}$ given above.

25. A project team identifies three quantitative variables (x_1, x_2, and x_3) to be used in an experiment intended to find an optimal way to run a vibratory burnishing process. The project budget will allow 18 runs of the process to be made. Suppose that the process variables have been coded in such a way that it is plausible to expect optimum conditions to satisfy

$$-2 \leq x_1 \leq 2, \quad -2 \leq x_2 \leq 2, \quad \text{and} -2 \leq x_3 \leq 2.$$

(a) Make and defend a recommendation as to how the 18 runs should be allocated. That is, make up a data collection sheet giving 18 combinations (x_1, x_2, x_3) that you would use. List these in the order you recommend running the experiment, and defend your proposed order.

(b) Suppose that after running the experiment and fitting a quadratic response surface, $\hat{y}$ appears to be maximized at the point (x_1^*, x_2^*, x_3^*). What are some circumstances in which you would be willing to recommend immediate use of the conditions (x_1^*, x_2^*, x_3^*) for production? What are some circumstances in which immediate use of the conditions (x_1^*, x_2^*, x_3^*) would be ill advised?

APPENDIX A
TABLES

See Table A.1.
See Table A.2.
See Table A.3.
See Tables A.4.1, A.4.2, and A.4.3.
See Table A.5.
See Table A.6.1.
See Table A.6.2.

TABLE A.1. Standard normal cumulative probabilities

$$\Phi(z) = \int_{-\infty}^{z} \frac{1}{\sqrt{2\pi}} \exp\left(-\frac{t^2}{2}\right) dt$$

z	.00	.01	.02	.03	.04	.05	.06	.07	.08	.09
−3.4	.0003	.0003	.0003	.0003	.0003	.0003	.0003	.0003	.0003	.0002
−3.3	.0005	.0005	.0005	.0004	.0004	.0004	.0004	.0004	.0004	.0003
−3.2	.0007	.0007	.0006	.0006	.0006	.0006	.0006	.0005	.0005	.0005
−3.1	.0010	.0009	.0009	.0009	.0008	.0008	.0008	.0008	.0007	.0007
−3.0	.0013	.0013	.0013	.0012	.0012	.0011	.0011	.0011	.0010	.0010
−2.9	.0019	.0018	.0018	.0017	.0016	.0016	.0015	.0015	.0014	.0014
−2.8	.0026	.0025	.0024	.0023	.0023	.0022	.0021	.0021	.0020	.0019
−2.7	.0035	.0034	.0033	.0032	.0031	.0030	.0029	.0028	.0027	.0026
−2.6	.0047	.0045	.0044	.0043	.0041	.0040	.0039	.0038	.0037	.0036
−2.5	.0062	.0060	.0059	.0057	.0055	.0054	.0052	.0051	.0049	.0048
−2.4	.0082	.0080	.0078	.0075	.0073	.0071	.0069	.0068	.0066	.0064
−2.3	.0107	.0104	.0102	.0099	.0096	.0094	.0091	.0089	.0087	.0084
−2.2	.0139	.0136	.0132	.0129	.0125	.0122	.0119	.0116	.0113	.0110
−2.1	.0179	.0174	.0170	.0166	.0162	.0158	.0154	.0150	.0146	.0143
−2.0	.0228	.0222	.0217	.0212	.0207	.0202	.0197	.0192	.0188	.0183
−1.9	.0287	.0281	.0274	.0268	.0262	.0256	.0250	.0244	.0239	.0233
−1.8	.0359	.0351	.0344	.0336	.0329	.0322	.0314	.0307	.0301	.0294
−1.7	.0446	.0436	.0427	.0418	.0409	.0401	.0392	.0384	.0375	.0367
−1.6	.0548	.0537	.0526	.0516	.0505	.0495	.0485	.0475	.0465	.0455
−1.5	.0668	.0655	.0643	.0630	.0618	.0606	.0594	.0582	.0571	.0559
−1.4	.0808	.0793	.0778	.0764	.0749	.0735	.0721	.0708	.0694	.0681
−1.3	.0968	.0951	.0934	.0918	.0901	.0885	.0869	.0853	.0838	.0823
−1.2	.1151	.1131	.1112	.1093	.1075	.1056	.1038	.1020	.1003	.0985
−1.1	.1357	.1335	.1314	.1292	.1271	.1251	.1230	.1210	.1190	.1170
−1.0	.1587	.1562	.1539	.1515	.1492	.1469	.1446	.1423	.1401	.1379
−0.9	.1841	.1814	.1788	.1762	.1736	.1711	.1685	.1660	.1635	.1611
−0.8	.2119	.2090	.2061	.2033	.2005	.1977	.1949	.1922	.1894	.1867
−0.7	.2420	.2389	.2358	.2327	.2297	.2266	.2236	.2206	.2177	.2148
−0.6	.2743	.2709	.2676	.2643	.2611	.2578	.2546	.2514	.2483	.2451
−0.5	.3085	.3050	.3015	.2981	.2946	.2912	.2877	.2843	.2810	.2776
−0.4	.3446	.3409	.3372	.3336	.3300	.3264	.3228	.3192	.3156	.3121
−0.3	.3821	.3783	.3745	.3707	.3669	.3632	.3594	.3557	.3520	.3483
−0.2	.4207	.4168	.4129	.4090	.4052	.4013	.3974	.3936	.3897	.3859
−0.1	.4602	.4562	.4522	.4483	.4443	.4404	.4364	.4325	.4286	.4247
−0.0	.5000	.4960	.4920	.4880	.4840	.4801	.4761	.4721	.4681	.4641

This table was generated using Minitab

TABLE A.1. Standard normal cumulative probabilities *(continued)*

z	.00	.01	.02	.03	.04	.05	.06	.07	.08	.09
0.0	.5000	.5040	.5080	.5120	.5160	.5199	.5239	.5279	.5319	.5359
0.1	.5398	.5438	.5478	.5517	.5557	.5596	.5636	.5675	.5714	.5753
0.2	.5793	.5832	.5871	.5910	.5948	.5987	.6026	.6064	.6103	.6141
0.3	.6179	.6217	.6255	.6293	.6331	.6368	.6406	.6443	.6480	.6517
0.4	.6554	.6591	.6628	.6664	.6700	.6736	.6772	.6808	.6844	.6879
0.5	.6915	.6950	.6985	.7019	.7054	.7088	.7123	.7157	.7190	.7224
0.6	.7257	.7291	.7324	.7357	.7389	.7422	.7454	.7486	.7517	.7549
0.7	.7580	.7611	.7642	.7673	.7704	.7734	.7764	.7794	.7823	.7852
0.8	.7881	.7910	.7939	.7967	.7995	.8023	.8051	.8078	.8106	.8133
0.9	.8159	.8186	.8212	.8238	.8264	.8289	.8315	.8340	.8365	.8389
1.0	.8413	.8438	.8461	.8485	.8508	.8531	.8554	.8577	.8599	.8621
1.1	.8643	.8665	.8686	.8708	.8729	.8749	.8770	.8790	.8810	.8830
1.2	.8849	.8869	.8888	.8907	.8925	.8944	.8962	.8980	.8997	.9015
1.3	.9032	.9049	.9066	.9082	.9099	.9115	.9131	.9147	.9162	.9177
1.4	.9192	.9207	.9222	.9236	.9251	.9265	.9279	.9292	.9306	.9319
1.5	.9332	.9345	.9357	.9370	.9382	.9394	.9406	.9418	.9429	.9441
1.6	.9452	.9463	.9474	.9484	.9495	.9505	.9515	.9525	.9535	.9545
1.7	.9554	.9564	.9573	.9582	.9591	.9599	.9608	.9616	.9625	.9633
1.8	.9641	.9649	.9656	.9664	.9671	.9678	.9686	.9693	.9699	.9706
1.9	.9713	.9719	.9726	.9732	.9738	.9744	.9750	.9756	.9761	.9767
2.0	.9773	.9778	.9783	.9788	.9793	.9798	.9803	.9808	.9812	.9817
2.1	.9821	.9826	.9830	.9834	.9838	.9842	.9846	.9850	.9854	.9857
2.2	.9861	.9864	.9868	.9871	.9875	.9878	.9881	.9884	.9887	.9890
2.3	.9893	.9896	.9898	.9901	.9904	.9906	.9909	.9911	.9913	.9916
2.4	.9918	.9920	.9922	.9925	.9927	.9929	.9931	.9932	.9934	.9936
2.5	.9938	.9940	.9941	.9943	.9945	.9946	.9948	.9949	.9951	.9952
2.6	.9953	.9955	.9956	.9957	.9959	.9960	.9961	.9962	.9963	.9964
2.7	.9965	.9966	.9967	.9968	.9969	.9970	.9971	.9972	.9973	.9974
2.8	.9974	.9975	.9976	.9977	.9977	.9978	.9979	.9979	.9980	.9981
2.9	.9981	.9982	.9983	.9983	.9984	.9984	.9985	.9985	.9986	.9986
3.0	.9987	.9987	.9987	.9988	.9988	.9989	.9989	.9989	.9990	.9990
3.1	.9990	.9991	.9991	.9991	.9992	.9992	.9992	.9992	.9993	.9993
3.2	.9993	.9993	.9994	.9994	.9994	.9994	.9994	.9995	.9995	.9995
3.3	.9995	.9995	.9996	.9996	.9996	.9996	.9996	.9996	.9996	.9997
3.4	.9997	.9997	.9997	.9997	.9997	.9997	.9997	.9997	.9997	.9998

TABLE A.2. t Distribution quantiles

ν	$Q(.9)$	$Q(.95)$	$Q(.975)$	$Q(.99)$	$Q(.995)$	$Q(.999)$	$Q(.9995)$
1	3.078	6.314	12.706	31.821	63.657	318.317	636.607
2	1.886	2.920	4.303	6.965	9.925	22.327	31.598
3	1.638	2.353	3.182	4.541	5.841	10.215	12.924
4	1.533	2.132	2.776	3.747	4.604	7.173	8.610
5	1.476	2.015	2.571	3.365	4.032	5.893	6.869
6	1.440	1.943	2.447	3.143	3.707	5.208	5.959
7	1.415	1.895	2.365	2.998	3.499	4.785	5.408
8	1.397	1.860	2.306	2.896	3.355	4.501	5.041
9	1.383	1.833	2.262	2.821	3.250	4.297	4.781
10	1.372	1.812	2.228	2.764	3.169	4.144	4.587
11	1.363	1.796	2.201	2.718	3.106	4.025	4.437
12	1.356	1.782	2.179	2.681	3.055	3.930	4.318
13	1.350	1.771	2.160	2.650	3.012	3.852	4.221
14	1.345	1.761	2.145	2.624	2.977	3.787	4.140
15	1.341	1.753	2.131	2.602	2.947	3.733	4.073
16	1.337	1.746	2.120	2.583	2.921	3.686	4.015
17	1.333	1.740	2.110	2.567	2.898	3.646	3.965
18	1.330	1.734	2.101	2.552	2.878	3.610	3.922
19	1.328	1.729	2.093	2.539	2.861	3.579	3.883
20	1.325	1.725	2.086	2.528	2.845	3.552	3.849
21	1.323	1.721	2.080	2.518	2.831	3.527	3.819
22	1.321	1.717	2.074	2.508	2.819	3.505	3.792
23	1.319	1.714	2.069	2.500	2.807	3.485	3.768
24	1.318	1.711	2.064	2.492	2.797	3.467	3.745
25	1.316	1.708	2.060	2.485	2.787	3.450	3.725
26	1.315	1.706	2.056	2.479	2.779	3.435	3.707
27	1.314	1.703	2.052	2.473	2.771	3.421	3.690
28	1.313	1.701	2.048	2.467	2.763	3.408	3.674
29	1.311	1.699	2.045	2.462	2.756	3.396	3.659
30	1.310	1.697	2.042	2.457	2.750	3.385	3.646
40	1.303	1.684	2.021	2.423	2.704	3.307	3.551
60	1.296	1.671	2.000	2.390	2.660	3.232	3.460
120	1.289	1.658	1.980	2.358	2.617	3.160	3.373
∞	1.282	1.645	1.960	2.326	2.576	3.090	3.291

This table was generated using Minitab

TABLE A.3. Chi-square distribution quantiles

ν	$Q(.005)$	$Q(.01)$	$Q(.025)$	$Q(.05)$	$Q(.1)$	$Q(.9)$	$Q(.95)$	$Q(.975)$	$Q(.99)$	$Q(.995)$
1	0.000	0.000	0.001	0.004	0.016	2.706	3.841	5.024	6.635	7.879
2	0.010	0.020	0.051	0.103	0.211	4.605	5.991	7.378	9.210	10.597
3	0.072	0.115	0.216	0.352	0.584	6.251	7.815	9.348	11.345	12.838
4	0.207	0.297	0.484	0.711	1.064	7.779	9.488	11.143	13.277	14.860
5	0.412	0.554	0.831	1.145	1.610	9.236	11.070	12.833	15.086	16.750
6	0.676	0.872	1.237	1.635	2.204	10.645	12.592	14.449	16.812	18.548
7	0.989	1.239	1.690	2.167	2.833	12.017	14.067	16.013	18.475	20.278
8	1.344	1.646	2.180	2.733	3.490	13.362	15.507	17.535	20.090	21.955
9	1.735	2.088	2.700	3.325	4.168	14.684	16.919	19.023	21.666	23.589
10	2.156	2.558	3.247	3.940	4.865	15.987	18.307	20.483	23.209	25.188
11	2.603	3.053	3.816	4.575	5.578	17.275	19.675	21.920	24.725	26.757
12	3.074	3.571	4.404	5.226	6.304	18.549	21.026	23.337	26.217	28.300
13	3.565	4.107	5.009	5.892	7.042	19.812	22.362	24.736	27.688	29.819
14	4.075	4.660	5.629	6.571	7.790	21.064	23.685	26.119	29.141	31.319
15	4.601	5.229	6.262	7.261	8.547	22.307	24.996	27.488	30.578	32.801
16	5.142	5.812	6.908	7.962	9.312	23.542	26.296	28.845	32.000	34.267
17	5.697	6.408	7.564	8.672	10.085	24.769	27.587	30.191	33.409	35.718
18	6.265	7.015	8.231	9.390	10.865	25.989	28.869	31.526	34.805	37.156
19	6.844	7.633	8.907	10.117	11.651	27.204	30.143	32.852	36.191	38.582
20	7.434	8.260	9.591	10.851	12.443	28.412	31.410	34.170	37.566	39.997
21	8.034	8.897	10.283	11.591	13.240	29.615	32.671	35.479	38.932	41.401
22	8.643	9.542	10.982	12.338	14.041	30.813	33.924	36.781	40.290	42.796
23	9.260	10.196	11.689	13.091	14.848	32.007	35.172	38.076	41.638	44.181
24	9.886	10.856	12.401	13.848	15.659	33.196	36.415	39.364	42.980	45.559
25	10.520	11.524	13.120	14.611	16.473	34.382	37.653	40.647	44.314	46.928
26	11.160	12.198	13.844	15.379	17.292	35.563	38.885	41.923	45.642	48.290
27	11.808	12.879	14.573	16.151	18.114	36.741	40.113	43.195	46.963	49.645
28	12.461	13.565	15.308	16.928	18.939	37.916	41.337	44.461	48.278	50.994
29	13.121	14.256	16.047	17.708	19.768	39.087	42.557	45.722	49.588	52.336
30	13.787	14.953	16.791	18.493	20.599	40.256	43.773	46.979	50.892	53.672
31	14.458	15.655	17.539	19.281	21.434	41.422	44.985	48.232	52.192	55.003
32	15.134	16.362	18.291	20.072	22.271	42.585	46.194	49.480	53.486	56.328
33	15.815	17.074	19.047	20.867	23.110	43.745	47.400	50.725	54.775	57.648
34	16.501	17.789	19.806	21.664	23.952	44.903	48.602	51.966	56.061	58.964
35	17.192	18.509	20.569	22.465	24.797	46.059	49.802	53.204	57.342	60.275
36	17.887	19.233	21.336	23.269	25.643	47.212	50.998	54.437	58.619	61.581
37	18.586	19.960	22.106	24.075	26.492	48.364	52.192	55.668	59.893	62.885
38	19.289	20.691	22.878	24.884	27.343	49.513	53.384	56.896	61.163	64.183
39	19.996	21.426	23.654	25.695	28.196	50.660	54.572	58.120	62.429	65.477
40	20.707	22.164	24.433	26.509	29.051	51.805	55.759	59.342	63.691	66.767

This table was generated using Minitab. For $\nu > 40$, the approximation $Q(p) \approx \nu \left(1 - \frac{2}{9\nu} + Q_z(p)\sqrt{\frac{2}{9\nu}}\right)^3$ can be used

TABLE A.4.1. F distribution .90 quantiles

ν_2 Denominator Degrees of Freedom	\multicolumn{19}{c}{ν_1 (Numerator degrees of freedom)}
	1	2	3	4	5	6	7	8	9	10	12	15	20	24	30	40	60	120	∞
1	39.86	49.50	53.59	55.84	57.24	58.20	58.90	59.44	59.85	60.20	60.70	61.22	61.74	62.00	62.27	62.53	62.79	63.05	63.33
2	8.53	9.00	9.16	9.24	9.29	9.33	9.35	9.37	9.38	9.39	9.41	9.42	9.44	9.45	9.46	9.47	9.47	9.48	9.49
3	5.54	5.46	5.39	5.34	5.31	5.28	5.27	5.25	5.24	5.23	5.22	5.20	5.18	5.18	5.17	5.16	5.15	5.14	5.13
4	4.54	4.32	4.19	4.11	4.05	4.01	3.98	3.95	3.94	3.92	3.90	3.87	3.84	3.83	3.82	3.80	3.79	3.78	3.76
5	4.06	3.78	3.62	3.52	3.45	3.40	3.37	3.34	3.32	3.30	3.27	3.24	3.21	3.19	3.17	3.16	3.14	3.12	3.10
6	3.78	3.46	3.29	3.18	3.11	3.05	3.01	2.98	2.96	2.94	2.90	2.87	2.84	2.82	2.80	2.78	2.76	2.74	2.72
7	3.59	3.26	3.07	2.96	2.88	2.83	2.78	2.75	2.72	2.70	2.67	2.63	2.59	2.58	2.56	2.54	2.51	2.49	2.47
8	3.46	3.11	2.92	2.81	2.73	2.67	2.62	2.59	2.56	2.54	2.50	2.46	2.42	2.40	2.38	2.36	2.34	2.32	2.29
9	3.36	3.01	2.81	2.69	2.61	2.55	2.51	2.47	2.44	2.42	2.38	2.34	2.30	2.28	2.25	2.23	2.21	2.18	2.16
10	3.28	2.92	2.73	2.61	2.52	2.46	2.41	2.38	2.35	2.32	2.28	2.24	2.20	2.18	2.16	2.13	2.11	2.08	2.06
11	3.23	2.86	2.66	2.54	2.45	2.39	2.34	2.30	2.27	2.25	2.21	2.17	2.12	2.10	2.08	2.05	2.03	2.00	1.97
12	3.18	2.81	2.61	2.48	2.39	2.33	2.28	2.24	2.21	2.19	2.15	2.10	2.06	2.04	2.01	1.99	1.96	1.93	1.90
13	3.14	2.76	2.56	2.43	2.35	2.28	2.23	2.20	2.16	2.14	2.10	2.05	2.01	1.98	1.96	1.93	1.90	1.88	1.85
14	3.10	2.73	2.52	2.39	2.31	2.24	2.19	2.15	2.12	2.10	2.05	2.01	1.96	1.94	1.91	1.89	1.86	1.83	1.80
15	3.07	2.70	2.49	2.36	2.27	2.21	2.16	2.12	2.09	2.06	2.02	1.97	1.92	1.90	1.87	1.85	1.82	1.79	1.76
16	3.05	2.67	2.46	2.33	2.24	2.18	2.13	2.09	2.06	2.03	1.99	1.94	1.89	1.87	1.84	1.81	1.78	1.75	1.72
17	3.03	2.64	2.44	2.31	2.22	2.15	2.10	2.06	2.03	2.00	1.96	1.91	1.86	1.84	1.81	1.78	1.75	1.72	1.69
18	3.01	2.62	2.42	2.29	2.20	2.13	2.08	2.04	2.00	1.98	1.93	1.89	1.84	1.81	1.78	1.75	1.72	1.69	1.66
19	2.99	2.61	2.40	2.27	2.18	2.11	2.06	2.02	1.98	1.96	1.91	1.86	1.81	1.79	1.76	1.73	1.70	1.67	1.63
20	2.97	2.59	2.38	2.25	2.16	2.09	2.04	2.00	1.96	1.94	1.89	1.84	1.79	1.77	1.74	1.71	1.68	1.64	1.61
21	2.96	2.57	2.36	2.23	2.14	2.08	2.02	1.98	1.95	1.92	1.87	1.83	1.78	1.75	1.72	1.69	1.66	1.62	1.59
22	2.95	2.56	2.35	2.22	2.13	2.06	2.01	1.97	1.93	1.90	1.86	1.81	1.76	1.73	1.70	1.67	1.64	1.60	1.57
23	2.94	2.55	2.34	2.21	2.11	2.05	1.99	1.95	1.92	1.89	1.84	1.80	1.74	1.72	1.69	1.66	1.62	1.59	1.55
24	2.93	2.54	2.33	2.19	2.10	2.04	1.98	1.94	1.91	1.88	1.83	1.78	1.73	1.70	1.67	1.64	1.61	1.57	1.53
25	2.92	2.53	2.32	2.18	2.09	2.02	1.97	1.93	1.89	1.87	1.82	1.77	1.72	1.69	1.66	1.63	1.59	1.56	1.52

26	2.91	2.52	2.31	2.17	2.08	2.01	1.96	1.92	1.88	1.86	1.81	1.76	1.71	1.68	1.65	1.61	1.58	1.54	1.50
27	2.90	2.51	2.30	2.17	2.07	2.00	1.95	1.91	1.87	1.85	1.80	1.75	1.70	1.67	1.64	1.60	1.57	1.53	1.49
28	2.89	2.50	2.29	2.16	2.06	2.00	1.94	1.90	1.87	1.84	1.79	1.74	1.69	1.66	1.63	1.59	1.56	1.52	1.48
29	2.89	2.50	2.28	2.15	2.06	1.99	1.93	1.89	1.86	1.83	1.78	1.73	1.68	1.65	1.62	1.58	1.55	1.51	1.47
30	2.88	2.49	2.28	2.14	2.05	1.98	1.93	1.88	1.85	1.82	1.77	1.72	1.67	1.64	1.61	1.57	1.54	1.50	1.46
40	2.84	2.44	2.23	2.09	2.00	1.93	1.87	1.83	1.79	1.76	1.71	1.66	1.61	1.57	1.54	1.51	1.47	1.42	1.38
60	2.79	2.39	2.18	2.04	1.95	1.87	1.82	1.77	1.74	1.71	1.66	1.60	1.54	1.51	1.48	1.44	1.40	1.35	1.29
120	2.75	2.35	2.13	1.99	1.90	1.82	1.77	1.72	1.68	1.65	1.60	1.55	1.48	1.45	1.41	1.37	1.32	1.26	1.19
∞	2.71	2.30	2.08	1.94	1.85	1.77	1.72	1.67	1.63	1.60	1.55	1.49	1.42	1.38	1.34	1.30	1.24	1.17	1.00

This table was generated using Minitab

TABLE A.4.2. F distribution .95 quantiles

ν_2 Denominator Degrees of Freedom	ν_1 (Numerator degrees of freedom)																			
	1	2	3	4	5	6	7	8	9	10	12	15	20	24	30	40	60	120	∞
1	161.44	199.50	215.69	224.57	230.16	233.98	236.78	238.89	240.55	241.89	243.91	245.97	248.02	249.04	250.07	251.13	252.18	253.27	254.31
2	18.51	19.00	19.16	19.25	19.30	19.33	19.35	19.37	19.39	19.40	19.41	19.43	19.45	19.45	19.46	19.47	19.48	19.49	19.50
3	10.13	9.55	9.28	9.12	9.01	8.94	8.89	8.85	8.81	8.79	8.74	8.70	8.66	8.64	8.62	8.59	8.57	8.55	8.53
4	7.71	6.94	6.59	6.39	6.26	6.16	6.09	6.04	6.00	5.96	5.91	5.86	5.80	5.77	5.75	5.72	5.69	5.66	5.63
5	6.61	5.79	5.41	5.19	5.05	4.95	4.88	4.82	4.77	4.74	4.68	4.62	4.56	4.53	4.50	4.46	4.43	4.40	4.36
6	5.99	5.14	4.76	4.53	4.39	4.28	4.21	4.15	4.10	4.06	4.00	3.94	3.87	3.84	3.81	3.77	3.74	3.70	3.67
7	5.59	4.74	4.35	4.12	3.97	3.87	3.79	3.73	3.68	3.64	3.57	3.51	3.44	3.41	3.38	3.34	3.30	3.27	3.23
8	5.32	4.46	4.07	3.84	3.69	3.58	3.50	3.44	3.39	3.35	3.28	3.22	3.15	3.12	3.08	3.04	3.01	2.97	2.93
9	5.12	4.26	3.86	3.63	3.48	3.37	3.29	3.23	3.18	3.14	3.07	3.01	2.94	2.90	2.86	2.83	2.79	2.75	2.71
10	4.96	4.10	3.71	3.48	3.33	3.22	3.14	3.07	3.02	2.98	2.91	2.85	2.77	2.74	2.70	2.66	2.62	2.58	2.54
11	4.84	3.98	3.59	3.36	3.20	3.09	3.01	2.95	2.90	2.85	2.79	2.72	2.65	2.61	2.57	2.53	2.49	2.45	2.40
12	4.75	3.89	3.49	3.26	3.11	3.00	2.91	2.85	2.80	2.75	2.69	2.62	2.54	2.51	2.47	2.43	2.38	2.34	2.30
13	4.67	3.81	3.41	3.18	3.03	2.92	2.83	2.77	2.71	2.67	2.60	2.53	2.46	2.42	2.38	2.34	2.30	2.25	2.21
14	4.60	3.74	3.34	3.11	2.96	2.85	2.76	2.70	2.65	2.60	2.53	2.46	2.39	2.35	2.31	2.27	2.22	2.18	2.13
15	4.54	3.68	3.29	3.06	2.90	2.79	2.71	2.64	2.59	2.54	2.48	2.40	2.33	2.29	2.25	2.20	2.16	2.11	2.07
16	4.49	3.63	3.24	3.01	2.85	2.74	2.66	2.59	2.54	2.49	2.42	2.35	2.28	2.24	2.19	2.15	2.11	2.06	2.01
17	4.45	3.59	3.20	2.96	2.81	2.70	2.61	2.55	2.49	2.45	2.38	2.31	2.23	2.19	2.15	2.10	2.06	2.01	1.96
18	4.41	3.55	3.16	2.93	2.77	2.66	2.58	2.51	2.46	2.41	2.34	2.27	2.19	2.15	2.11	2.06	2.02	1.97	1.92
19	4.38	3.52	3.13	2.90	2.74	2.63	2.54	2.48	2.42	2.38	2.31	2.23	2.16	2.11	2.07	2.03	1.98	1.93	1.88
20	4.35	3.49	3.10	2.87	2.71	2.60	2.51	2.45	2.39	2.35	2.28	2.20	2.12	2.08	2.04	1.99	1.95	1.90	1.84
21	4.32	3.47	3.07	2.84	2.68	2.57	2.49	2.42	2.37	2.32	2.25	2.18	2.10	2.05	2.01	1.96	1.92	1.87	1.81
22	4.30	3.44	3.05	2.82	2.66	2.55	2.46	2.40	2.34	2.30	2.23	2.15	2.07	2.03	1.98	1.94	1.89	1.84	1.78
23	4.28	3.42	3.03	2.80	2.64	2.53	2.44	2.37	2.32	2.27	2.20	2.13	2.05	2.01	1.96	1.91	1.86	1.81	1.76
24	4.26	3.40	3.01	2.78	2.62	2.51	2.42	2.36	2.30	2.25	2.18	2.11	2.03	1.98	1.94	1.89	1.84	1.79	1.73
25	4.24	3.39	2.99	2.76	2.60	2.49	2.40	2.34	2.28	2.24	2.16	2.09	2.01	1.96	1.92	1.87	1.82	1.77	1.71

26	4.23	3.37	2.98	2.74	2.59	2.47	2.39	2.32	2.27	2.22	2.15	2.07	1.99	1.95	1.90	1.85	1.80	1.75	1.69
27	4.21	3.35	2.96	2.73	2.57	2.46	2.37	2.31	2.25	2.20	2.13	2.06	1.97	1.93	1.88	1.84	1.79	1.73	1.67
28	4.20	3.34	2.95	2.71	2.56	2.45	2.36	2.29	2.24	2.19	2.12	2.04	1.96	1.91	1.87	1.82	1.77	1.71	1.65
29	4.18	3.33	2.93	2.70	2.55	2.43	2.35	2.28	2.22	2.18	2.10	2.03	1.94	1.90	1.85	1.81	1.75	1.70	1.64
30	4.17	3.32	2.92	2.69	2.53	2.42	2.33	2.27	2.21	2.16	2.09	2.01	1.93	1.89	1.84	1.79	1.74	1.68	1.62
40	4.08	3.23	2.84	2.61	2.45	2.34	2.25	2.18	2.12	2.08	2.00	1.92	1.84	1.79	1.74	1.69	1.64	1.58	1.51
60	4.00	3.15	2.76	2.53	2.37	2.25	2.17	2.10	2.04	1.99	1.92	1.84	1.75	1.70	1.65	1.59	1.53	1.47	1.39
120	3.92	3.07	2.68	2.45	2.29	2.18	2.09	2.02	1.96	1.91	1.83	1.75	1.66	1.61	1.55	1.50	1.43	1.35	1.25
∞	3.84	3.00	2.60	2.37	2.21	2.10	2.01	1.94	1.88	1.83	1.75	1.67	1.57	1.52	1.46	1.39	1.32	1.22	1.00

This table was generated using Minitab

TABLE A.4.3. F distribution .99 quantiles

ν_2 Denominator Degrees of Freedom	\multicolumn{16}{c}{ν_1 (Numerator degrees of freedom)}															
	1	2	3	4	5	6	7	8	9	10	12	15	20	24	30	40	60	120	∞
1	4052	4999	5403	5625	5764	5859	5929	5981	6023	6055	6107	6157	6209	6235	6260	6287	6312	6339	6366
2	98.51	99.00	99.17	99.25	99.30	99.33	99.35	99.38	99.39	99.40	99.41	99.43	99.44	99.45	99.47	99.47	99.48	99.49	99.50
3	34.12	30.82	29.46	28.71	28.24	27.91	27.67	27.49	27.35	27.23	27.05	26.87	26.69	26.60	26.51	26.41	26.32	26.22	26.13
4	21.20	18.00	16.69	15.98	15.52	15.21	14.98	14.80	14.66	14.55	14.37	14.20	14.02	13.93	13.84	13.75	13.65	13.56	13.46
5	16.26	13.27	12.06	11.39	10.97	10.67	10.46	10.29	10.16	10.05	9.89	9.72	9.55	9.47	9.38	9.29	9.20	9.11	9.02
6	13.75	10.92	9.78	9.15	8.75	8.47	8.26	8.10	7.98	7.87	7.72	7.56	7.40	7.31	7.23	7.14	7.06	6.97	6.88
7	12.25	9.55	8.45	7.85	7.46	7.19	6.99	6.84	6.72	6.62	6.47	6.31	6.16	6.07	5.99	5.91	5.82	5.74	5.65
8	11.26	8.65	7.59	7.01	6.63	6.37	6.18	6.03	5.91	5.81	5.67	5.52	5.36	5.28	5.20	5.12	5.03	4.95	4.86
9	10.56	8.02	6.99	6.42	6.06	5.80	5.61	5.47	5.35	5.26	5.11	4.96	4.81	4.73	4.65	4.57	4.48	4.40	4.31
10	10.04	7.56	6.55	5.99	5.64	5.39	5.20	5.06	4.94	4.85	4.71	4.56	4.41	4.33	4.25	4.17	4.08	4.00	3.91
11	9.65	7.21	6.22	5.67	5.32	5.07	4.89	4.74	4.63	4.54	4.40	4.25	4.10	4.02	3.94	3.86	3.78	3.69	3.60
12	9.33	6.93	5.95	5.41	5.06	4.82	4.64	4.50	4.39	4.30	4.16	4.01	3.86	3.78	3.70	3.62	3.54	3.45	3.36
13	9.07	6.70	5.74	5.21	4.86	4.62	4.44	4.30	4.19	4.10	3.96	3.82	3.66	3.59	3.51	3.43	3.34	3.25	3.17
14	8.86	6.51	5.56	5.04	4.69	4.46	4.28	4.14	4.03	3.94	3.80	3.66	3.51	3.43	3.35	3.27	3.18	3.09	3.00
15	8.68	6.36	5.42	4.89	4.56	4.32	4.14	4.00	3.89	3.80	3.67	3.52	3.37	3.29	3.21	3.13	3.05	2.96	2.87
16	8.53	6.23	5.29	4.77	4.44	4.20	4.03	3.89	3.78	3.69	3.55	3.41	3.26	3.18	3.10	3.02	2.93	2.84	2.75
17	8.40	6.11	5.18	4.67	4.34	4.10	3.93	3.79	3.68	3.59	3.46	3.31	3.16	3.08	3.00	2.92	2.83	2.75	2.65
18	8.29	6.01	5.09	4.58	4.25	4.01	3.84	3.71	3.60	3.51	3.37	3.23	3.08	3.00	2.92	2.84	2.75	2.66	2.57
19	8.19	5.93	5.01	4.50	4.17	3.94	3.77	3.63	3.52	3.43	3.30	3.15	3.00	2.92	2.84	2.76	2.67	2.58	2.49
20	8.10	5.85	4.94	4.43	4.10	3.87	3.70	3.56	3.46	3.37	3.23	3.09	2.94	2.86	2.78	2.69	2.61	2.52	2.42
21	8.02	5.78	4.87	4.37	4.04	3.81	3.64	3.51	3.40	3.31	3.17	3.03	2.88	2.80	2.72	2.64	2.55	2.46	2.36
22	7.95	5.72	4.82	4.31	3.99	3.76	3.59	3.45	3.35	3.26	3.12	2.98	2.83	2.75	2.67	2.58	2.50	2.40	2.31
23	7.88	5.66	4.76	4.26	3.94	3.71	3.54	3.41	3.30	3.21	3.07	2.93	2.78	2.70	2.62	2.54	2.45	2.35	2.26
24	7.82	5.61	4.72	4.22	3.90	3.67	3.50	3.36	3.26	3.17	3.03	2.89	2.74	2.66	2.58	2.49	2.40	2.31	2.21
25	7.77	5.57	4.68	4.18	3.85	3.63	3.46	3.32	3.22	3.13	2.99	2.85	2.70	2.62	2.54	2.45	2.36	2.27	2.17

26	7.72	5.53	4.64	4.14	3.82	3.59	3.42	3.29	3.18	3.09	2.96	2.81	2.66	2.58	2.50	2.42	2.33	2.23	2.13
27	7.68	5.49	4.60	4.11	3.78	3.56	3.39	3.26	3.15	3.06	2.93	2.78	2.63	2.55	2.47	2.38	2.29	2.20	2.10
28	7.64	5.45	4.57	4.07	3.75	3.53	3.36	3.23	3.12	3.03	2.90	2.75	2.60	2.52	2.44	2.35	2.26	2.17	2.06
29	7.60	5.42	4.54	4.04	3.73	3.50	3.33	3.20	3.09	3.00	2.87	2.73	2.57	2.49	2.41	2.33	2.23	2.14	2.03
30	7.56	5.39	4.51	4.02	3.70	3.47	3.30	3.17	3.07	2.98	2.84	2.70	2.55	2.47	2.39	2.30	2.21	2.11	2.01
40	7.31	5.18	4.31	3.83	3.51	3.29	3.12	2.99	2.89	2.80	2.66	2.52	2.37	2.29	2.20	2.11	2.02	1.92	1.80
60	7.08	4.98	4.13	3.65	3.34	3.12	2.95	2.82	2.72	2.63	2.50	2.35	2.20	2.12	2.03	1.94	1.84	1.73	1.60
120	6.85	4.79	3.95	3.48	3.17	2.96	2.79	2.66	2.56	2.47	2.34	2.19	2.03	1.95	1.86	1.76	1.66	1.53	1.38
∞	6.63	4.61	3.78	3.32	3.02	2.80	2.64	2.51	2.41	2.32	2.18	2.04	1.88	1.79	1.70	1.59	1.47	1.32	1.00

This table was generated using Minitab

TABLE A.5. Control chart constants

m	d_2	d_3	c_4	A_2	A_3	B_3	B_4	B_5	B_6	D_1	D_2	D_3	D_4
2	1.128	0.853	0.7979	1.880	2.659		3.267		2.606		3.686		3.267
3	1.693	0.888	0.8862	1.023	1.954		2.568		2.276		4.358		2.575
4	2.059	0.880	0.9213	0.729	1.628		2.266		2.088		4.698		2.282
5	2.326	0.864	0.9400	0.577	1.427		2.089		1.964		4.918		2.114
6	2.534	0.848	0.9515	0.483	1.287	0.030	1.970	0.029	1.874		5.079		2.004
7	2.704	0.833	0.9594	0.419	1.182	0.118	1.882	0.113	1.806	0.205	5.204	0.076	1.924
8	2.847	0.820	0.9650	0.373	1.099	0.185	1.815	0.179	1.751	0.388	5.307	0.136	1.864
9	2.970	0.808	0.9693	0.337	1.032	0.239	1.761	0.232	1.707	0.547	5.394	0.184	1.816
10	3.078	0.797	0.9727	0.308	0.975	0.284	1.716	0.276	1.669	0.686	5.469	0.223	1.777
11	3.173	0.787	0.9754	0.285	0.927	0.321	1.679	0.313	1.637	0.811	5.535	0.256	1.744
12	3.258	0.778	0.9776	0.266	0.886	0.354	1.646	0.346	1.610	0.923	5.594	0.283	1.717
13	3.336	0.770	0.9794	0.249	0.850	0.382	1.618	0.374	1.585	1.025	5.647	0.307	1.693
14	3.407	0.763	0.9810	0.235	0.817	0.406	1.594	0.399	1.563	1.118	5.696	0.328	1.672
15	3.472	0.756	0.9823	0.223	0.789	0.428	1.572	0.421	1.544	1.203	5.740	0.347	1.653
20	3.735	0.729	0.9869	0.180	0.680	0.510	1.490	0.504	1.470	1.549	5.921	0.415	1.585
25	3.931	0.708	0.9896	0.153	0.606	0.565	1.435	0.559	1.420	1.806	6.056	0.459	1.541

This table was computed using Mathcad

TABLE A.6.1. Factors for two-sided tolerance intervals for normal distributions

n	95 % Confidence			99 % Confidence		
	$p = .90$	$p = .95$	$p = .99$	$p = .90$	$p = .95$	$p = .99$
2		36.519	46.944		182.720	234.877
3	8.306	9.789	12.647	18.782	22.131	28.586
4	5.368	6.341	8.221	9.416	11.118	14.405
5	4.291	5.077	6.598	6.655	7.870	10.220
6	3.733	4.422	5.758	5.383	6.373	8.292
7	3.390	4.020	5.241	4.658	5.520	7.191
8	3.156	3.746	4.889	4.189	4.968	6.479
9	2.986	3.546	4.633	3.860	4.581	5.980
10	2.856	3.393	4.437	3.617	4.294	5.610
11	2.754	3.273	4.282	3.429	4.073	5.324
12	2.670	3.175	4.156	3.279	3.896	5.096
13	2.601	3.093	4.051	3.156	3.751	4.909
14	2.542	3.024	3.962	3.054	3.631	4.753
15	2.492	2.965	3.885	2.967	3.529	4.621
16	2.449	2.913	3.819	2.893	3.441	4.507
17	2.410	2.868	3.761	2.828	3.364	4.408
18	2.376	2.828	3.709	2.771	3.297	4.321
19	2.346	2.793	3.663	2.720	3.237	4.244
20	2.319	2.760	3.621	2.675	3.184	4.175
25	2.215	2.638	3.462	2.506	2.984	3.915
30	2.145	2.555	3.355	2.394	2.851	3.742
35	2.094	2.495	3.276	2.314	2.756	3.618
40	2.055	2.448	3.216	2.253	2.684	3.524
50	1.999	2.382	3.129	2.166	2.580	3.390
60	1.960	2.335	3.068	2.106	2.509	3.297
80	1.908	2.274	2.987	2.028	2.416	3.175
100	1.875	2.234	2.936	1.978	2.357	3.098
150	1.826	2.176	2.859	1.906	2.271	2.985
200	1.798	2.143	2.816	1.866	2.223	2.921
500	1.737	2.070	2.721	1.777	2.117	2.783
1000	1.709	2.036	2.676	1.736	2.068	2.718
∞	1.645	1.960	2.576	1.645	1.960	2.576

This table was computed using Mathcad

TABLE A.6.2. Factors for one-sided tolerance intervals for normal distributions

	95 % Confidence			99 % Confidence		
n	$p = .90$	$p = .95$	$p = .99$	$p = .90$	$p = .95$	$p = .99$
2						
3	6.155	7.656	10.553	14.006	17.372	23.896
4	4.162	5.144	7.042	7.380	9.083	12.388
5	3.407	4.203	5.741	5.362	6.578	8.939
6	3.006	3.708	5.062	4.411	5.406	7.335
7	2.755	3.399	4.642	3.859	4.728	6.412
8	2.582	3.187	4.354	3.497	4.285	5.812
9	2.454	3.031	4.143	3.240	3.972	5.389
10	2.355	2.911	3.981	3.048	3.738	5.074
11	2.275	2.815	3.852	2.898	3.556	4.829
12	2.210	2.736	3.747	2.777	3.410	4.633
13	2.155	2.671	3.659	2.677	3.290	4.472
14	2.109	2.614	3.585	2.593	3.189	4.337
15	2.068	2.566	3.520	2.521	3.102	4.222
16	2.033	2.524	3.464	2.459	3.028	4.123
17	2.002	2.486	3.414	2.405	2.963	4.037
18	1.974	2.453	3.370	2.357	2.905	3.960
19	1.949	2.423	3.331	2.314	2.854	3.892
20	1.926	2.396	3.295	2.276	2.808	3.832
25	1.838	2.292	3.158	2.129	2.633	3.601
30	1.777	2.220	3.064	2.030	2.515	3.447
35	1.732	2.167	2.995	1.957	2.430	3.334
40	1.697	2.125	2.941	1.902	2.364	3.249
50	1.646	2.065	2.862	1.821	2.269	3.125
60	1.609	2.022	2.807	1.764	2.202	3.038
80	1.559	1.964	2.733	1.688	2.114	2.924
100	1.527	1.927	2.684	1.639	2.056	2.850
150	1.478	1.870	2.611	1.566	1.971	2.740
200	1.450	1.837	2.570	1.524	1.923	2.679
500	1.385	1.763	2.475	1.430	1.814	2.540
1000	1.354	1.727	2.430	1.385	1.762	2.475
∞	1.282	1.645	2.326	1.282	1.645	2.326

This table was computed using Mathcad

BIBLIOGRAPHY

[1] Anand, K. N., Bhadkamkar, S. M., & Moghe, R. (1994–1995). "Wet method of chemical analysis of cast iron: upgrading accuracy and precision through experimental design." *Quality Engineering, 7*(1), 187–208.

[2] Bisgaard, S. (1994). "Blocking generators for small 2^{k-p} designs." *Journal of Quality Technology, 26*(4), 288–296.

[3] Bisgaard, S., & Fuller, H. T. (1995–1996). "Reducing variation with two-level factorial experiments." *Quality Engineering, 8*(2), 373–377.

[4] Box, G. E. P., & Draper, N. R. (1969). *Evolutionary operation.* New York: Wiley.

[5] Box, G. E. P., & Draper, N. R. (1986). *Empirical model building and response surfaces.* New York: Wiley.

[6] Box, G. E. P., Hunter, W. G., & Hunter, J. S. (1978). *Statistics for experimenters.* New York: Wiley.

[7] Brassard, M. (Ed.). (1984). *Proceedings–case study seminar–Dr. Deming's management methods: How they are being implemented in the U.S. and abroad.* Andover, MA: Growth Opportunity Alliance of Lawrence.

[8] Brezler, P. (1986). "Statistical analysis: Mack Truck gear heat treat experiments." *Heat Treating, 18*(11), 26–29.

[9] Brown, D. S., Turner, W. R., & Smith, A. C. (1958). "Sealing strength of wax-polyethylene blends." *Tappi, 41*(6), 295–300.

[10] Burdick, R. K., & Graybill, F. A. (1992). *Confidence intervals on variance components.* New York: Marcel Dekker.

[11] Burns, W. L. (1989–90). "Quality control proves itself in assembly." *Quality Engineering, 2*(1), 91–101.

[12] Burr, I. W. (1953). *Engineering statistics and quality control.* New York: McGraw-Hill.

[13] Burr, I. W. (1979). *Elementary statistical quality control.* New York: Marcel Dekker.

[14] Champ, C. W., & Woodall, W. H. (1987). "Exact results for shewhart control charts with supplementary runs rules." *Technometrics, 29*(4), 393–399.

[15] Champ, C. W., & Woodall, W. H. (1990). "A program to evaluate the run length distribution of a shewhart control chart with supplementary runs rules." *Journal of Quality Technology, 22*(1), 68–73.

[16] Chau, K. W., & Kelley, W. R. (1993). "Formulating printable coatings via d-optimality." *Journal of Coatings Technology, 65*(821), 71–78.

[17] Currie, L. A. (1968). "Limits for qualitative detection and quantitative determination." *Analytical Chemistry, 40*(3), 586–593.

[18] Crowder, S. V., Jensen, K. L., Stephenson, W. R., & Vardeman, S. B. (1988). "An interactive program for the analysis of data from two-level factorial experiments via probability plotting." *Journal of Quality Technology, 20*(2), 140–148.

[19] Duncan, A. J. (1986). *Quality control and industrial statistics* (5th ed.). Homewood, IL: Irwin.

[20] Eibl, S., Kess, U., & Pukelsheim, F. (1992). "Achieving a target value for a manufacturing process: a case study." *Journal of Quality Technology, 24*(1), 22–26.

[21] Ermer, D. S., & Hurtis, G. M. (1995–1996). "Advanced SPC for higher-quality electronic card manufacturing." *Quality Engineering, 8*(2), 283–299.

[22] Grego, J. M. (1993). "Generalized linear models and process variation." *Journal of Quality Technology, 25*(4), 288–295.

[23] Hahn, J. G., & Meeker, W. Q. (1991). *Statistical intervals: a guide for practitioners.* New York: Wiley.

[24] Hendrix, C. D. (1979). "What every technologist should know about experimental design." *Chemical Technology, 9*(3), 167–174.

[25] Hill, W. J., & Demler, W. R. (1970). "More on planning experiments to increase research efficiency." *Industrial and Engineering Chemistry, 62*(10), 60–65.

[26] Kolarik, W. J. (1995). *Creating quality: concepts, systems, strategies and tools.* New York: McGraw-Hill.

[27] Lawson, J. S. (1990–1991). "Improving a chemical process through use of a designed experiment." *Quality Engineering, 3*(2), 215–235.

[28] Lawson, J. S., & Madrigal, J. L. (1994). "Robust design through optimization techniques." *Quality Engineering, 6*(4), 593–608.

[29] Leigh, H. D., & Taylor, T. D. (1990). "Computer-generated experimental designs." *Ceramic Bulletin, 69*(1), 100–106.

[30] Lochner, R. H., & Matar, J. E. (1990). *Designing for quality: an introduction to the best of Taguchi and Western methods of statistical experimental design.* London and New York: Chapman and Hall.

[31] Mielnik, E. M. (1993–1994). "Design of a metal-cutting drilling experiment: a discrete two-variable problem." *Quality Engineering, 6*(1), 71–98.

[32] Miller, A., Sitter, R. R., Wu, C. F. J., & Long, D. (1993–1994). "Are large taguchi-style experiments necessary? A reanalysis of gear and pinion data." *Quality Engineering, 6*(1), 21–37.

[33] Moen, R. D., Nolan, T. W., & Provost, L. P. (1991). *Improving quality through planned experimentation.* New York: McGraw-Hill.

[34] Myers, R. H. (1976). *Response surface methodology.* Ann Arbor: Edwards Brothers.

[35] Nair, V. N. (Ed.) (1992). "Taguchi's parameter design: a panel discussion." *Technometrics, 34*(2), 127–161.

[36] Nelson, L. S. (1984). "The Shewhart control chart-tests for special causes." *Journal of Quality Technology, 16*(4), 237–239.

[37] Neter, J., Kutner, M. H., Nachtsheim, C. J., Wasserman, W. (1996). *Applied linear statistical models* (4th ed.). Chicago: Irwin.

[38] Ophir, S., El-Gad, U., & Snyder, M. (1988). "A case study of the use of an experimental design in preventing shorts in nickel-cadmium cells." *Journal of Quality Technology, 20*(1), 44–50.

[39] Quinlan, J. (1985). "Product improvement by application of Taguchi methods." *American Supplier Institute News* (special symposium ed., pp. 11–16). Dearborn, MI: American Supplier Institute.

[40] Ranganathan, R., Chowdhury, K. K., & Seksaria, A. (1992). "Design evaluation for reduction in performance variation of TV electron guns." *Quality Engineering, 4*(3), 357–369.

[41] Schneider, H., Kasperski, W. J., & Weissfeld, L. (1993). "Finding significant effects for unreplicated fractional factorials using the n smallest contrasts." *Journal of Quality Technology, 25*(1), 18–27.

[42] Sirvanci, M. B., & Durmaz, M. (1993). "Variation reduction by the use of designed experiments." *Quality Engineering, 5*(4), 611–618.

[43] Snee, R. D. (1985). "Computer-aided design of experiments: some practical experiences." *Journal of Quality Technology, 17*(4), 222–236.

[44] Snee, R. D. (1985). "Experimenting with a large number of variables," in *Experiments in industry: design, analysis and interpretation of results* (pp. 25–35). Milwaukee: American Society for Quality Control.

[45] Snee, R. D., Hare, L. B., & Trout, J. R. (Eds.) (1985). *Experiments in industry: design, analysis and interpretation of results*. Milwaukee: American Society for Quality Control.

[46] Sutter, J. K., Jobe, J. M., & Crane, E. (1995). "Isothermal aging of polyimide resins." *Journal of Applied Polymer Science, 57*(12), 1491–1499.

[47] Taguchi, G., & Wu, Y. (1980). *Introduction to off-line quality control*. Nagoya: Japan Quality Control Organization.

[48] Tomlinson, W. J., & Cooper, G. A. (1986). "Fracture mechanism of Brass/Sn-Pb-Sb solder joints and the effect of production variables on the joint strength." *Journal of Materials Science, 21*(5), 1730–1734.

[49] Vander Wiel, S. A., & Vardeman, S. B. (1994). "A discussion of all-or-none inspection policies." *Technometrics, 36*(1), 102–109.

[50] Vardeman, S. B. (1986). "The legitimate role of inspection in modern SQC." *The American Statistician, 40*(4), 325–328.

[51] Vardeman, S. B. (1994). *Statistics for engineering problem solving*. Boston: PWS Publishing.

[52] Vardeman, S. B., & Jobe, J. M. (2001). *Basic engineering data collection and analysis*. Pacific Gove, CA: Duxbury/Thomsan Learning.

[53] Walpole, R. E., & Myers, R. H. (1993). *Probability and statistics for engineers and scientists* (5th ed.). New York: Macmillan.

[54] Wernimont, G. (1989–1990). "Statistical quality control in the chemical laboratory." *Quality Engineering, 2*(1), 59–72.

[55] Western Electric Company. (1984). *Statistical quality control handbook* (2nd ed.). New York: Western Electric Company.

[56] Zwickl, R. D. (1985). "An example of analysis of means for attribute data applied to a 2^4 factorial design." *ASQC electronics division technical supplement*, Issue 4. Milwaukee: American Society for Quality Control.

INDEX

A

A_2 (Shewhart charts), 116, 117, 160
A_3 (Shewhart charts), 117, 160
Accuracy, 18, 34, 35, 90, 240, 241
    of measurements, 34, 35
Aircraft wing three-surface
    configuration study,
    356–358
Alarm rules, 136–144, 148, 160,
    161, 167
    Shewhart control charts, 10,
    113, 128, 139, 160
Aliased main effects, 347
Alias structure, 336, 338, 340,
    344–348
    for fractional factorials
        2^{p-q} studies, 338, 345–348
        2^{p-1} studies, 338, 340–344
Alumina powder packing
    properties, 280–281
Analysis of variance (ANOVA)
    one-way methods contrasted,
    252–260
    two-way and gage R&R, 69,
    234
ANOVA. *See* Analysis of variance
    (ANOVA)
ARLs. *See* Average run lengths
As past data scenario, Shewhart
    control charts, 110
Assignable cause variation, 108
Attributes data, 128–135, 160
    Shewhart charts, 128–135, 160
Average run lengths (ARLs),
    144–150
    Shewhart charts, 109, 145, 147
Axial points, 364
    with 2^p factorial, 364

B

B_4 (Shewhart charts), 121, 122,
    124, 160
B_5 (Shewhart charts), 122, 124, 160
B_6 (Shewhart charts), 122, 124, 160
Balanced data, 272, 274, 275,
    297–301

Balanced data (cont.)
2^p factorials, 275, 297–299
Balanced data two-way factorials, 262–266
Baseline variation, 108
Benchmarking, 7, 8, 11
Between group, 56
Bimodality, 19
Black box process, 252
Blocking, 313, 379, 380
Blocks, 196, 313, 379, 380
Bolt shanks, 92, 93
Bond strength, integrated circuits, 393–394
Bonferroni's inequality, 259, 308, 314, 317, 318, 324
Box-Behnken design, 364
Box plots, 192, 194–199
Bridgeport numerically controlled milling machine, 240–241
Brinell hardness measurements, 35
Brush ferrules, 28, 318–320
Bunching, 140, 141
    in control charts, 140, 141

C

c_4 (Shewhart charts), 418,
c_5 (Shewhart charts), 418,
Calibration, 6, 10, 33, 35, 37–39, 75–79, 87, 89, 90, 140
Capability analysis sheet, 206
Capability index C_{pk}, 210, 212,
Capability measures, 207–211
Capability ration C_p, 207, 209, 211,
Carbon atmospheric blank, 188
Casehardening in red oak, 238–239
Cast iron carbon content, 188
Catalyst development fractional factorial study, 335, 343
Cause-and-effect diagrams, 14

c charts, 132, 133, 147, 148
    average run lengths, 144
    formulas, 144
Center line, 110, 114, 119–123, 129–134, 137, 141, 142
    Shewhart control charts, 110
Central composite designs, 365, 366, 375
Check sheets, 16–18
Chemical process improvement fractional factorial study, 137–138
Chemical process yield studies, 401–402
Chi-square distribution, 411
    quantiles, 411
Clean room process improvement, 4–6
CNC lathe turning, 29
Cold crack resistance fractional factorial study, 334
Collator machine stoppage rate, 328–331
Color index fractional factorial study, 338, 340
Common causes, 112
    variation, 112
Complete factorial, 262, 301, 379
Computer locks, 96–97
Concomitant variables, 379
Confidence interval, 10, 38, 40, 42, 46, 56, 58, 68, 70, 76, 79, 84, 207, 208, 210, 214, 218, 252, 256–260, 266, 271, 272, 274, 331, 341, 386
Confidence limits
    for C_p, 209, 210,
    for C_{pk}, 210–211,
    for a difference in means, 258
    for a mean, 45
    for a mean difference, 45, 87
    for 2^p factorial effect, 289, 292

for $p_1 - p_2$, 85,
for a process standard
deviation, 55
for a ratio of two standard
deviations, 46
for a standard deviation, 71
Conforming, 80–82, 84, 128
Confounded main effects, 336
Continual improvement, 8, 27, 372
Contour plots, 357–362, 367, 368, 375, 400
Control. *See* Engineering control; Statistical process monitoring
Control charts, 6, 8, 10, 102, 108–110, 113, 115–117, 119–126, 128, 130, 135–143, 146, 149, 160, 161, 170–174, 176–178, 186–188, 243, 245, 403, 418. *See also* Shewhart control charts
  constants, 116, 117, 418
  count data Shewhart charts, 129
Control factors, 383
Control gain, PID control, 158
Control limits, 109–125, 129–134, 136–138, 142–146, 148. *See also* Three sigma control limits
  engineering specifications contrasted, 111
  Shewhart control charts, 128, 139, 143, 160
Corporate culture, TQM efforts at changing, 8
Counts, 85, 113, 128–135, 281, 282, 293
  Shewhart charts, 128–143
C_p, 207–212, 214, 242, 243, 245, 246

C_{pk}, 207, 210–212, 214–216, 233, 234, 242–246
Crack inspection, by magnaflux, 175
Cube plots, 285–287
Cumulative frequency ogive, 195
Curvature, 156, 363, 364, 367
  in response surfaces, 363
Customer orientation, 5
  of modern quality assurance, 4, 5
  of TQM, 7, 8
Cut-off machine, 20

D

D_1 (Shewhart charts), 418
D_2 (Shewhart charts), 418
d_2 (Shewhart charts), 418
D_3 (Shewhart charts), 418
d_3 (Shewhart charts), 418
D_4 (Shewhart charts), 418
Data collection, 10, 12, 15–18, 24, 25, 40, 47, 53, 54, 75, 111, 252, 260, 364, 379, 381
  principles, 6, 10
Dead time, 151–153, 155, 156
  in PID control, 151, 153
Deciles, 196
Defects, defining what constitutes, 15
Defining relations, 339–341, 345–347, 349
  for fractional factorials
    2^{p-q} studies, 334–352
    2^{p-1} studies, 337
Delay, in PID control, 151
Derivative gain, in PID control, 156
Destructive, 48
Device, 2, 34, 107, 281, 372
  bias, 42–46, 48, 50
  variability, 54
Difference in means, 258

Dimethyl phenanthrene
    atmospheric blank, 401
Discrete time PID control, 150–157
DMAIC, 9, 10
Documentation, in data collection, 15
Dot plots, 192–194, 199, 205, 239
Drilling depths, 244–245
Drilling experiment response
    surface study, 355
Dual in-line package bond pullouts, 281, 282, 293, 295
Dummy variables, 275, 297
Duncan's alarm rules, 142

E
Effect sparsity principle, 192, 335
Eigenvalues, 369–371
    of quadratic response function, 369–371
    principle, 370
Electrical discharge machining (EDM)
    hole angles, 211
    prediction/tolerance intervals, 213–220
    process capability analysis, 204
Electrical switches, 174–176
Electronic card assemblies, 186–187
Empowerment, 30
Engineering control, 107, 150, 151, 157–160
    statistical process monitoring contrasted, 157–158
Engineering specifications
    control limits contrasted, 111
    marking on histograms to communicate improvement needs, 21
Error bars, 264, 265, 308, 311, 319, 324, 329, 330

Error propagation formulas, 225–227
Evolutionary operation (EVOP) studies, 372–374
EVOP. *See* Evolutionary operation (EVOP) studies
Experimental design. *See also*
    Factorial designs;
    One-way methods;
    Response surface studies
    black box processes, 252
    issues in, 261
    sequential strategy, 364
    Taguchi methods, 381

F
Factorial designs. *See also* 2^p factorials; 2^{p-q} factorials;
    Threeway factorials;
    Two-way factorials
    blocks, 379
    central composite designs, 365–367
    fractional designs, 299
    qualitative considerations, 378–381
Farm implement hardness, 98
F distribution, 412–417
Fiber angle, carpets, 60, 90
First quartile, 196, 205
Fishbone diagrams, 14
Fitted interactions, 270, 273, 275, 288
    three-way factorials, 280, 281, 283–286, 288
    two-way factorials, 261–279, 283, 285, 286, 297
Fitted main effects, 267, 268, 270, 295
    three-way factorials, 284
    two-way factorials, 267, 268
Fitted response values, 254

Index 429

Fitted three-factor interactions, 288
Fitted two-factor interaction, 286
Flowcharts, 5, 6, 12, 13
Formal multiplication, 338, 345, 346
Fractional factorials, 10, 299, 334–354, 384, 389, 390, 393, 394, 396, 402
    limitations, 335
Fraction nonconforming, 128–132, 386, 387
    Shewhart charts for, 128–132
Freaks, 140
    in control charts, 140
Full factorial in p factors, 280, 301

G

Gauge capability ratio, 65, 73, 88, 89, 91, 93, 99, 100, 104, 105
Gauge R&R studies, 62–74, 83, 116
    and change detection, 65
Gear heat treatment, 193
General quadratic relationships, 363
Generators, 221, 339–341, 345–347, 350
    of fractional factorial designs
        2^{p-q} studies, 340, 345, 347, 348
        2^{p-1} studies, 340
go/no-go inspection, 80–86
Graphical methods
    for process characterization, 191–230
    for quality assurance, 15–18
    for surface response studies, 354–374
Graphics, 207, 355–362
Grit composition, 205, 207, 213, 220
Grouping, 140, 194
    in control charts, 140

H

Half fractions, 331, 336–345, 348
    2^p factorials, 337–344
    $1/2^q$ fraction, 334, 336
Half-normal plots, 293–296, 313, 317, 319, 328, 342, 344, 351, 352, 386, 388, 389, 392, 396
Hardness measurement method comparison, 198
Histograms, 19–21, 24, 31, 140, 161, 192, 194, 205, 221, 223, 230, 236
    dot plots/stem-and-leaf diagrams contrasted, 192–194
Hose cleaning, 242–244
Hose skiving, 27–28, 179
Hypothesis testing, 64, 83, 141, 179, 227, 268, 269, 273, 337, 339, 341, 382, 404

I

$I \times J$ complete factorial, 262
Impedance in micro-circuits, 247
Implement hardness, 98–99
Individual observation control charts, 109, 139, 140
Individuals chart, 124, 125
Injection molding cause-effect diagram, 14
Inner array, 383
    experimental design, 383
Instability, 6, 120, 121, 123, 130, 132, 133, 135, 138, 159
    in control charts, 6, 120, 246
Integral gain, in PID control, 179
Interactions, 270–273, 275, 286, 289, 291, 335, 336, 339–342, 349
    aliasing in fractional factorials, 342

Interactions (cont.)
   plots, 264, 265, 269, 270, 277, 279, 287, 291, 303, 308, 311, 319, 321, 323, 324, 329, 330
   parallelism, 269, 270
   three-way factorials, 280, 281, 283–286, 288
   two-way factorials, 261–279, 283–286, 297
Interquartile range, 197, 205
Ishikawa diagrams, 5, 14, 387

J

Jet engine visual inspection, 172–174
JMP, 58–60, 70, 72, 78
Journal diameters, 161–163, 245

L

Laboratory carbon blank, 30–31, 188
Laser metal cutting, 90
Lathes. *See* Turning operations
Level changes, in control charts, 111
Lift-to-drag ratio wing configuration study, 356–358
Linear combination of means, 257, 289
   in one-way methods, 251–261
   in 2^p factorials, 279–301
   in two-way factorials, 261–279, 283, 285, 286, 297
Linearity, 37, 45, 78, 79, 200, 251
Long-term variation, 108
Lower control limits. *See* Control limits
Lower specification, for variables acceptance sampling, 208

M

Machined steel lever tongue thickness, 192–193
Machined steel sleeves, inside diameter monitoring, 88
Machined steel slugs, 29–30
Magnaflux crack inspection, 175
Main effects, 262, 267–271, 273, 274, 283–288, 290, 291, 293, 295, 335, 336, 339–342, 347, 349–351
   aliasing in fractional factorials, 342
   three-way factorials, 284
   two-way factorials, 261–279, 283–286, 297
Maximum, in quadratic response surface, 367
Mean, 1, 35, 45, 46, 109, 194, 202, 222, 252, 258, 335. *See also* Linear combinations of means; specific Quality assurance methods
   difference, 45, 81, 84, 86
   nonconformities per unit, 132–136, 328
   Shewhart charts for, 132–136
   square error, 77
Measurand, 36–51, 54, 56, 57, 62–65, 75, 76, 78, 79
Measurements
   accuracy, 34, 90
   bias, 36, 53
   error, 10, 33, 36–38, 40–53, 63, 64, 66, 87
   gauge R&R studies, 62–75
   noise, 59
   precision, 6, 10, 37, 38, 60
      and change detection, 60
   Shewhart charts, 113–128
   validity, 90

Median, 113, 114, 118, 160, 162, 166, 186, 196, 197, 205
  chart, 113, 162
  formulas, 160
Metrology, 5, 6, 34–39, 77, 81
Micro-circuit impedance, 247–248
Micrometer caliper gauge R&R, 62–63
Milling operation, 187
Minimum, in quadratic response surface, 388
Mixtures, 139
  in control charts, 139
Monitoring chart, 108
Moving range, 124, 125
Multimodality, 19
Multiple regression analysis, 262, 355, 357, 361

N

NASA polymer
  iron content, 317
  percent weight loss, 315–317
NASA polymer, 323–331
National Institute of Standards and Technology (NIST), 18, 30, 35
Nelson's alarm rules, 143
Nickel-cadmium cells, 387–388, 390
Noise factors, 383
Noise variables, 379, 383, 384
Non-conforming, 81, 82, 84–86, 128–132, 135, 384
Nonparametric limits, 217
Nonrandom variation, 108
Normal distribution
  confidence intervals, 10
  expected value of sample range, 90
Normal plot(ing), 10, 201–204, 214, 217, 229–230, 254, 255, 275, 276, 290, 292–295, 342, 343, 351, 352

Normal probability plots, 198–204
  formulas, 202
  np charts, 128, 129, 131
np chart, 128, 129, 131, 132

O

Observed variation, 37, 38, 108, 138
Offline quality control, 381
Oil field production, 231–235
Oil viscosity measurement uncertainty, 225
One-way ANOVA, 56–58, 69
One-way methods, 252–260
  linear combinations of means, 257–261
  pooled variance estimator, 253–256
One-way normal model, 252–256, 260, 275
Operational definitions, 15, 128, 142
  of measurable quantities, 15
Optimization, surface response studies for, 366–369
Outer array, 383
  experimental design, 383
Out of control point, 109, 120, 130

P

Packaging, box size choice, 223–225
Paint coat thickness, 390–392
Paper burst strength measurement, 102–103
Paper dry weight, 153, 154, 180–184, 189
Paper tensile strength measurement, 104–105
Paper thickness measurement, 100–102
Paper weight
  measurement, 99
  PID control, 153
Parallelism, 269, 270, 273, 287

Parallelism (cont.)
    in interaction plots, 270
Parameter design experiments, 381
Pareto diagrams, 21, 25
Pareto principle
    formulas, 292
    p charts, 19
Part-to-part variation, 42
Past data, 110
p chart, 129, 141
Pellet densification, 39, 51, 52
Pelletizing process, fraction nonconforming, 129–130
Perspective plots, 358
Physical variation, 40, 53
PID control, 150–158, 179, 185
    algorithm, 152, 153, 155
Pipe compound mixture study, 235
Plastic packaging, 26, 176–178
Poisson processes, 132
Polymer density, 375–376, 397
Pooled estimator of σ^2, 256
Pooled fraction nonconforming, 131
Pooled mean nonconformities per unit, 133
Pooled-sample variance estimator, in one-way methods, 253–256
Pooled variance estimator, 253–256
Potentiometers, 30
p quantile, 194, 195, 202, 257
Precision, 6, 10, 21, 34–38, 44, 60, 73, 192, 225, 226, 228, 264, 268, 270, 373
    of measurements (*see* Measurement precision)
Precision-to-tolerance ratio, 65
Prediction intervals, 214–218, 240
    nonparametric limits, 217
Prediction limits in SLR, 76–78
Principle of effect sparsity, 335
Printing process flowchart, 13

Probabilistic simulations, 221
Probabilistic tolerancing, 6, 10, 220–239
Probability plotting, 201, 202, 292, 293, 342
Process and characterization capability, 191–230
Process capability, 6, 191, 207–213, 229, 243, 245
    analysis, 207, 208
    measures, 207–213
    ratios, 208
Process characterization, 192–250
Process identification, 12–15
Process improvement, 3, 4, 6–11, 17, 21, 22, 25, 137, 251–301, 333–384
Process location charts, 113–119
Process mapping, 5, 25
Process monitoring, 16, 107–160. *See also* Control charts
    change detection, 108, 111
    statistical, 150
Process-oriented quality assurance, 4–6, 9, 26, 27
Process spread charts, 119–123
Process stability, 108, 112, 178, 242, 243
    control charts for monitoring, 118
Process standard deviation, 54–56, 124, 125, 146
Process variability, 121, 211
Product array, 383
    experimental design, 383
Production outcomes matrix, 26
Product-oriented inspection, 4
Product-oriented quality assurance, 4, 6
Propagation of error, 6, 225–227, 230
    formulas, 225–227
Proportional gain, in PID control, 156

Proportional-integral-derivative
  (PID) control, 150–158,
  179, 185
Pump end cap depths, 153
Pump housing drilling depths, 244

Q

Q-Q plots, 192, 198–202
Quadratic relationship, 363–364,
  369
Quadratic response surfaces
  adequately determined, 371
  analytical interpretation,
    369–372
  mixtures, 369
Quality, 1–7, 10–12, 14–17, 19–25,
    27, 28, 30, 34, 48–50, 56,
    58, 70, 74, 87, 99, 122,
    126, 133, 138, 142, 143,
    147, 148, 157, 174, 175,
    185, 186, 207, 212, 213,
    238, 245, 247, 248, 296,
    300, 326, 327
  assurance
    cycle, 4–6, 8
    data collection, 15–18
    modern philosophy of, 3–6
    process-oriented cycle, 4–6
    process-*versus*
      product-oriented, 4
    role of inspection in, 4
    statistical graphics for, 6,
      19–25
  of conformance, 2, 3, 25, 26,
    296
  uniformity and, 5, 24
  consultants, 381
  of design, 2, 25
  improvement, 3–7, 19, 20, 25,
    378–384
Quantile plots, 192, 195, 196, 236
Quantiles, 192, 194–198, 201, 204,
  208, 214, 293, 410–412,
  414, 416

chi-square distribution, 411
standard normal, 201–203
studentized range distribution,
  5, 34, 67, 121
t-distribution, 410

R

Radiator leak monitoring, 135
Random effects, 56–60, 62–65, 70,
  72, 81
  models, 52–65
    two-way and gage studies,
    62–65
Randomization, 101, 379, 380
Randomized block model, 313
Random variation, 254, 292, 372
Range charts, 119, 124. *See* R
  charts
Rational sampling, 111
Rational subgrouping, 111, 141
  formulas, 381
  R charts, 120, 122
  Taguchi emphasis and,
    381–384
R chart, 119–122, 138, 382
Reamed holes, surface finish, 114
Red oak casehardening, 238–239
Regression analysis, 10, 75, 79, 80,
  151, 262, 274, 275, 333,
  355–357, 361, 384, 392
Reliability, 1, 63
Remeasurement, replication
  contrasted, 381
Repeatability, 44, 60–68, 70, 73,
  76–78, 81–84, 292
Replication, 279, 289–296, 312,
  316, 324, 326, 340, 342,
  348, 351, 356, 361, 377,
  380, 381, 393, 396, 399,
  404
Reproducibility, 44, 60, 62–68,
  71–74, 81–84

Residuals, 254, 255, 274–277, 279, 290, 297–299, 357, 366, 367, 382
   balanced data 2^p factorials, 279–280
   balanced data two-way factorials, 274–276
   one-way methods, 252
Resistors
   resistance distribution, 221–223
   resistance measurements, 325–326
Response optimization strategies, 372–374
Response surface studies, 354–377
   adequately determined, 371
   graphics for, 355
   optimization strategies, 390–393
   quadratic, 362–364
Retrospective, 110, 115–118, 120–123, 130–135, 160
   c charts, 132
   median chart, 162
   np charts, 131, 132
   p charts, 388, 389
   R chart, 120, 121
   scenario, Shewhart control charts, 128
   s charts, 122, 123, 162, 164
   u charts, 134, 135
   x chart, 115, 118, 162, 164
Reverse Yates algorithm, 298, 299, 320
Ridge geometry, in quadratic response surface, 370
Robust design, 247, 382
Rolled paper, 163, 165
Rolled products, 29
R&R standard deviation, 65
Rule of 10, 28
Run charts, 19, 23, 24, 139–141
Run length, 107, 144–148
   distribution, 144

S

Saddle geometry, in quadratic response surface, 370
Sample medians, 118
Sample standard deviation, 38, 39, 42, 51, 52, 54, 66, 86, 88, 116, 123, 166, 185, 186, 208, 213, 232, 234, 245, 247, 256, 260, 278, 281, 282, 290, 293, 296, 306, 308, 314, 316, 319, 366, 396, 403
Scatterplots, 16, 19, 22, 23, 77, 200
   formulas, 16
   s charts, 122
s chart, 122, 123
Seal strength optimization study, 366–369
Sequential experimental strategy, 364–369
Service industries, statistical control applications, 12
Sheet metal saddle diameters, 235–236
Sheet metal slips, 235
Shewhart control charts. 8, 108–112, 128, 160. *See also specific* Shewhart charts
   alarm rules, 136–144
   average run lengths, 144
   constants, 128
   data hugging center line, 110
   formula summary, 119, 120
   fraction nonconforming charts, 129
   limitations, 211

mean nonconformities per unit
chart, 132, 133
patterns, 136–144
process location charts,
113–119
process spread charts, 119, 120
Short-term variation, 108, 178
Signal-to-noise ratios, 383
Simple linear regression, 75–79
Single sourcing, for reducing
product variation, 19
Six Sigma, 7–10, 112
Solder joint strength, 259, 263–266,
271, 276
Solder thickness, 395–396
Sorting operations, histogram
shapes, 20
Sparsity of effects principle, 292
Special cause, 137, 159
variation, 112, 159
Specifications, 2, 21, 65, 111, 192,
203, 208, 209, 211. See
also Engineering
specifications
Speedometer cable, 391–392
Standard
normal, 145, 194, 201, 202,
210, 211, 293, 408, 409
normal cumulative
probabilities, 408, 409
normal quantiles, 201, 231,
236, 293
Standard deviation, 35–46, 49, 52,
56, 60, 61, 66–71, 73, 74,
76, 78, 80, 84, 86–92, 94,
101, 103, 110, 113, 114,
116–119, 123–125, 127,
129, 135, 145, 146,
162–167, 169, 172, 177,
178, 185–188, 202, 205,
207, 208, 212, 213, 221,
223, 226–229, 232, 234,
241, 242, 245–249, 254,
256, 258, 260, 261, 263,
264, 278, 281, 282, 290,
293, 295, 296, 298, 300,
305, 306, 308, 309, 314,
316, 319, 323, 327–329,
331, 357, 366, 382, 383,
386, 388, 396, 403
Standardized residuals, 254
Standards given, 110, 113–122,
124, 129–133, 145, 146,
160
c charts, 132, 133
median charts, 110
np charts, 131
p charts, 129
R charts, 121
scenario, Shewhart control
charts, 110
s charts, 133
u charts, 133
x charts, 115
Star points, 364–367, 369, 374
with 2^p factorial, 292
Stationary point, 370, 371
in quadratic response surface,
372
Statistical graphics methods. See
Graphical methods
Statistical process monitoring,
150–159
engineering control contrasted,
150–159
Statistical quality assurance. See
Quality assurance
Statistics, 1–3, 8, 10, 17, 19–24,
33–87, 114, 142, 143,
159, 162, 166, 167, 191,
195, 201, 213, 214, 253,
260, 264, 341, 345, 355,
357, 371, 393, 394, 403
Steel heat treatment, 15, 29
Steel machined sleeves, inside
diameter monitoring, 225
Stem-and-leaf diagrams, 192–194
Stratification, 141–143
in control charts, 141
Sum of squares, 57, 69, 70, 72

Surface finish
    machined bars, 397–398
    reamed holes, 114
Surface plots, 358–361
Systematic variation/cycles, 137
Systematic variation, in con, 137
System causes, 108

T

Tablet hardness, 323–324
Tabular alumina powder packing properties, 280–281
Taguchi methods, 381
Tar yield, 376–377, 400
Taylor expansion, 225
t-distribution, 48, 86, 214, 215, 410
    quantiles, 410
Third quartile, 196, 205
Three-factor interactions, 288, 339, 340, 347
Three sigma control limits, 143, 144
    Shewhart control charts, 139
Three-way factorials
    3^p factorials, fitting quadratic response functions with, 364
    in three-way factorials, 285, 286
Tile manufacturing, 384–386
Time constant, PID control, 151, 155
Tolerance intervals, 191, 213–218, 230, 240, 419, 420
    nonparametric limits, 217
Total quality management (TQM), 7–8, 10.
    elements of, 7
    limitations, 361, 378
Training, need for in data collection, 9

Transmission gear measurement, 94–96
Trends, 19, 31, 137, 139, 140, 179, 275, 276
    in control charts, 139, 140
Turning operations
    with CNC lathe, 29
    CUSUM-controlled CNC lathe, 29
    run chart for, 23
    and surface roughness, 402–403
TV electron guns, 262–263
2^p factorials, 279–300
    algorithm for producing, 344
    balanced data, 298, 299
    effect detection without replication, 292, 312–317
    effect detection with replication, 289–291, 308–312
    for fitting quadratic response surfaces, 364
    half fraction, 334
    2^{p-1} fractional factorials, 340
    2^p half fractions, 337–344
    plus repeated center point, 364, 382
    2^{p-q} factorials, 334–336
    reverse Yates algorithm for, 298
    with star or axial points, 364
    Yates algorithm for, 318
Two-factor interactions, 285–288, 339, 347
    in three-way factorials, 285
2^{p-1} *fractional factorial,* 337, 340
Two-way ANOVA, 69
Two-way factorials, 261–276, 283, 285, 286, 297
    balanced data, 275

effect definition and
estimation, 58, 66
Two-way model, 64, 264
Two-way random effects models,
62–65
for gauge R&R studies, 62–65

U
U-bolt threads
formulas, 246
u charts, 133
u chart, 132–135, 147
Uniformity, and quality of
conformance, 2
Unit of analysis, 381
problem, 381
Upper control limits. *See* Control
limits

V
Validity, 34, 35
of measurements, 245
Valve airflow, 327–328
Variables data, 113, 161
Shewhart charts, 113–128
Variance. *See* specific quality
assurance methods
Variance components, 63, 70
random effects model, 56–60
Variation, 2, 36, 108, 193, 252, 341
Voltage regulator, for alternator,
248, 249

W
Wax-polyethylene blends seal
strength optimization
study, 366
Weld pulls, 230
Western electric alarm rules, 142,
143, 148
Window frame sampling, 29
Within group, 56

X
x charts, 112–115, 118, 119, 124,
127, 138, 141, 145, 146,
148, 149, 161–162, 164,
165, 178, 179, 239, 382
symbol)

Y
Yates algorithm, 296–300, 312,
316, 319, 320, 324, 328,
341, 342, 351, 354, 364,
386, 388, 389, 391, 394,
396, 410
reverse, 298, 299, 320
Yates standard order, 296–298, 341,
342, 351, 386, 387, 389,
391, 394, 396, 409

Z
Zones, 142, 143

CPSIA information can be obtained
at www.ICGtesting.com
Printed in the USA
LVOW02s1452080916